Spon's Asia-Pacific Construction Costs Handbook

Fourth edition

Spon's Asia-Pacific Construction Costs Handbook

Fourth edition

Davis Langdon & Seah International

LONDON AND NEW YORK

First edition published 1993
by Spon Press

Second edition published 1996
by Spon Press

Third edition published 2000
By Spon Press
This edition published 2010 by Spon Press

2 Park Square, Milton Park, Abingdon, Oxon OX14 4RN

Simultaneously published in the USA and Canada by Spon Press
270 Madison Avenue, New York, NY 10016, USA

Spon Press is an imprint of the Taylor & Francis Group, an informa business

Typeset in Times New Roman
Printed and bound in Singapore by Markono Print Media Pte Ltd

This publication presents material of a broad scope and applicability. Despite stringent efforts by all concerned in the publishing process, some typographical or editorial errors may occur, and readers are encouraged to bring these to our attention where they represent errors of substance. The publisher and author disclaim any liability, in whole or in part, arising from information contained in this publication. The reader is urged to consult with an appropriate licensed professional prior to taking any action or making any interpretation that is within the realm of a licensed professional practice.

British Library Cataloguing in Publication Data
A catalogue record for this book is available from the British Library

Library of Congress Cataloging-in-Publication Data
A catalog record has been requested for this book

ISBN13: 978-0-415-46565-6 (hbk)
ISBN13: 978-0-203-85534-8 (ebk)

Contents

Preface

In 1994, the Editors published the first Asia Pacific Construction Costs Handbook which covered fifteen countries. Three countries in the Asia Pacific region – Sri Lanka, Taiwan and Vietnam – were added in the second edition which was published in 1997. Two other countries – Canada and India – were added in the third edition which was published in 2000. In this edition, two countries – Cambodia and Pakistan – not previously covered in the last volume have been included. Future volumes will no doubt add to the list of countries.

This book is designed to be a convenient reference. Its purpose is to present coherent snapshots of the economies and construction industries of the Asia Pacific region; it also places this information in an international context with the inclusion of the United Kingdom and United States of America. It is not a substitute for local knowledge and professional advice. It will, however, be extremely useful as an introduction to a country and its construction industry for clients, consultants, contractors, manufacturers of construction materials and equipment and others concerned with development, property and construction in the region.

Davis Langdon & Seah International
2010

Acknowledgements

The contents of this book have been gathered together from a variety of sources – individuals, organizations and publications. Construction cost data and general background information on local construction industries are based on contributions from a network of professional colleagues and associates worldwide. These include:

- Australia — Davis Langdon
- Brunei Darussalam — Davis Langdon & Seah Brunei Darussalam
- Cambodia — Davis Langdon & Seah (Thailand) Ltd
- China — Davis Langdon & Seah China Limited
- Hong Kong — Davis Langdon & Seah Hong Kong Limited
- India — Davis Langdon & Seah Consulting India Pvt Ltd
- Indonesia — PT. Davis Langdon & Seah Indonesia
- Japan — Davis Langdon & Seah Japan Limited; Mr David Yip (Takenaka Corporation) and Taisei Corporation
- Malaysia — Davis Langdon & Seah (M) Sdn Bhd
- New Zealand — Davis Langdon New Zealand Ltd
- Pakistan — Davis Langdon & Seah Pakistan (Pte) Ltd
- Philippines — Davis Langdon & Seah Philippines, Inc
- Singapore — Davis Langdon & Seah Singapore Pte Ltd
- South Korea — Davis Langdon & Seah Korea Co Ltd
- Sri Lanka — Overseas Realty [Ceylon] Limited
- Taiwan — BRAVO Project Management
- Thailand — Davis Langdon & Seah (Thailand) Ltd
- United Kingdom — Davis Langdon LLP
- United States of America — Davis Langdon
- Vietnam — Davis Langdon & Seah Vietnam Co Ltd

Much of the statistical data is from World Bank Development Reports, the *Economist World in Figures* and official published statistics. The background on individual countries has come from local sources, national yearbooks, annual reports and *Economist Intelligence Unit Reports.*

Important sources of general and construction industry data have been various embassies, high commissions, trade missions, statistical offices, and government departments in the UK and overseas. Information on international contracting is largely based on surveys undertaken by *Engineering News Record* magazine. Data on exchange rates and consumer price indices come mainly from the *Financial Times* or International Monetary Fund publications.

Specific acknowledgements and sources are given where appropriate in each country. The research and compilation of this book were undertaken by Davis Langdon & Seah International.

How to use this book

This book is in three parts – *Part One: Regional Overview*; *Part Two: Individual Countries*; and *Part Three: Comparative Data*. The twenty countries covered in the book are listed in the Contents. The United Kingdom and USA are outside the region but are included for comparative purposes.

Part One: Regional Overview

Part One comprises an essay – *The construction industry in the Asia Pacific region* – which describes the current situation and main trends in the construction industries covered in this publication.

Part Two: Individual Countries

In Part Two the twenty countries are arranged in alphabetical order, and each is presented in a similar format under the following main headings:

- *Key data* provides main national, economic and construction indicators.
- *The construction industry* outlines the structure of the industry, tendering and contract procedures plus the regulations and standards.
- *Construction cost data* includes data on labour and material costs, measured rates for items of construction work and approximate estimating costs per square metre for different building types.
- *Exchange rates and inflation* presents data on exchange rates with the pound sterling, US$, euro and Japanese yen, and includes data on the main indices of price movements for retail prices and construction.
- *Useful addresses* gives the names and addresses of public and private organizations associated with the construction industry.

Part Three: Comparative Data

To allow comparison between countries covered in the book, Part Three brings together data from Part Two and presents them under three main headings:

- *Key national indicators* including financial and demographic data.
- *Construction output indicators* including output per capita.
- *Construction cost data* including labour and material costs and costs per square metre.

Abbreviations

LENGTH	
kilometre	km
metre	m
decimetre	dm
millimetre	mm
yard	yd
foot	ft
inch	in
AREA	
hectare	ha
VOLUME	
kilolitre	kl
hectolitre	hl
litre	l
millilitre	ml
WEIGHT (MASS)	
tonne	t
kilogram	kg
gram	g
hundredweight	cwt
pound	lb
ounce	oz
FORCE	
kilonewton	kN
newton	N
not available	N/A

Conversion factors

LENGTH

Metric		**Imperial equivalent**
1 kilometre	1000 metres	0.6214 miles
		1093.6 yards
1 metre	100 centimetres	1.0936 yards
	1000 millimetres	3.2808 feet
		39.370 inches
1 centimetre	10 millimetres	0.3937 inches
1 millimetre		0.0394 inches
Imperial		**Metric equivalent**
1 mile	1760 yards	1.6093 kilometres
	5280 feet	1609.3 metres
1 yard	3 feet	0.9144 metres
	36 inches	914.40 millimetres
1 foot	12 inches	0.3048 metres
		304.80 millimetres
1 inch		25.400 millimetres

AREA

Metric		**Imperial equivalent**
1 square kilometre	100 hectares	0.3861 square miles
	10^6 square metres	247.11 acres
1 hectare	10 000 square metres	2.4711 acres
		11 960 square yards
1 square metre	10 000 square centimetres	1.1960 square yards
		10.764 square feet
1 square centimetre	100 square millimetres	0.1550 square inches
1 square millimetre		0.0016 square inches
Imperial		**Metric equivalent**
1 square mile	640 acres	2.5900 square kilometres
		259.00 hectares
1 acre	4840 square yards	0.4047 hectares
		4046.9 square metres
1 square yard	9 square feet	0.8361 square metres
1 square inch		6.4516 square centimetres
		645.16 square millimetres

VOLUME

Metric		Imperial equivalent
1 cubic metre or	10 hectolitres	1.3080 cubic yards
1 kilolitre	1000 cubic decimetres	35.315 cubic feet
	1000 litres	
1 hectolitre	100 litres	3.5315 cubic feet
		21.997 gallons
1 cubic decimetre	1000 cubic centimetres	61.023 cubic inches
or 1 litre	1000 millilitres	0.2200 gallons
		1.7598 pints
		0.2642 US gallons
		2.1134 US pints
1 cubic centimetre	1000 cubic millimetres	0.0610 cubic inches
or 1 millilitre		

Imperial		Metric equivalent
1 cubic yard	9 cubic feet	0.7646 cubic metres
1 cubic foot	1728 cubic inches	28.317 litres
	6.2288 gallons	
	7.4805 US gallons	
1 cubic inch		16.387 cubic centimetres
1 gallon	8 pints	4.5461 litres
1 pint		0.5683 litres

US		
1 barrel	42 gallons	158.99 litres
1 gallon	8 pints	3.7854 litres
1 pint		0.4732 litres

WEIGHT (MASS)

Metric		Imperial equivalent
1 tonne	1000 kilograms	0.9842 tons
		1.1023 US tons
		2204.6 pounds
1 kilogram	1000 grams	2.2046 pounds
		35.274 ounces
1 gram		0.0353 ounces

Imperial		**Metric equivalent**
1 ton	20 hundredweights	1.0160 tonnes
	2240 pounds	1016.0 kilograms
1 hundredweight	112 pounds	50.802 kilograms
1 pound	16 ounces	0.4536 kilograms
		453.59 grams
1 ounce		28.350 grams

US		
1 ton	20 hundredweights	0.9072 tonnes
	2000 pounds	907.18 kilograms
1 hundredweight	100 pounds	45.359 kilograms

FORCE

Metric		**Imperial equivalent**
1 kilonewton	1000 newtons	0.1004 tons force
		0.1124 US tons force
1 newton		0.2248 pounds force

Imperial		**Metric equivalent**
1 ton force	2240 pounds force	9.9640 kilonewtons
1 pound force		4.4482 newtons

US		
1 ton force	2000 pounds force	8.8964 kilonewtons

PRESSURE

Metric	**Imperial equivalent**
1 newton per square millimetre	145.04 pounds force per square inch
1 kilonewton per square metre	20.885 pounds force per square foot

Imperial	**Metric equivalent**
1 pound force per square inch	6.8948 kilonewtons per square metre
	0.0069 newtons per square millimetre
1 ton force per square inch	107.25 kilonewtons per square metre
	0.1073 newtons per square millimetre

US	
1 ton force per square foot	95.761 kilonewtons per square metre
	0.9576 newtons per square millimetre

Davis Langdon & Seah International

Practice Profile

Davis Langdon & Seah International (DLSI) is a worldwide organization which provides professional quantity surveying, cost engineering and construction cost consultancy services, project management, value management and quality management consultancy. DLSI operates throughout the Asia Pacific region, Australasia, Europe, Africa and America and has associations with firms in Kenya and Uganda. The DLSI group employs over 5,500 staff in 106 offices located in more than 28 countries around the world.

Asia Pacific region and Australasia:

- Singapore : Davis Langdon & Seah Singapore Pte Ltd
  Davis Langdon & Seah Project Management Pte Ltd
  DLS Contract Advisory & Dispute Management Services
- Brunei Darussalam : Davis Langdon & Seah Brunei Darussalam
  Juruukur Bahan Utama-DLS
- China : Davis Langdon & Seah China Limited
  Davis Langdon & Seah (Beijing) Construction Consultants Co Ltd
  Davis Langdon & Seah Consultancy (Chengdu) Co Ltd
  Davis Langdon & Seah China Limited Chongqing Representative Office
  Davis Langdon & Seah Consultancy (Shenzhen) Co Ltd
  Davis Langdon & Seah Consultancy (Shenzhen) Co Ltd, Guangzhou Branch
  Davis Langdon & Seah Consultancy (Shanghai) Co Ltd
  Davis Langdon & Seah (Beijing) Construction Consultancy Co Ltd, Shenyang Branch
  Davis Langdon & Seah Consultancy (Shenzhen) Co Ltd
  Davis Langdon & Seah Consultancy Co Ltd, Tianjin Branch
  Davis Langdon & Seah Consultancy (Shanghai) Co Ltd, Wuhan Branch
- Hong Kong : Davis Langdon & Seah Hong Kong Limited
  Davis Langdon & Seah China Limited
  DLS Management Limited
  DLS Specifications China Ltd
  Davis Langdon & Seah Macau Limited

- India : Davis Langdon & Seah Consulting India Pvt Ltd with offices at Bangalore, Chennai, Delhi, Hyderabad and Mumbai
- Indonesia : PT. Davis Langdon & Seah Indonesia with offices at Bali, Jakarta and Surabaya
- Japan : Davis Langdon & Seah Japan Limited
- Malaysia : Davis Langdon & Seah Malaysia Sdn Bhd in association with JUBM Sdn Bhd DLS Management (M) Sdn Bhd with offices at Johor Bahru, Kota Kinabalu, Kuala Lumpur, Kuching and Penang
- Pakistan : Davis Langdon & Seah Pakistan (Private) Limited
- Philippines : Davis Langdon & Seah Philippines Inc
- South Korea : Davis Langdon & Seah Korea Co Ltd
- Thailand : Davis Langdon & Seah (Thailand) Ltd LECE Thailand Ltd
- Vietnam : Davis Langdon & Seah Vietnam Co Ltd with offices at Hanoi and Ho Chi Minh City
- Australasia : Davis Langdon with offices at Adelaide, Brisbane, Cairns, Canberra, Darwin, Hobart, Melbourne, Perth, Sunshine Coast, Sydney and Townsville
- New Zealand : Davis Langdon with offices at Auckland, Christchurch and Wellington

Europe:

- United Kingdom : Davis Langdon LLP Davis Langdon Schumann Smith Davis Langdon Mott Green Wall Davis Langdon Crosher & James with offices at Birmingham, Bristol, Cambridge, Cardiff, Edinburgh, Glasgow, Heathrow, Leeds, Liverpool, London, Maidstone, Manchester, Milton Keynes, Norwich, Oxford, Peterborough, Plymouth and Southampton
- Ireland : Davis Langdon PKS with offices at Cork, Dublin, Galway and Limerick

- Spain (Barcelona) : Davis Langdon Edetco
- Russia (Moscow) : Ruperti Project Services International

Middle East:

- Bahrain (Manama) : Davis Langdon
- Lebanon (Beirut) : Davis Langdon
- United Arab Emirates (UAE) : Davis Langdon with offices at Abu Dhabi and Dubai
- Qatar (Doha) : Davis Langdon

Africa:

- South Africa : Davis Langdon with offices at Bloemfontein, Botswana, Cape Town, Durban, George, Johannesburg, Klerksdorp, Mozambique, Pietermaritzburg, Port Shepstone, Pretoria, Richards Bay, Stellenbosch and Vanderbijlpark

United States of America:

- United States of America : Davis Langdon with offices at Boston, Honolulu, New York, Philadelphia, San Francisco, Santa Monica, Sacramento and Seattle

In addition to the financial management of construction projects, DLSI also undertakes varied construction industry research and consultancy assignments worldwide, ensuring a broadly based and truly international information service to their clients.

The value of the organization's international experience, research and information is distilled into the strategic advice and services offered to DLSI's individual clients and also provides data for publications, such as this current Asia Pacific Construction Costs Handbook.

Professional Services

DLSI specializes in the financial management of construction projects, from inception to completion. Their range of services includes:

- Investment Appraisals
- Brief Development
- Construction Cost Management
- Cost and Time Planning

- Strategic Procurement Advice and Management
- Tender and Contract Documentation
- Project Management
- Dispute Resolution
- Development Economics and Appraisals
- Risk Analysis and Management
- Value Analysis and Management
- Legal Support Services
- Capital Allowances Taxation Assistance
- Quality Management Consultancy and Training
- Research and Consultancy

The organization's experience covers a wide range of construction projects, such as:

- Airports and Airport Buildings
- Arts and Cultural Buildings
- Business Park Developments
- Civic Buildings
- Civil Engineering and Infrastructure Works
- Educational Buildings
- Health and Hospital Buildings
- Historic and Gazetted Buildings
- Hotels
- Internet Data Centres
- Industrial/Warehouse Developments
- Leisure Projects
- Office Buildings and Interior Fit-out Works
- Petro-Chemical Projects
- Power Generation Projects
- Public Buildings
- Residential Developments
- Retail Developments
- Sports Centres
- Transportation
- Water and Waste Projects

Practice Statement

Davis Langdon & Seah International (DLSI) is committed to giving the best possible professional service in meeting the needs of each client – whether large or small, local, multi-national or international.

The strategic and integrated management of cost, time and quality – the client's 'risk' areas of a contract – are essential functions, which are necessary to ensure the satisfactory planning, procurement, execution and operation of construction projects.

DLSI specializes in the financial management of construction projects and their risk areas, from project inception to completion.

The organization employs highly qualified and skilled professional staff, with specialist experience in all sectors of the construction industry, including international cost variables, procurement options and management structures.

It operates a sophisticated information support system, based on the latest computer technology, enabling large-scale capture and retrieval of cost and relevant market data.

The highest operational standards are observed to ensure quality of product and are quality assured in respect of the services in those countries where formal registration is available.

DLSI draws upon their international network of offices but works in manageable teams under direct partner or director leadership, and maintains personal client contact at all stages of a project.

The organization's approach is:

- to be positive and creative in their advice, rather than simply reactive;
- to concentrate on value for money and value engineering rather than on superficial cost-cutting;
- to give advice that is matched to the client's own criteria, rather than to impose standard or traditional solutions;
- to see cost as one component of a successful design solution, which needs to be balanced with many others, and to work as an integrated member of a design team in achieving that balance;
- to pay attention to the life-long costs of owning and operating a facility, rather than to the initial capital cost only.

The overall objective is to manage client requirements, control risk, manage cost and maximise value for money, throughout the course of construction projects, always aiming to be – and to deliver – the best.

As a company, DLSI believes that the protection of the environment is a responsibility that everyone carries on his or her shoulders. The firm believes that a healthy environment is the main catalyst in determining the well-being of our society and our people and business, and is the foundation for a sustainable and strong economy.

With this vision, DLSI has setup a Sustainable Consultancy Group to study into the various issues to aid our clients in developing the necessary guidelines to minimize the environment impact of their projects.

DAVIS LANGDON & SEAH INTERNATIONAL ASIA PACIFIC

BRUNEI DARUSSALAM

DAVIS LANGDON & SEAH BRUNEI DARUSSALAM

JURUUKUR BAHAN UTAMA – DLS

25, BT Complex Kg. Jaya Setia
Mukim Berakas 'A' BB2713
P O Box 313
Bandar Seri Begawan BS 8670
Negara Brunei Darussalam
Tel : (673) 233 2833
Fax: (673) 233 2933
E-mail: dlsbsb@dls.com.bn

CHINA

DAVIS LANGDON & SEAH CHINA LIMITED

BEIJING
Suite 1225 – 1238
Junefield Plaza Central
Tower South
No. 10 Xuan Wu Men Wai St.
Beijing 100 052
Tel : (86-10) 6310 1136
Fax: (86-10) 6310 1143
E-mail: dlsbj@dlsbj.com

CHONGQING
Room 3408
International Trade Centre
No. 38 Qing Nian Road,
Central District
Chongqing 400 010
Tel : (86-23) 8655 1333
Fax: (86-23) 8655 1616
E-mail: dlscq@dlscq.com

CHENGDU
Room 807 Block A, Times Plaza
No. 2 Zongfu Road
Chengdu 610 016
Tel : (86-28) 8671 8373
Fax: (86-28) 8671 8535
E-mail: dlscd@dlscd.com

FOSHAN
Unit 1803 Room 2
18/F Hua Hui Mansion
46 Zu Miao Road
Foshan 528 000
Shenyang 110 013
Tel : (86-757) 8203 0028
Fax: (86-757) 8203 0029
E-mail: dlsgz@dlsgz.com

GUANGZHOU
Unit 2711-2712,
Bank of America Plaza
No. 555 Ren Min Zhong Road
Guangzhou 510 145
Tel : (86-20) 8130 3813
Fax: (86-20) 8130 3812
E-mail: dlsgz@dlsgz.com

SHANGHAI
Room 1582 Tower B
City Centre of Shanghai
No. 100 Zun Yi Road
Shanghai 200 051
Tel : (86-21) 6091 2800
Fax: (86-21) 6091 2999
E-mail: dlssh@dlssh.com

SHENYANG
Room 8-9, 11/F
E Tower of Fortune Plaza
No. 59 Beizhan Road
Shenhe District
Shenyang 110 013
Tel : (86-24) 3128 6678
Fax: (86-24) 3128 6983
E-mail: dlssy@dlssy.com

SHENZHEN
Room E & F
42/F World Finance Centre
Block A, 4003 East Shennan Rd
Shenzhen 518 001
Tel : (86-755) 8269 0642
Fax: (86-755) 8269 0641
E-mail: dlssz@dlssz.com

TIANJIN
Suite 1-1-2103
Tianjin Harbour Centre
No. 240 Zhang Zizhong Road
Heping District
Tianjin 300 041
Tel : (86-22) 8331 1618
Fax: (86-22) 2319 3186
E-mail: dlstj@dlstj.com

WUHAN
Room B, 5th Floor
2-1 Building, Wuhan Tiandi
No. 68 Lu Gou Qiao Road
Wuhan 430 010
Tel : (86-27) 5920 9299
Fax: (86-27) 5920 9298
E-mail: dlswh@dlswh.com

HONG KONG

DAVIS LANGDON & SEAH HONG KONG LIMITED

HONG KONG
21st Floor, Leighton Centre
77 Leighton Road
Tel : (852) 2830 3500
Fax: (852) 2576 0416
E-mail: dlshk@dlshk.com

MACAU
Avenida da Praia Grande No 599
Edifacio Commercial Rodrigues
14 Andar B
Macau
Tel : (853) 2833 1710
Fax: (853) 2833 1532
E-mail: dlsmacau@dlsmacau.com

INDIA

DAVIS LANGDON & SEAH CONSULTING INDIA PVT LTD

BANGALORE
3rd Floor, Raheja Chancery Building
No. 133 Brigade Road
Bangalore 560 025
Tel : 91 (80) 4123 9141
Fax: 91 (80) 4123 8922
E-mail: dlsindia@dls.co.in

CHENNAI
New No. 125 (Old No. 63)
Jammi Building, 1st Floor
Royapettah High Road
Mylapore
Chennai 600 004
Tel : 91 (44) 2498 8141
Fax: 91 (44) 2498 8137
E-mail: dlsindia@dls.co.in

DELHI
2nd Floor, Unit No. 465
Phase V, Udyog Vihar
Gurgaon Haryana 122 016
Tel : 91 (12) 4430 8790
Fax: 91 (12) 4430 8793
E-mail: dlsindia@dls.co.in

DAVIS LANGDON & SEAH INTERNATIONAL
ASIA PACIFIC

HYDERABAD
2nd Floor, Hitex Exhibition Center
Izzat Nagar
Hyderabad 500 034
Tel : 91 (40) 2311 4942
Fax: 91 (40) 2311 2942
E-mail: dlsindia@dls.co.in

MUMBAI
1204/5/6, Maithili's Signet
Plot No. 39/4
Sector No. 30A, Vashi
Navi Mumbai 400 703
Tel : 91 (22) 4156 8686
Fax: 91 (22) 4156 8615
E-mail: dlsindia@dls.co.in

INDONESIA

PT. DAVIS LANGDON & SEAH INDONESIA

BALI
Kuta Poleng Blok B/3A
Jalan Setiabudi
Kuta, Badung, Bali 80361
Tel : 62 (361) 766 260
Fax: 62 (361) 750 312
E-mail: dlsbali@dls.co.id

JAKARTA
18th Floor,
Ratu Plaza Office Tower
Jalan Jenderal Sudirman 9
Jakarta 10270
Tel : (62-21) 739 7550
Fax: (62-21) 739 7846
E-mail: dlsjkt@dls.com.id

SURABAYA
Room 601-A
Bank Mandiri Building
Jalan Basuki Rahmand 129-137
Surabaya 60271
Tel : (62-31) 546 5857
Fax: (62-31) 531 6579
E-mail: dlssby@dls.co.id

KOREA

DAVIS LANGDON & SEAH KOREA CO LTD
#429 G-Five Central Plaza
1685-8 Seocho 4-Dong
Seocho-Gu
Seoul, Korea 137-882
Tel : (82-2) 543 3888
Fax: (82-2) 543 0234
E-mail: dlsk@dlskorea.com

JAPAN

DAVIS LANGDON & SEAH JAPAN LIMITED
5F Kowa Building
2-8-16 Akasaka Minato-ku
Tokyo 107-0052
Tel : 81 (3) 6459 1277
Fax: 81 (3) 6459 1278
E-mail: dlsj@dlsjapan.com

MALAYSIA

DAVIS LANGDON & SEAH (M) SDN BHD

JUBM SDN BHD

JOHOR BAHRU
49-01 Jalan Tun Abdul Razak
Susur 1/1 Medan Cahaya
Johor Bahru 80000
Tel : (60-7) 223 6229
Fax: (60-7) 223 5975
E-mail: jubmjb@dlsjubm.com.my

KOTA KINABALU
Suite 8A, 8th Floor
Wisma Pendidikan
Jalan Madang, P O Box 11598
Kota Kinabalu 88817
Tel : (60-88) 223 369
Fax: (60-88) 216 537
E-mail: jubmkk@dlsjubm.com.my

KUALA LUMPUR
2 Jalan PJU 5/15
Kota Damansara
Petaling Jaya 47810
Selangor
Tel : (60-3) 6156 9000
Fax: (60-3) 6157 8660
E-mail: info@dlsjubm.com.my

KUCHING
No. 2 (3rd Floor)
Jalan Song Thian Cheok
Kuching 93100
Tel : (60-82) 232 212
Fax: (60-82) 232 198
E-mail: kuching@dlsjubm.com.my

PENANG
Suite 3A-3
Level 3A Wisma Great Eastern
No. 25 Lebuh Light
Penang 10200
Tel : (60-4) 264 2071
Fax: (60-4) 264 2068
E-mail: penangoffice@dlsjubm.com.my

PAKISTAN

DAVIS LANGDON & SEAH PAKISTAN (PTE) LTD
18C, Nishat Commercial Lane 4
Khayaban-e-Bukhari, Phase 6
D.H.A Karachi
75500 Pakistan
Tel : (92-21) 3524 0191 to 94
Fax : (92-21) 3524 0195

PHILIPPINES

DAVIS LANGDON & SEAH PHILIPPINES INC
4th Floor, Kings Court 1
2129 Pasong Tamo
Makati City
Manila 1231
Tel : (63-2) 811 2971
Fax: (63-2) 811 2071
E-mail: manila@dls.com.ph

DAVIS LANGDON & SEAH INTERNATIONAL ASIA PACIFIC

SINGAPORE

DAVIS LANGDON & SEAH SINGAPORE PTE LTD
1 Magazine Road
#05-01Central Mall
Singapore 059567
Tel : (65) 6222 3888
Fax: (65) 6224 7089
E-mail: dlssp3@dls.com.sg

THAILAND

DAVIS LANGDON & SEAH (THAILAND) LTD
10th Floor, Kian Gwan Building II
140/1 Wireless Road, Lumpini, Pratumwan
Bangkok 10330
Tel : (66-2) 253 1438
Fax: (66-2) 253 4977
E-mail: general@dls.co.th

VIETNAM

DAVIS LANGDON & SEAH VIETNAM CO LTD

HANOI
#706 7th Floor
North Star Building
4 Da Tuong Street
Hoan Kiem District
Hanoi
Tel : (84-4) 942 7525
Fax: (84-4) 942 7526
E-mail: dlshan@netnam.vn

HO CHI MINH CITY
9th Level, Unit E,
OSIC Building
8 Nguyen Hue Street
District 1
Ho Chi Minh City
Tel : (848) 823 8297
Fax: (848) 823 8197
E-mail: dlsvietnam@hcm.vnn.vn

DAVIS LANGDON & SEAH INTERNATIONAL ASIA PACIFIC

AUSTRALASIA

DAVIS LANGDON AUSTRALIA

ADELAIDE
Level 12
25 Grenfell Street
Adelaide SA 5000
Tel : (61-8) 8410 4044
Fax: (61-8) 8410 4166

BRISBANE
Level 13
324 Queen Street
Brisbane QLD 4000
Tel : (61-7) 3221 1788
Fax: (61-7) 3221 3417

CAIRNS
Suite 8
78 Mulgrave Road
PO Box 751 Cairns QLD 4870
Tel : (61-7) 4051 7511
Fax: (61-7) 4051 7611

CANBERRA
Suite 702
54 Marcus Clarke Street
Canberra Act 2600
Tel : (61-2) 6257 4428
Fax: (61-2) 6247 1468

DARWIN
PO Box 3419
Darwin NT 0801
Suite 1A, Level 1 CML Building
59 Smith Street
Darwin NT 0800
Tel : (61-8) 8981 8020
Fax: (61-8) 8941 1092

HOBART
53 Salamanca Place
Hobart TAS 7000
Tel : (61-3) 6234 8788
Fax: (61-3) 6231 1429

MELBOURNE
Level 20, 350 Queen Street
Melbourne VIC 3000
Tel : (61-3) 9933 8800
Fax: (61-3) 9933 8801

PERTH
Level 8, 251 Adelaide Terrace
Perth WA 6000
Tel : (61-8) 9221 8870
Fax: (61-8) 9221 8871

SUNSHINE COAST
Suite 6, 94 Memorial Avenue
Maroochydore QLD 4558
Tel : (61-7) 5479 2005
Fax: (61-7) 5479 5949

SYDNEY
Level 5
100 Pacific Highway
North Sydney NSW 2060
Tel : (61-2) 9956 8822
Fax: (61-2) 9956 8848

TOWNSVILLE
Unit 1, 2 Mcllwraith Street
South Townsville QLD 4810
Tel : (61-7) 4721 2788
Fax: (61-7) 4721 3766

DAVIS LANGDON NEW ZEALAND LTD

AUCKLAND
Level 10, Citigroup Centre
23 Customs Street East
Auckland 1010
Postal Address:
P O Box 935
Shortland Street
Auckland 1140
Tel : (64-9) 379 9903
Fax: (64-9) 309 9814

CHRISTCHURCH
Level 1, 93 – 95 Cambridge Terrace
Christchurch 8013
Postal Address:
P O Box 3166
Christchurch Mail Centre
Christchurch 8140
Tel : (64-3) 366 2669
Fax: (64-3) 366 9231

WELLINGTON
Level 15, 49 Boulcott Street
Wellington 6011
Postal Address:
P O Box 358
Wellington 6140
Tel : (64-4) 472 7505
Fax: (64-4) 473 3778

MIDDLE EAST AND LEBANON

DAVIS LANGDON

BAHRAIN - MANAMA
Al Saffar House
Unit 22B, Building No. 1042
Block 436, Road 3621, Seef District
P O Box 640, Manama
Kingdom of Bahrain
Tel : (973-17) 588 796
Fax: (973-17) 581 288

LEBANON - BEIRUT
1st Floor, Chatilla Building
Australia Street
Rawche, Shouran
P O Box 13/5422-Shouran
Beirut, Lebanon
Tel (96-11) 780 111
Fax: (96-11) 809 045

UAE – ABU DHABI
Villa 213/3
25th Street, Mushrif
P O Box 113971
Abu Dhabi, United Arab Emirates
Tel (971-2) 444 2040
Fax: (971-2) 444 2039

UAE - DUBAI
Level 7, Building C/P 54
Dubai Healthcare City
P O Box 7856
Dubai, United Arab Emirates
Tel (971-4) 324 3690
Fax: (971-4) 324 3691

QATAR - DOHA
Salwa Commercial Complex Building
1st Floor, Behind Seal Building,
Salwa Road
P O Box 3206
Doha, State of Qatar
Tel : (974) 458 0150
Fax: (974) 469 7905

DAVIS LANGDON & SEAH INTERNATIONAL

EUROPE

DAVIS LANGDON EDETCO

SPAIN - BARCELONA
C/Muntaner, 479, 1-2
Barcelona 08021, Spain
Tel : (34-93) 418 6899
Fax: (34-93) 211 0003

RUPERTI PROJECT SERVICES INTERNATIONAL

RUSSIA - MOSCOW
8th March Street
Bld. 6A Block 1
Moscow 127083
Russia
Tel : (7-495) 983 0850
Fax: (7-495) 983 0851

SOUTH AFRICA

DAVIS LANGDON

BLOEMFONTEIN
4 Langeberg Avenue
Bainsviei
Bloemfontein
Tel : (27-51) 451 1548
Fax: (27-51) 451 1832

BOTSWANA
Plot 20620 Acts House – Unit 9
Samedupe Road
Broadhurst Industrial
Gaborone, Botswana
Tel : (267) 390 0711
Fax: (267) 395 7550

CAPE TOWN
45 Buitengracht Street
Cape Town 8000
Tel : (27-21) 423 7840
Fax: (27-21) 423 7841

DURBAN
First Floor, 17 The Boulevard
Westway Office Park
Westville 3630
Tel : (27-31) 275 4200
Fax: (27-31) 265 0038

GEORGE
97 Mitchell Street
George 6529
Tel : (27-44) 873 5070
Fax: (27-44) 873 3931

JOHANNESBURG
3rd Floor, MPF House
32 Princess of Wales Terrace
Sunnyside Office Park
Parktown
Tel : (27-11) 484 2330
Fax: (27-11) 484 2361

KLERKSDORP
Second Floor
22 Boom Street
Klerksdorp 2570
Tel : (27-18) 464 1641
Fax: (27-18) 464 1644

MOZAMBIQUE
Rua D. Esterao de Ataide No. 38/48
Sommerschield 1
Maputo Mozambique
Tel : (258-21) 490 696
Fax: (258-21) 490 699

PIETERMARTIZBURG
300 Loop Street
Pietermartizburg 3201
Tel : (27-33) 345 8371
Fax: (27-33) 394 9201

PORT SHEPSTONE
Suite 7 Portston Centre
Aiken Street
Port Shepstone 4240
Tel : (27-39) 682 4114
Fax: (27-39) 682 4114

PRETORIA
First Floor, Lakeview 2
Nieuw Muckleneuk
Pretoria
Tel : (27-12) 460 5100
Fax: (27-12) 460 5677

RICHARDS BAY
Zululand Chamber of Business
Foundation Office Park Alton
South Central Arterial
Alton Buscom Building
Richards Bay
Tel : (27-35) 797 3039
Fax: (27-35) 797 3977

STELLENBOSCH
Town Square
9 Electron Street
Technopark
Stellenbosch 7599
Tel : (27-21) 880 8300
Fax: (27-21) 880 2984

VANDERBIJLPARK
Delfos Boulevard
Iscor Works (Engineering Building)
Vanderbijlpark
Tel : (27-16) 889 4159
Fax: (27-16) 889 4159

DAVIS LANGDON & SEAH INTERNATIONAL

UNITED STATES OF AMERICA

DAVIS LANGDON

BOSTON
211 Congress Street
Suite 202
Boston
MA 02110
Tel : (1-617) 357 1496
Fax: (1-617) 357 1812

HONOLULU
841 Bishop Street
Suite 1560
Honolulu
HI 96813
Tel : (1-808) 536 6100
Fax: (1-808) 536 6135

LOS ANGELES - SANTA MONICA
301 Arizona Avenue
Suite 301
Santa Monica
California 90401
Tel : (1-310) 393 9411
Fax: (1-310) 393 7493

NEW YORK
370 Lexington Avenue,
Suite 2500
New York
NY 10017
Tel : (1-212) 697 1340
Fax: (1-212) 697 1344

PHILADELPHIA
1717 Arch Street
Suite 3430
Philadelphia
PA 19103
Tel : (1-215) 564 3104
Fax: (1-215) 564 3109

SACRAMENTO
1331 Garden Highway
Suite 310
Sacramento
CA 95833
Tel : (1-916) 925 8335
Fax: (1-916) 925 0989

SAN FRANCISCO
343 Sansome Street
Suite 1050
San Francisco
CA 94104
Tel : (1-415) 981 1004
Fax: (1-415) 981 1419

SEATTLE
719 2nd Avenue
Suite 400
Seattle
WA 98104
Tel : (1-206) 343 8119
Fax: (1-206) 343 8541

DAVIS LANGDON & SEAH INTERNATIONAL

UNITED KINGDOM

DAVIS LANGDON LLP

BIRMINGHAM
Colmore Plaza
Colmore Circus Queensway
Birmingham
B4 6AT
Tel : (44-121) 710 1100
Fax: (44-121) 710 1399

BRISTOL
St Lawrence House
29/31 Broad House
Bristol
BS1 2HF
Tel : (44-117) 927 7832
Fax: (44-117) 925 1350

CAMBRIDGE
36 Storey's Way
Cambridge
CB3 ODT
Tel : (44-1223) 351 258
Fax: (44-1223) 321 002

CARDIFF
4 Pierhead Street
Capital Waterside
Cardiff
CF10 4QP
Tel : (44-29) 2049 7497
Fax: (44-29) 2049 7111

EDINBURGH
5th Floor
40 Princes Street
Edinburgh
EH2 2BY
Tel : (44-131) 550 9440
Fax: (44-131) 550 9499

GLASGOW
7th Floor
Aurora
120 Bothwell Street
Glasgow
G2 7JS
Tel : (44-141) 248 0300
Fax: (44-141) 248 0303

HEATHROW
East Wing Level 5
3 World Business Centre
1208 Newall Road
Heathrow
TW6 2TA
Tel : (44-20) 8564 6640
Fax: (44-20) 8759 3243

LEEDS
No 4 The Embankment
Victoria Wharf
Sovereign Street
Leeds
LS1 4BA
Tel : (44-113) 243 2481
Fax: (44-113) 242 4601

LIVERPOOL
Cunard Building
Water Street
Liverpool
L3 1JR
Tel : (44-151) 236 1992
Fax: (44-151) 227 5401

LONDON
MidCity Place
71 High Holborn
London
WC1V 6QS
Tel : (44-20) 7061 7400
Fax :(44-20) 7061 7061

MAIDSTONE
42 Kings Hill Avenue
Kings Hill
West Malling
Kent
ME19 4AJ
Tel : (44-1732) 840 429
Fax: (44-1732) 842 305

MANCHESTER
Cloister House
Riverside
New Bailey Street
Manchester
M3 5AG
Tel : (44-161) 819 7600
Fax: (44-161) 819 1818

MILTON KEYNES
Level 5
Metropolitan House
321 Avebury Boulevard
Milton Keynes
MK9 2GA
Tel : (44-1908) 304 700
Fax: (44-1908) 660 059

NORWICH
63 Thorpe Road
Norwich
NR1 1UD
Tel : (44-1603) 628 194
Fax: (44-1603) 615 928

OXFORD
Avalon House
Marcham Road, Abingdon
Oxon
OX14 1TZ
Tel : (44-1235) 555 025
Fax: (44-1235) 554 909

PETERBOROUGH
Clarence House
Minerva Business Park
Lynchwood
Peterborough
PE2 6FT
Tel : (44-1733) 362 000
Fax: (44-1733) 230 875

PLYMOUTH
1 Ensign House
Parkway Court
Longbridge Road
Plymouth
PL6 8LR
Tel : (44-1752) 827 444
Fax: (44-1752) 221 219

SOUTHAMPTON
Brunswick House
Brunswick Place
Southampton
S015 2AP
Tel : (44-23) 8033 3438
Fax: (44-23) 8022 6099

DAVIS LANGDON & SEAH INTERNATIONAL

IRELAND

DAVIS LANGDON PKS

CORK
Hibernian House
80A South Mall
Cork
Tel : (353-21) 422 2800
Fax: (353-21) 422 2801

DUBLIN
24 Lower Hatch Street
Dublin 2
Ireland
Tel : (353-1) 676 3671
Fax: (353-1) 676 3672

GALWAY
Heritage Hall
Kirwan's Lane
Galway
Tel : (353-91) 530 199
Fax: (353-91) 530 198

LIMERICK
Mezzanine Suite
Riverpoint
Lower Mallow Street
Limerick
Tel : (353-61) 318 870
Fax: (353-61) 318 871

PART ONE: REGIONAL OVERVIEW

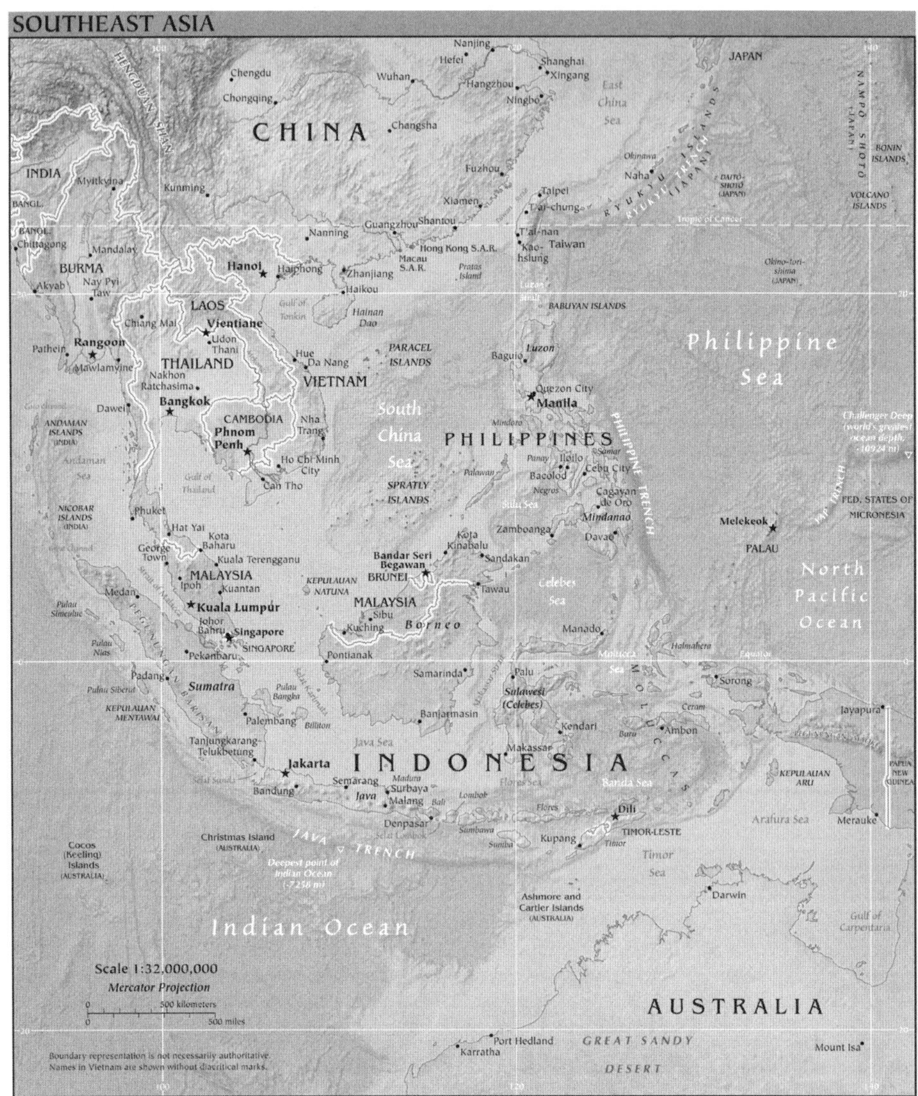

Source: www.cia.gov

1. The construction industry in the Asia Pacific region

INTRODUCTION

This book covers twenty major countries bordering the Pacific Ocean. It also includes the UK for comparative purposes. The region includes the two major industrialized powers of the USA and Japan, various Asian countries varying from the vast area of China and India to tiny Brunei Darussalam and Singapore as well as Australia and New Zealand. The levels of development range from some of the poorest countries of the world to the very richest and from long established market economies to centrally planned and transition economies.

Year 2007/2008 have seen dramatic changes in the world economic landscape. The turbulence in financial markets started from the USA sub-prime crisis has caused rippling effects beyond the USA, affecting Europe, India, China, Japan and many other countries. It has resulted in uncertainties in the global economic outlook with many countries entering into recession.

As a result of the global economic crisis, several fiscal stimulus packages were announced by many countries, in the hope that such packages would be able to provide a fillip to the countries' economy amid this recession crisis. So far, more than US$1 trillion (S$1.5 trillion) has been pledged by various countries in a bid to jump start their waning economies.

This introduction describes the international groupings of the countries, their key characteristics and various measures of their standard of living. The section on Construction Output and the Economy relates value of construction output to Gross Domestic Product (GDP) and to levels of investment for each country. It includes some data on housing stock. Finally there is a section on the Organization of the Construction Sector.

The main active regional grouping is the Association of South East Asian Nations (ASEAN) established in 1967 with the aims of accelerating economic growth, social progress and cultural development; the promotion of collaboration and mutual assistance in matters of common interest; and the continuing stability of the region.

The selection of the countries included in this book is in part based on their importance and in part on the availability of data both published and unpublished. Table A lists the countries, indicating their membership of international groupings and whether they have formal Davis Langdon & Seah International (DLSI) offices or are represented by a firm associated with the DLSI Group.

Table A: MEMBERSHIP OF INTERGOVERNMENTAL ORGANIZATIONS AND DLSI REPRESENTATION

Country	*ASEAN*	*OECD*	*DLSI offices*
Australia		X	X
Brunei	X		X
Cambodia	X		
China			X
Hong Kong			X
India			X
Indonesia	X		X
Japan		X	X
Malaysia	X		X
New Zealand		X	X
Pakistan			X
Philippines	X		X
Singapore	X		X
South Korea		X	X
Sri Lanka			
Taiwan			
Thailand	X		X
UK		X	X
USA		X	X
Vietnam	X		X

Sources: OECD, Organization for Economic Co-operation and Development
ASEAN, Association of South East Asian Nations

Table B, in summarising the key characteristics of the countries included in this volume highlights their diversity. Populations range from around 400,000 in Brunei to more than 1.3 billion in China. The smallest country in terms of area is Singapore, followed by Hong Kong and the largest, China, followed by the USA. Density of population is high in the smallest countries but only 3 persons per square kilometre in Australia, 33 persons in the USA, 140 in China and 350 in India. Definitions of the statistical terms in Table B and others used in this book are discussed in Part Two Section 2.

Because of the need to use monetary values in a common currency, GDP at nominal exchange rates is not always the best indicator of standard of living. Table C compares the money value of GDP per capita with estimates of GDP per capita on a purchasing power parity (PPP) basis, that is an approximation to what the GDP per capita will actually buy in the respective countries. In Table C, the countries are arranged as in Table B but in fact it is clear that on a PPP basis the order changes noticeably.

Table B: KEY CHARACTERISTICS OF COUNTRIES

Country	*Population mn*	*Land area 000 km²*	*Population per km²*	*GDP US$ bn 2007*	*GDP per capita US$ 2007*
Australia	21.2	7,686.85	2.76	712.93	33,628.95
Brunei	0.4	5.77	67.65	12.77	32,735.86
Cambodia	14.4	181.04	79.54	8.49	591.37
China	1,321.0	9,425.29	140.15	3,681.99	2,772.18
Hong Kong	7.0	1.08	6,441.15	208.52	30,110.71
India	1,149.0	3,287.59	349.50	3,100.00	2,600.00
Indonesia	225.6	1,890.75	119.32	360.52	1,410.48
Japan	127.7	377.84	337.98	5,804.37	32,250.97
Malaysia	27.2	330.25	82.36	133.26	6,446.91
New Zealand	4.3	268.67	15.97	78.50	18,280.92
Pakistan	164.7	803.94	204.92	143.65	2,600.00
Philippines	88.6	300.00	295.33	154.78	1,746.94
Singapore	4.6	0.71	6,506.36	153.76	35,566.44
South Korea	48.6	99.68	487.57	801.47	18,250.66
Sri Lanka	20.0	65.61	304.83	32.58	1,629.94
Taiwan	23.0	36.19	636.66	374.13	16,238.28
Thailand	66.0	514.00	128.48	242.38	3,670.29
UK	61.0	244.82	249.08	2,257.81	35,604.69
USA	304.1	9,161.63	33.19	14,264.60	47,395.00
Vietnam	86.2	331.69	259.76	86.00	998.08

Table C: GDP PER CAPITA ON A PPP AND MONEY BASIS, 2007

Country	*GDP per capita on PPP basis* US$*	*Index USA=100*	*Rank Order*	*GDP per capita US$*	*Rank Order*
Australia	37,700	80.56	16	33,628.95	17
Brunei	53,700	114.74	20	32,735.86	16
Cambodia	1,900	4.06	1	591.37	1
China	5,500	11.75	8	2,772.18	8
Hong Kong	42,900	91.67	17	30,110.71	14
India	2,700	5.77	3	2,600.00	6
Indonesia	3,700	7.91	6	1,410.48	3
Japan	34,300	73.29	14	32,250.97	15
Malaysia	14,800	31.62	10	6,446.91	10
New Zealand	28,200	60.26	12	18,280.92	13
Pakistan	2,500	5.34	2	2,600.00	6
Philippines	3,300	7.05	5	1,746.94	5
Singapore	51,200	109.40	19	35,566.44	18
South Korea	25,500	54.49	11	18,250.66	12
Sri Lanka	4,100	8.76	7	1,629.94	4
Taiwan	31,100	66.45	13	16,238.28	11
Thailand	8,200	17.52	9	3,670.29	9
UK	36,500	77.99	15	35,604.69	19
USA	46,800	100.00	18	47,395.00	20
Vietnam	2,700	5.77	3	998.08	2

*Source: * – www.cia.gov*

Other factors are also relevant in assessing the standard of living and quality of life. Table D, overleaf, shows some non-monetary indicators of development.

Table D: SELECTED NON-MONETARY INDICATORS OF DEVELOPMENT

Country	*Population* *mn*	*Internet users* *2007*	*Internet users as a % of population* *2007*	*Infant mortality rate (per 1,000 live births)* *2009*
Australia	21.2	11,240,000	53.0	4.75
Brunei	0.4	199,532	49.9	12.27
Cambodia	14.4	70,000	0.5	54.79
China	1,321.0	253,000,000 *	19.2	20.25
Hong Kong	7.0	3,961,000	56.6	2.92
India	1,149.0	80,000,000	7.0	30.15
Indonesia	225.6	13,000,000	5.8	29.97
Japan	127.7	88,110,000	69.0	2.79
Malaysia	27.2	15,868,000	58.3	15.87
New Zealand	4.3	3,360,000	78.1	4.92
Pakistan	164.7	17,500,000	10.6	65.14
Philippines	88.6	5,300,000	6.0	20.56
Singapore	4.6	3,105,000	67.5	2.31
South Korea	48.6	35,590,000	73.2	4.26
Sri Lanka	20.0	771,700	3.9	18.57
Taiwan	23.0	14,760,000	64.2	5.35
Thailand	66.0	13,416,000	20.3	17.63
UK	61.0	40,200,000	65.9	4.85
USA	304.1	223,000,000 *	73.3	6.26
Vietnam	86.2	17,870,000	20.7	22.88

Source: www.cia.gov
** Figures are for years other than those specified*

None of the factors reviewed above indicates the financial viability of the economies. Table E, overleaf, shows external debt and gross domestic investment, both as a percentage of GDP, for the countries for which information is available.

Table E: FOREIGN DEBT AND INVESTMENT

Country	*External debt as a % of GDP * 2007/2008*	*Gross domestic investment as a % of GDP*
Australia	144.8	18.0
Brunei	0.0	8.8
Cambodia	50.8	20.8
China	11.4	42.3
Hong Kong	37.8	20.3
India	5.3	33.8
Indonesia	42.1	24.9
Japan	25.7	29.6
Malaysia	40.6	23.4
New Zealand	75.3	22.0
Pakistan	30.1	23.0
Philippines	34.6	5.3
Singapore	16.6	45.0
South Korea	47.5	25.0
Sri Lanka	39.9	29.0
Taiwan	29.3	16.6
Thailand	26.7	26.5
UK	462.8	17.1
USA	85.9	13.7
Vietnam	27.6	43.1

*Source: * – www.cia.gov*

CONSTRUCTION OUTPUT AND THE ECONOMY

Table F, overleaf, shows the relationships of investment, gross construction output and net construction output to GDP and to each other. Gross output is the total value of construction produced; net output is gross output minus the inputs from other industries (see Statistical Notes in Section 2). These inputs are mainly materials but also plant and equipment and other goods or services. Thus net output consists mainly of labour, management costs and profits.

The gross and net output figures are those contained in the country sections. For some countries a considerable amount of estimation is involved. Generally the national accounts contain estimates of the contribution of construction to GDP which is net output. It is often difficult to obtain estimates of gross construction output for developing countries. Because of the difficulty in gathering data on construction any figures of gross and net output are subject to considerable margins of error. These are a result of large numbers of small projects particularly in renovation and repair and maintenance; the wide geographical distribution of construction activity; the fact that the price of a construction project is not always determined in one operation and changes may not be recorded; and the large number of construction clients and construction firms.

Table F: INVESTMENT AND CONSTRUCTION OUTPUT RELATED TO GDP

Country	*Net construction output as a % of GDP*	*Gross construction output as a % of GDP*	*Net construction output as a % of gross output*	*Gross domestic investment as a % of GDP*
Australia	-	10.9	-	18.0
Brunei	-	7.5	-	8.8
Cambodia	6.7	-	-	20.8
China	5.6	20.3	27.4	42.3
Hong Kong	2.3	5.8	40.8	20.3
India	-	-	-	33.8
Indonesia	1.7	7.7	22.1	24.9
Japan	-	8.8	-	29.6
Malaysia	1.6	-	-	23.4
New Zealand	4.6	9.7	47.7	22.0
Pakistan	-	1.3	-	23.0
Philippines	3.4	8.1	42.1	5.3
Singapore	3.7	7.8	47.2	45.0
South Korea	18.3	27.4	66.6	25.0
Sri Lanka	4.0	7.4	54.1	29.0
Taiwan	2.2	2.3	94.1	16.6
Thailand	2.9	8.5	34.2	26.5
UK	-	8.4	-	17.1
USA	-	8.1	-	13.7
Vietnam	6.1	6.5	-	43.1

It is possible however, to make estimates of gross output especially if data for net output and investment are available. The relationship between gross construction output and net construction output depends on:

- the work mix. Some work is more labour intensive than other, for example repair and maintenance in the UK probably involves about three times more labour input than new civil engineering work.
- the sophistication of construction including the extent of use of capital equipment.
- wage rates and productivity.

Thus, a country which has low wage rates and high productivity would be expected to have low net output in relation to gross output but in fact low wage rates are often combined with low productivity so that the effects to some extent cancel each other out. A country with a sophisticated construction product probably uses a high level of equipment and expensive materials so that the tendency would be for net output to be a low proportion of gross output. Such a country however probably also has high wage rates and high productivity thus compensating to some extent for the high input costs. In general the proportion of net output of gross output is around 50%.

Another factor to be taken into account is the relationship between construction new work output and total investment. In most countries, construction accounts for about half of all investment and is likely to be higher in less developed countries than in developed ones because the construction industry provides much of the very basic infrastructure.

Where the authors have estimated gross output they have generally done so by estimating the percentage which it is likely to take of GDP bearing in mind the proportion accounted for by net output and by total investment.

Considering the individual countries in Table F the preceding general statements may be seen reflected in the figures.

TOP 225 INTERNATIONAL CONSTRUCTION COMPANIES

Every year, the *Engineering News Record* (ENR) ranks the 225 largest construction contracting companies from around the world. The table below shows the number of construction companies that are listed in the ENR's Top 225 International Construction Companies for the following countries. The companies are ranked according to construction revenue generated outside of each company's home country in 2007 in US dollars.

Table G: NUMBER OF CONSTRUCTION COMPANIES IN TOP 225 INTERNATIONAL COMPANIES, 2008

Country	*Number of construction companies in top 225 companies*
Australia	4
Brunei	-
Cambodia	-
China	51
Hong Kong	-
India	2
Indonesia	-
Japan	16
Malaysia	-
New Zealand	-
Pakistan	1
Philippines	-
Singapore	-
South Korea	11
Sri Lanka	-
Taiwan	1
Thailand	-
UK	4
USA	35
Vietnam	-

Source: www.enr.construction.com

It can be seen that China, USA, and Japan have a large number of construction companies that operate internationally.

PART TWO: INDIVIDUAL COUNTRIES

2. Introductory notes to country sections

INTRODUCTION

In this part of the book, twenty countries are arranged alphabetically, and each country is presented as far as possible in a similar format, under five main headings – Key data, The construction industry, Construction cost data, Exchange rates and inflation, and Useful addresses. These notes introduce the five main sets of information presented on the individual countries and provide, in one place, general notes, definitions and explanations, in order to keep the individual country sections as succinct as possible. A final heading, Statistical notes, discusses and explains the statistical definitions and concepts adopted in the book.

KEY DATA

The key data sheet at the start of each country lists main population, geographic, economic and construction indicators and thus provides a brief statistical overview of that country. In many cases data produced by national statistical offices have been used; in other cases, UN or World Bank sources have been relied on. Some estimates are included for construction data especially for gross construction output. The methods are discussed in Part One – The Construction Industry in the Asia Pacific Region. In Part Three, Comparative Data, international agency data have been used throughout in order to ensure consistency. Further notes on economic indicators are provided below in the Statistical notes.

THE CONSTRUCTION INDUSTRY

The main topics covered in this section are the contribution of the industry to the economy; the structure of the industry; the availability of and constraints on construction labour and materials; tendering and contract procedures and standards.

Although construction is often fragmented and tends to be labour intensive with low capital investment, it is invariably the single largest industry in a country. In most countries the net output of construction contributes between 2% and 18% to Gross Domestic Product (GDP) and a similar percentage to direct construction employment (indirect employment – in the construction materials industries and other related activities – can more than double the contribution). Gross construction output including materials and plant and equipment is normally around twice net output but the range is quite wide and the reasons are discussed in Part One – The Construction Industry in the Asia Pacific Region.

CONSTRUCTION COST DATA

This section includes both construction costs incurred by contractors and the costs they charge their clients. The costs of labour and materials are input costs of construction, i.e. the costs incurred by contractors.

Unit costs, measured rates for construction work and approximate estimating costs per square metre are output costs, i.e. the costs contractors charge their clients. Problems of definition make meaningful and consistent presentation of unit rates extremely difficult. For unit rates to be useful it is essential to be clear what is included and what is excluded. Notes are provided in each country section, for example, on the treatment of preliminary items and on the methods of measurement adopted for approximate estimating rates per square metre.

Cost of Labour and Materials

Typical costs for construction labour and materials are given in most country sections. Two figures are generally given for each grade of labour. The wage rate is the basis of an employee's income – his basic weekly wage will be the number of hours worked multiplied by his wage rate. The cost of labour, on the other hand, is the cost to the employer of employing that employee; it is also based on the wage rate but includes (where applicable) allowances for:

- incentive payments
- traveling time and fares
- lodging and subsistence
- public and annual holidays with pay
- training levies
- employer's liability and third party insurances
- health insurance
- payroll taxes
- other mandatory and voluntary payments.

The costs of main construction materials are given as delivered to site in quantities appropriate to a reasonably substantial building project. It is presumed that there are no particular difficulties of access which would significantly affect costs. Generally tax, and particularly any value added tax, is excluded from material costs mainly because the rate of tax to be levied may depend on the type of work in which the material is to be incorporated.

Unit Rates

Rates for a variety of commonly occurring construction items are provided for most countries. They are usually based on a major, if not the capital, city and the relevant date is always fourth quarter of 2008. Rates generally include all necessary labour, materials, plant and equipment and, where appropriate, allowances for contractors' overheads and profit, preliminary and general items associated with site set-up, etc. and contractors' profit and attendance on specialist trades. Where the basis of rates is different from this, notes are provided in the text in each country section. Value added tax and other taxes are excluded. The rates are appropriate to a reasonably substantial building project.

In the country sections abbreviated descriptions are given for each work item; a full description of each work item is presented in section 4.

Approximate Estimating

Approximate estimating costs per unit area (square metre and square feet) are given for most countries for a variety of building types. Notes on the method of measurement and what is or is not included in unit rates are provided in each country section. Areas generally are measured on all floors inside external walls and with no deduction for internal walls, columns, etc. Where this is not the case it is noted. Generally tax, and particularly value added tax, is excluded in approximate estimating costs.

When making comparisons of construction costs between countries it is important to be clear about what is being compared. There are two main methods of comparison: first the comparison of identical buildings in each country and, second, the comparison of functionally similar buildings in each country. In the country sections, the approximate estimating rates given are for the standard of building of each type normally built in that country. Rates are therefore closer to the 'functionally similar' approach. The rate per square metre given for an office building, for example, or a warehouse in any particular country refers to the normal type of office building or warehouse built in that country. In country sections they are presented in national currencies. A selection of approximate estimating costs are also presented in pound sterling, US dollar and 100 Japanese yen equivalents in Part Three thus enabling comparisons on a common currency basis to be made.

EXCHANGE RATES AND INFLATION

Exchange Rates

Currency exchange rates are important when comparing costs between one country and another. While it is most useful to consider costs within a country in that country's currency, it is necessary, from time to time, to use a common currency in order to compare one country's costs with another. But exchange rates can fluctuate dramatically and few currencies (even those considered strong) can be considered really stable. It can be risky to think in terms, for example, of one country being consistently a set percentage more or less expensive than another.

Different rates of internal inflation affect the relative values of currencies and, therefore, the rates of exchange between them. However, the reasons behind exchange rate fluctuations are complex and often political as much as economic; they include such factors as interest rates, balance of payments, trade figures and, of course, government intervention in the foreign exchange markets, and, for that matter, other government actions.

Graphs of exchange rates since 1998 against the Pound Sterling, Euro, the US Dollar and the Japanese Yen are included for most countries. They have been calculated by averaging the published weekly values in each quarter. The values given are therefore smoothed – the most dramatic peaks and troughs have been ironed out. They are, however, useful for indicating long term trends. As far as possible, the form of the graph is kept the same; hence the vertical scale is adjusted to accommodate different currencies. It should always be checked whether marked movement in a graph is a result of erratic exchange rates or merely the selected vertical scale.

If a line moves up from left to right (for example, the Vietnam Dong against the Pound Sterling – see below) it indicates that the subject currency (the Vietnam Dong) is declining in value against the currency of the line (the Pound Sterling). The higher the line is, the more subject currency is required to purchase the line currency. If, on the other hand, a line moves down from left to right (for example, the Australian Dollar against the Pound Sterling) it indicates the subject currency is strengthening against the line currency. Where there is virtually no movement at all, that is the line is horizontal, this usually indicates a currency effectively 'tied' or 'pegged' to the line currency.

EXCHANGE RATE GRAPH

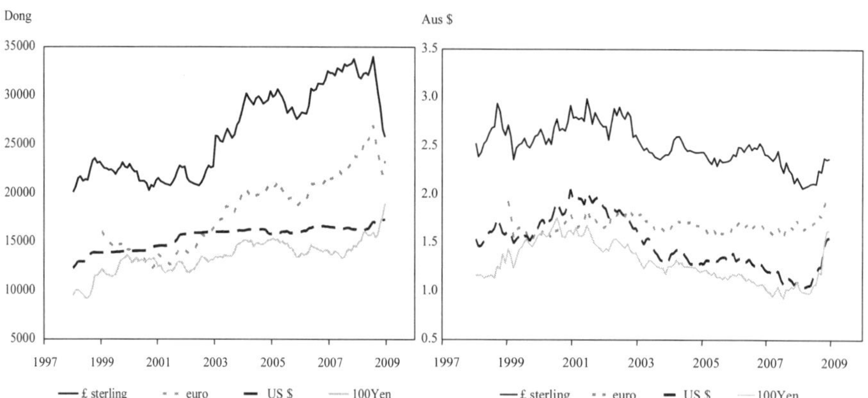

Inflation

General inflation has been measured using consumer or retail price indices. These reflect price changes in a basket of goods and services weighted according to the spending patterns of a typical family. Weights are changed periodically, and new items inserted. General inflation indices usually rise and, in so doing erode the purchasing power of a given currency unit. Other measures of inflation tend to be related to specific items. The two most commonly prepared for the construction industry are discussed below.

Cost and tender price indices measure different types of inflation which occur within the construction industry. Building costs are the costs actually incurred by a contractor in the course of his business, the major ones being labour and materials; tender prices are the prices for which a contractor offers to erect a building. Tender prices include building costs but also take into account the prevailing market situation. When there is plenty of construction work tender prices may increase at a greater rate than building costs while, when work is scarce, tender prices may actually fall even if building costs are rising.

Most countries have building cost indices – the method of compilation is generally relatively simple, basically comprising a weighted basket of the main inputs to construction. Rather fewer countries have tender price indices – their method of compilation is more complex usually involving a detailed analysis of accepted tenders for construction work and comparing these with a common base schedule of prices. When construction indices are described as price indices it is not always clear what these are.

USEFUL ADDRESSES

At the end of each country section, a list of addresses is given. This usually comprises main government, contracting, professional, standards and research organizations involved in the construction industry.

STATISTICAL NOTES

Gross Domestic Product (GDP) is the total value of all the goods and services produced in a country. Thus it shows the wealth generated within a country. Gross National Product (GNP) is the total value of all the goods and services produced in a country plus or minus net income from outside. Thus it represents the total amount of income available to the population. Reasons why GNP can be greater than GDP include that nationals abroad send back money, that the country receives aid or that the country has an income from investments abroad. Debt repayment and payment of interest can make GNP less than GDP.

It is appropriate to use GNP as a measure of wealth when income is being considered, e.g. in allocation for various purposes. GDP is more appropriate where productive capacity is being considered. Because a primary focus of this book is on the productive capacity of the construction industry in the key data sheets for each country the emphasis is on GDP, although there are a few countries where only statistics of GNP are available.

In considering expenditure the data for private and public consumption and investment are expressed as a proportion of GDP. This is partly because the main source for this expresses it this way and use of this one source gives a consistent picture. Because expenditure is made out of GNP the three percentages do not always total to 100.

Data on construction output for most countries are available in the form of net output or value added, that is, broadly, gross value of construction output minus the value of the material input, and the cost of plant and equipment.

However, the method of arriving at these data and gross value of construction output varies from country to country and is sometimes so indirect that it is of dubious reliability/quality. Both gross and net construction output are given where possible. The authors have made estimates based on relationships to other indicators and past data of gross output where reliable data are not available and have for some countries also estimated net output.

The exchange rates given in the key data are those which are appropriate for use with the cost data. For conversion of figures for a year, e.g. GDP in 2008, the mid-year exchange rate has been used. Because a single yen has a small value compared to the US dollar or the pound sterling, a rate for 100 yen is given in each case. Purchasing power parity (PPP) is the exchange rate which would be appropriate to express an income in one country in terms of its purchasing power in another country.

All the statistics are subject to considerable margins of error but particularly so for the less developed countries. As soon as they are converted from national currencies to US$ in order to permit comparison, the difficulty arises that the exchange rate may not reflect the purchasing power parity (PPP). In using exchange rates to convert value of construction output the difficulty is greater because of the greater specificity of production. The statistics sometimes do not reflect the real situation, and this is the problem which exists in a greater or lesser degree for all countries. Indeed even taking authoritative sources, variations of a factor of eight are possible. Table C in section 1 shows a comparison of money and purchasing power parity GNP or GDP. The statistics in this volume are those considered by the editors to be as accurate and as representative of the real situation as possible.

3. Individual countries

DAVIS LANGDON

Davis Langdon manages client requirements, controls risk, manages cost and maximises value for money, throughout the course of construction projects, always aiming to be – and to deliver – the best.

TYPICAL PROJECT STAGES, INTEGRATED SERVICES AND THEIR EFFECT

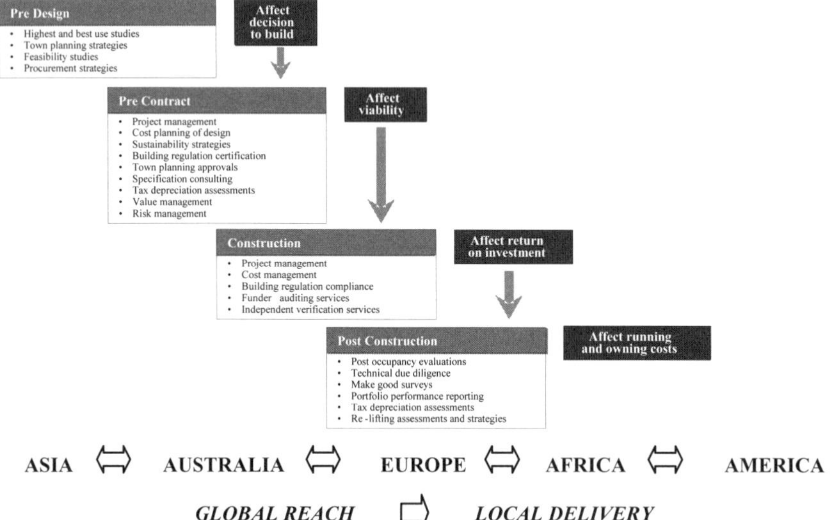

ASIA ⇔ AUSTRALIA ⇔ EUROPE ⇔ AFRICA ⇔ AMERICA

GLOBAL REACH ⇨ LOCAL DELIVERY

ADELAIDE OFFICE
Level 12, 25 Grenfell Street
Adelaide 5000
South Australia
Tel : (61-8) 8410 4044
Fax: (61-8) 8410 4166

BRISBANE OFFICE
Level 13, 324 Queen Street
Brisbane, Queensland 4000
Australia
Tel : (61-7) 3221 1788
Fax: (61-7) 3221 3417

CAIRNS OFFICE
Suite 8, 78 Mulgrave Road
PO Box 751,
Cairns, Queensland 4870
Australia
Tel : (61-7) 4051 7511
Fax: (61-7) 4051 7611

CANBERRA OFFICE
Suite 702, 54 Marcus Clarke Street
Canberra Australia
Capital Territory 2600
Australia
Tel : (61-2) 6257 4428
Fax: (61-2) 6247 1468

DARWIN OFFICE
Suite 1A, Level 1, CML Building
59 Smith Street, NT 0800, Australia
PO Box 3419, Darwin
NT 0801 Australia
Tel : (61-8) 8981 8020
Fax: (61-8) 8941 1092

HOBART OFFICE
53 Salamanca Place
Hobart Tasmania 7000
Australia
Tel : (61-3) 6234 8788
Fax: (61-3) 6231 1429

MELBOURNE OFFICE
Level 20, 350 Queen Street
Melbourne Victoria 3000
Australia
Tel : (61-3) 9933 8800
Fax: (61-3) 9933 8801

PERTH OFFICE
Level 8, 251 Adelaide Terrace
Perth Western Australia 6000
Australia
Tel : (61-8) 9221 8870
Fax: (61-8) 9221 8871

SUNSHINE COAST OFFICE
Suite 6, 94 Memorial Avenue
Maroochydore Queensland 4558
Australia
Tel : (61-7) 5479 2005
Fax: (61-7) 5479 5949

SYDNEY OFFICE
Level 5, 100 Pacific Highway
North Sydney
New South Wales 2060
Australia
Tel : (61-2) 9956 8822
Fax: (61-2) 9956 8848

TOWNSVILE OFFICE
Unit 1, 2 Mcllwraith Street
South Townsville
Queensland 4810
Australia
Tel : (61-7) 4721 2788
Fax: (61-7) 4721 3766

DAVIS LANGDON & SEAH INTERNATIONAL
www.davislangdon.com

AUSTRALIA

All data relate to 2008 unless otherwise indicated.

Population	
Population	21.2 mn
Urban population (2005)	75%
Population under 15 (2006)	20%
Population 65 and over (2006)	13%
Average annual growth rate (2006 to 2007)	1.5%
Geography	
Land area	7,686,850 km^2
Agricultural area	60%
Capital city	Canberra
Population of capital city (2006)	0.32 mn
Largest city	Sydney
Population of largest city (2006)	4.1 mn
Economy	
Monetary unit	Australian Dollar (A$)
Exchange rate (average fourth quarter 2008) to:	
the pound sterling	A$ 2.37
the US dollar	A$ 1.52
the euro	A$ 1.93
the yen x 100	A$ 1.58
Average annual inflation (1999 to 2008)	3.3%
Inflation rate	3.7%
Gross Domestic Product (GDP)	A$ 1,083.66 bn
GDP per capita	A$ 51,116
Average annual real change in (GDP) (1999 to 2008)	3.6%
Private consumption as a proportion of GDP	57.0%
Public consumption as a proportion of GDP	22.0%
Investment as a proportion of GDP	18.0%
Construction	
Gross value of construction output (2007 to 2008)	A$ 117.89 bn

THE CONSTRUCTION INDUSTRY

Construction Output

Gross construction output was A$117.89 billion for the year 2007-08, equivalent to US$78 billion. Construction output is divided into three broad sectors: residential building, non-residential building and engineering construction. Table below shows the level of output in each sector at average 2007-08 prices.

CONSTRUCTION ACTIVITY BY SECTOR, 2007-2008

Type of Work	*A$ billion*	*%*
Public and Private		
Residential building	39.15	33
Non-residential building	27.05	23
Engineering construction	51.69	44
Total	117.89	100

Source: Australian Bureau of Statistics

The private sector accounts for more than half of the construction activity mainly for residential building with the remaining public sector largely involved in engineering construction. The public sector plays a small role in housing construction but provides a third of non-residential building.

The table below shows a breakdown of the value of building work completed for 2007-08.

VALUE OF BUILDING WORK DONE, 2007-2008 (CURRENT PRICES)

Type of building	*Value (A$ million)*	*Percentage of total building work (%)*
New residential	35,503	51
Alterations and additions to residential buildings	6,620	10
Non-residential building		
public sector	5,651	8
private sector	21,490	31
Total non-residential building	27,142	39
Total building	69,265	100

Source: Australian Bureau of Statistics

Residential building has been relatively flat due to general economic uncertainty and there remains an ongoing nationwide shortfall of 25,000 to 30,000 dwellings per annum (relative to demand).

The value of new residential building reached A$35.5 billion in 2007-08 (or A$42.1 billion including alterations and additions), accounting for 60.8% of all building work done and down 4% on 2006-07 figures.

Non-residential building increased 18.3% to reach A$27.1 billion in 2007-08, up from A$22.9 billion in 2006-07.

The overall value for building work recorded an increase of 9% between 2006-07 and 2007-08.

VALUE OF ENGINEERING CONSTRUCTION WORK DONE, 2007-2008 (CURRENT PRICES)

Type of work	*Total of Private and Public sectors (A$ million)*	*Percentage of total engineering construction (%)*
Roads, highways and subdivisions	12,074.1	20.5
Bridges, railways and harbours	5,665.0	9.6
Electricity generation, transmission and distribution	8,967.4	15.3
Water storage and supply, sewerage and drainage	6,422.3	10.9
Telecommunications	4,810.3	8.2
Heavy industry	18,550.1	31.6
Recreation and other	2,297.0	3.9
Total	58,786.2	100

Source: Engineering Construction Activity, Australia

Growth in the engineering construction industry has been significant across Australia as a result of Government commitment to spending and has focused mainly on roads, highways and heavy industry – accounting for 52% of the total value of engineering construction work done during 2007-08. The trend is expected to continue with the construction of new tollways, tunnels rail links and shipping channels.

As more services are privatised, the value of work done in the private sector is expected to increase. Simultaneously, as large scale power generation begins to shift into newer climate friendly technologies the value of work done in this sector is expected to continue rising.

Characteristics and Structure of the Industry

After years of experiencing a skills shortage, the Australian workforce is now starting to see unemployment rates rise – albeit modestly. Increasing to 4.9% in February 2009.

With 9% of the national labour force in construction which represents 7% of Australia's GDP. This is leading to a backlog in work and consequently higher wages which is reflected with an increase in apprenticeships over the past years.

Since the housing downturn in 2001, the number of working days lost due to industrial disputes per 1,000 employees in the construction industry has dropped from 389 to just 12 over the 2007-08 financial year adding to the profitability and attitudes toward industrial relations in the sector.

The major contractors and some of their characteristics are shown on the table below. Total assets and revenue are for Australian based activities only.

MAJOR CONTRACTORS IN AUSTRALIA, 2006-2007

Contractor	*Total Assets (A$ million)*	*Revenue (A$ million)*
Lend Lease	9,336.2	14,281.9
Leighton Holdings	4,745.2	11,891.5
Multiplex Group	7,914.7	3,552.3
Mirvac Group	7,352.6	2,220.9
Watpac	507.5	640.2
AV Jennings	528.7	632.2
Hutchinson Builders	224.6	584.4
Devine	454.1	549.4
Hansen Yuncken	142.5	540.0
FKP Property Group	3,512.5	417.5
Henley Properties	-	380.0
Probuild Constructions	141.6	280.8
LU Simon Builders	-	265.0
Becton Property Group	897.7	259.6
Construction Engineering	64.3	229.8

Source: Business Review Weekly – The Top 1000 Companies (Australia)

Architects usually undertake a complete design service. On large projects they are seldom responsible for supervision of work on site, generally considered the responsibility of the developer on large scale projects or a project manager for lesser scale projects.

Engineering services are increasingly becoming a considerable portion of a building's cost. The quest for sustainable buildings is driving new and emerging technologies with engineering services typically representing 40% to 50% of a project's building costs. Specialist engineering services cost advice is also on increased demand to achieve best value outcomes for the client in capital cost and also in recurrent operational costs.

Clients and Finance

Construction activity rose to A$117.89 billion, increasing 11.6% on the previous year or A$12.25 billion in real terms across 2007-08. Engineering and construction work contributed 43.85% of the total value, while residential building contributed

33.21% and non-residential contributed 22.95% with the public sector contributing 20.8% compared to the private sectors 79.2%.

Selection of Design Consultants

Within Australia the three main options for Design Team procurement can be considered as (a) separate appointment for each consultant; (b) single appointments for teams of consultants; or (c) single appointment for a lead consultant responsible for the provision of all design services through sub-consultant agreements.

There is an emerging pattern that certain clients are moving away from single appointment procurement and towards separate appointments or team appointments. There are benefits and disadvantages associated in each procurement method that should be fully evaluated by clients before deciding the most beneficial route. For example, in single appointment procurement clients gain increased certainty in performance through a defined scope of services specific to each consultant contract, whereas clients do not have the same level of visibility via the lead consultant approach.

To ensure transparency and uniformity in consultant procurement clients may adopt questionnaires, competitions or similar techniques to select organisations for tender to provide a thorough assessment tool and a clear audit trail. This will also serve to provide a uniform format for response which makes proposals more easily comparable. A questionnaire or competition approach, tailored to the specific project will also serve to focus the consultants and provide an early opportunity to assess their willingness and ability to perform.

Clients must decide upon what weighting criteria is to be given more priority in consultant selection assessing project needs against consultant's proven track record, ability to perform, resource commitment, price, reputation and previous references amongst other things.

Some states continue to have more formal procurement systems than others. In Tasmania, for example, the State Government keeps a register of consultants and has standard conditions of engagement and there are *State Government Purchasing Guidelines* for Queensland.

Professional bodies all publish fee scales but they are not mandatory and are not usually used except in Tasmania.

Contractual Arrangements

Contractual arrangements are influenced both by State and by Federal Governments and there are some 20 contract types in use.

Recent years have seen an escalation in public private partnerships (PPP) to deliver public infrastructure projects. The common approach starts with a public sector cost benefit analysis to determine that any particular infrastructure project is suitable for a PPP delivery process. Clarity of output requirement is paramount along with realistic recognition and allocation of project and operational risk and responsibilities. Project requirements may also include a predetermined ownership phase.

Project alliancing is a partnership-type model of procurement, where risks and responsibilities are collectively shared between government and service providers, in the delivery of major capital works projects. This form of contractual arrangement is more dependent on developing trust and strong relationships to drive performance where risk, profit or loss is equitably shared in pre-agreed ownership ratios.

Liability and Insurance

Construction projects are increasingly subject to more stringent regulations as a result of changing legislation, Occupational Health and Safety (OH&S) settlement claims and the adoption of potentially higher risk design and construction contracts. Numerous professional and construction based insurance is available and should be reviewed on a regular basis to ensure coverage against the continuously evolving sector and its influences.

Development Control and Standards

Certification of building projects may be undertaken by a Local Authority (Council or State Government) or by a private accredited building practitioner. The building approval process is similar among Australian States and Territories. One of the key issues faced by the development industry is the lengthy delays in the planning approval process. The appointment of private certifiers is becoming increasingly common to undertake the role of the local government to certify compliance with appropriate standards and grant building approval.

Although the Disability Discrimination Act (1992) is not encompassed within the legislative requirements of the Building Code, Local Councils are now very active in enforcing the Act as part of the planning approval for developments to avoid future litigating circumstances regarding accessibility issues.

The new Occupational Health and Safety Act is about the safety of workplaces throughout the life-cycle of a building. It is considered an uninsurable offense if a designer does not consider future safe occupation of a building. This legislation refers directly to the designer and therefore cannot be contracted out.

When the construction of a commercial building is complete, it is now a legislative requirement that the building owner maintains the essential services. In addition, all owners of buildings built post 1 July 1994, have an obligation to produce and display an essential services report on the buildings safety measures.

Standards for design and construction of buildings are set by the Building Code of Australia which is produced as a performance based regulatory document and administered by local authorities. Generally, building approval is necessary in order to erect a structure in Australia.

Environmentally Sustainable Development

Australia's recent ratification of the Kyoto Protocol, in addition to increasing energy costs, prolonged drought conditions and climate awareness, has resulted in the inclusion of Section J – Energy Efficiency in the building code. Buildings designs need to demonstrate compliance with these minimum standards which broadly include: building fabric, external glazing, building sealing, hot water supply, air-conditioning and ventilation systems, lighting and electrical power, and maintenance.

Environmental rating tools have been developed in order to compare buildings that aim to achieve national and world leading design practices. These include the emissions focused Australian Building Greenhouse Rating (ABGR), the National Australian Built Environment Rating System (NABERS) and the holistic approach of the Green Building Council of Australia's Green Star rating tool.

CONSTRUCTION COST DATA

Cost of Labour

The figures that follow are typical of labour costs in Melbourne as at the fourth quarter of 2008. The wage rate is the basis of an employee's income, while the cost of labour indicates the cost to a contractor of employing that employee. The difference between the two covers a variety of mandatory and voluntary contributions – a list of items which could be included is given in section 2.

	Wage rate (per week) A$	*Cost of labour (per hour) A$*	*Number of hours worked per year*
Site operatives			
Mason/bricklayer	3,400	85	1,552
Carpenter	3,400	85	1,552
Plumber	3,600	90	1,552
Electrician	3,800	95	1,552
Structural steel erector	3,400	85	1,552
HVAC installer	3,800	95	1,552
Semi-skilled worker	3,320	83	1,552
Unskilled labourer	3,200	80	1,552
Equipment operator	3,280	82	1,552
Watchman/security	1,600	40	1,552
Site supervision			
General foreman	4,200	105	1,552
Trades foreman	4,000	100	1,552

	Wage rate (per year) A$	*Cost of labour (per hour) A$*	*Number of hours worked per year*
Contractors' personnel			
Project manager	185,000	155	1,552
Site manager	165,000	140	1,552
Contract administrator/QS	150,000	100	1,552
Junior coordinator	100,000	80	1,552
Junior administrator	110,000	85	1,552
Planner	170,000	135	1,552
Consultants' personnel			
Senior architect	170,000	250	1,552
Senior engineer	170,000	250	1,552
Senior surveyor	170,000	250	1,552
Qualified architect	140,000	170	1,552
Qualified engineer	140,000	170	1,552
Qualified surveyor	120,000	150	1,552

Cost of Materials

The figures that follow are the costs of main construction materials, delivered to site in the Melbourne area, as incurred by contractors in the fourth quarter of 2008. These assume that the materials would be in quantities as required for a medium sized construction project and that the location of the works would be neither constrained nor remote.

	Unit	*Cost A$*
Concrete		
Ready mixed concrete (Grade 40MPa)	m³	159.00
Ready mixed concrete (Grade 25MPa)	m³	140.00
Steel		
Mild steel reinforcement	tonne	1,250.00
Structural steel sections	tonne	1,395.00
Structural steel RHS	tonne	1,900.00
Bricks and blocks		
Common bricks (230 x 110 x 76mm)	1000	515.00
Good quality facing bricks (230 x 110 x 76mm)	1000	920.00
Hollow concrete blocks (400 x 200 x 100mm)	1000	158.00
Solid concrete blocks (400 x 200 x 100mm)	1000	190.00
Precast concrete cladding units with exposed aggregate finish	m²	120.00
Precast concrete cladding units with smooth finish	m²	150.00

	Unit	Cost A$
Timber and insulation		
Softwood sections for carpentry – scantlings	m^3	590.00
Softwood for joinery – dressed	m^3	1,450.00
Hardwood for joinery – dressed	m^3	2,100.00
Exterior quality plywood (6mm)	m^2	30.00
Plywood for interior joinery (6mm)	m^2	25.00
Softwood strip flooring (17mm)	m^2	27.00
Chipboard sheet flooring (19mm)	m^2	18.00
100mm thick quilt insulation	m^2	9.00
Softwood internal door complete with frames and ironmongery	each	350.00
Glass and ceramics		
Float glass (6mm)	m^2	49.00
Plaster and paint		
Good quality ceramic wall tiles (150 x 150mm)	m^2	40.00
Plaster in 25 kg bags	tonne	950.00
Plasterboard (13mm thick)	m^2	8.00
Acrylic paint in 4 litre tins	litre	15.00
Gloss enamel paint in 4 litre tins	litre	19.00
Tiles and paviors		
Quarry floor tiles (150 x 150mm)	m^2	55.00
Sheet vinyl 2mm thick	m^2	20.00
Precast concrete paving slabs (400 x 400 x 25mm)	m^2	50.00
Terracotta roof tiles	m^2	24.00
Precast concrete roof tiles	m^2	18.00
Drainage		
WC suite complete	each	800.00
Lavatory basin complete	each	550.00
150mm diameter cast iron drain pipes	m	80.00

Unit Rates

The descriptions on the next page are generally shortened versions of standard descriptions listed in full in section 4. Where an item has a two digit reference number (e.g. 05 or 33), this relates to the full description against that number in section 4. Where an item has an alphabetic suffix (e.g. 12A or 34B) this indicates that the standard description has been modified. Where a modification is major the complete modified description is included here and the standard description should be ignored; where a modification is minor (e.g. the insertion of a named hardwood) the shortened description has been modified here but, in general, the full description in section 4 prevails.

The unit rates below are for main work items on a typical construction project in the Melbourne area in the fourth quarter of 2008. The rates include all necessary labour, materials, equipment and contractors' overheads and profit. An allowance has been added to cover preliminary and general items.

		Unit	*Rate A$*
Excavation			
01	Mechanical excavation of foundation trenches	m^3	85.00
02	Hardcore filling making up levels	m^3	80.00
03	Earthwork support	m^2	20.00
Concrete work			
04	Plain in situ concrete in strip foundations in trenches	m^3	250.00
05	Reinforced in situ concrete in beds	m^3	250.00
06	Reinforced in situ concrete in walls	m^3	320.00
07	Reinforced in situ concrete in suspended floors or roof slabs	m^3	280.00
08	Reinforced in situ concrete in columns	m^3	310.00
09	Reinforced in situ concrete in isolated beams	m^3	280.00
10	Precast concrete slab	m^2	180.00
Formwork			
11	Formwork to concrete walls	m^2	140.00
12	Formwork to concrete columns	m^2	140.00
13	Formwork to horizontal soffits of slabs	m^2	140.00
Reinforcement			
14	Reinforcement in concrete walls	tonne	2,800.00
15	Reinforcement in suspended concrete slabs	tonne	2,800.00
16	Fabric reinforcement in concrete beds	m^2	22.00
Steelwork			
17	Fabricate, supply and erect steel framed structure	tonne	6,800.00
18	Framed structural steelwork in universal joist sections	tonne	7,000.00
19	Structural steelwork lattice roof trusses	tonne	8,000.00
Brickwork and blockwork			
20	Precast lightweight aggregate hollow concrete block walls	m^2	120.00
21A	Solid (perforated) concrete blocks	m^2	125.00
22	Sand lime bricks	m^2	130.00
23	Facing bricks	m^2	120.00
Roofing			
24	Concrete interlocking roof tiles 430 x 380mm	m^2	35.00
25A	Terracotta roof tiles 260 x 160mm	m^2	20.00
26	Fibre cement roof slates 600 x 300mm	m^2	145.00

		Unit	*Rate A$*
29	3 layers glass-fibre based bitumen felt roof covering	m^2	125.00
30	Bitumen based mastic asphalt roof covering	m^2	80.00
31A	Glass-fibre mat roof insulation 100mm thick	m^2	25.00
33	Troughed galvanised steel roof cladding	m^2	65.00
Woodwork and metalwork			
34	Preservative treated sawn softwood 50 x 100mm	m	10.40
35	Preservative treated sawn softwood 50 x 150mm	m	14.00
36	Single glazed casement window in local hardwood	m^2	600.00
37A	Two panel glazed door in local hardwood, size 820 x 2040mm with frame and hardware	each	1,500.00
38A	Solid core half hour fire resisting hardwood internal flush doors, size 820 x 2040mm with frame and hardware	each	1,300.00
39	Aluminium double glazed window	m^2	700.00
40A	Aluminium single glazed door, size 850 x 2100mm	each	2,000.00
41	Hardwood skirtings	m	18.00
Plumbing			
42A	Steel quadrant eaves gutter	m	23.00
43	UPVC rainwater pipes	m	27.00
44A	Light gauge copper cold water tubing 20mm diameter	m	35.00
45	High pressure plastic pipes for cold water supply	m	35.00
46	Low pressure plastic pipes for cold water distribution	m	18.40
47A	UPVC soil and vent pipes 50mm diameter	m	46.00
48	White vitreous china WC suite	each	812.00
49	White vitreous china lavatory basin	each	560.00
50A	Porcelain enamelled shower tray	each	513.00
51	Stainless steel single bowl sink and double drainer	each	800.00
Electrical work			
52	PVC insulated and copper sheathed cable	m	3.00
53A	10 amp power point	each	58.00
54	Flush mounted, 1 way light switch	each	65.00
Finishings			
55	2 coats gypsum based plaster on brick walls	m^2	55.00
56	White glazed tiles on plaster walls	m^2	130.00
57	Red clay quarry tiles on concrete floors	m^2	115.00
58	Cement and sand screed to concrete floors	m^2	40.00
60	Mineral fibre tiles on concealed suspension system	m^2	55.00
Glazing			
61A	Glazing to wood 6mm	m^2	300.00

		Unit	Rate A$
Painting			
62A	Acrylic on plaster walls	m²	12.00
63	Oil paint on timber	m²	35.00

Approximate Estimating

The building costs per unit area given below are averages incurred by building clients for typical buildings in the Melbourne area as at the fourth quarter of 2008. They are based upon the total floor area of all storeys, measured between external walls and without deduction for internal walls.

Approximate estimating costs generally include mechanical and electrical installations but exclude furniture, loose or special equipment, and external works; they also exclude fees for professional services. The costs shown are for specifications and standards appropriate to Australia and this should be borne in mind when attempting comparisons with similarly described building types in other countries. A discussion of this issue is included in section 2. Comparative data for countries covered in this publication, including construction cost data, are presented in Part Three.

Approximate estimating costs must be treated with caution; they cannot provide more than a rough guide to the probable cost of building.

	Cost m^2 A$	Cost ft^2 A$
Industrial buildings		
Factories for letting	450	42
Factories for owner occupation (light industrial use)	500	46
Factories for owner occupation (heavy industrial use)	650	60
Factory/office (high-tech) for letting (ground floor shell, first floor offices)	1,100	102
Factory/office (high tech) for owner occupation (controlled environment, fully finished)	1,500	139
High tech laboratory workshop centres (air-conditioned)	5,750	534
Warehouses, low bay (6 to 8m high) for letting	450	42
Warehouses, low bay for owner occupation (no heating)	500	46
Warehouses, high bay for owner occupation (no heating)	550	51
Administrative and commercial buildings		
Civic offices, fully air-conditioned	2,200	204
Offices for letting, 5 to 10 storeys, air-conditioned	2,400	223
Offices for letting, high rise, air-conditioned	2,800	260
Offices for owner occupation, high rise, air-conditioned	2,800	260
Prestige/headquarters office, 5 to 10 storeys, air-conditioned	2,900	269
Prestige/headquarters office, high rise, air-conditioned	3,200	297

	Cost m² A$	*Cost ft² A$*
Health and education buildings		
General hospitals (150 beds)	3,500	325
Teaching hospitals (200 beds)	4,000	372
Private hospitals (100 beds)	3,500	325
Health centres	1,900	177
Nursery school	1,700	158
Primary/junior schools	1,800	167
Secondary/middle schools	2,000	186
University (arts) buildings	2,800	260
University (science) buildings	4,400	409
Recreation and arts buildings		
Theatres (over 500 seats) including seating and stage equipment	3,200	297
Theatres (less than 500 seats) including seating and stage equipment	3,300	307
Sports halls including changing and social facilities	2,100	195
Swimming pools (international standard) excluding changing and social facilities	each	2.5m
Swimming pools (schools standard) excluding changing facilities	each	1.5m
National museums including full air-conditioning and standby generator	4,200	390
Local museums including air-conditioning	3,500	325
Branch/local libraries	2,600	242
Residential buildings		
Social/economic single family housing (multiple units)	1,000	93
Private/mass market single family housing 2 storey detached/semi detached (multiple units)	1,300	121
Purpose designed single family housing 2 storey detached (single unit)	1,600	149
Social/economic apartment housing, low rise (no lifts)	1,600	149
Social/economic apartment housing, high rise (with lifts)	2,100	195
Private sector apartment building (standard specification)	2,700	251
Private sector apartment buildings (luxury)	3,600	334
Student/nurses halls of residence	2,000	186
Homes for the elderly (shared accommodation)	1,600	149
Homes for the elderly (self contained with shared communal facilities)	1,600	149
Hotel, 5 star, city centre	3,450	321
Hotel, 3 star, city/provincial	2,800	260
Motel	2,000	186

Regional Variations

The approximate estimating costs are based on projects in Melbourne. For other parts of Australia, adjust these costs by the following factors:

Sydney	:	-2%	Brisbane	:	-1%
Perth	:	+8%	Adelaide	:	-6%
Hobart	:	-8%			
Darwin	:	-6%			

Goods and Services Tax (GST)

A Goods and Services Tax (GST) was introduced on 1 July 2000. Generally, all goods delivered and services provided are subject to GST at the rate of 10%.

EXCHANGE RATES AND INFLATION

The combined effect of exchange rates and inflation on prices within a country and price comparisons between countries is discussed in section 2.

Exchange Rates

The graph on the next page plots the movement of the Australian dollar against the sterling, the euro, the US dollar and 100 Japanese yen since 2000. The values used for the graph are quarterly and the method of calculating these is described and general guidance on the interpretation of the graph provided in section 2. The average exchange rate in the fourth quarter of 2008 was A$2.37 to pound sterling, A$1.93 to euro, A$1.52 to US dollar and A$1.58 to 100 Japanese yen.

THE AUSTRALIAN DOLLAR AGAINST STERLING, EURO, US DOLLAR AND 100 JAPANESE YEN

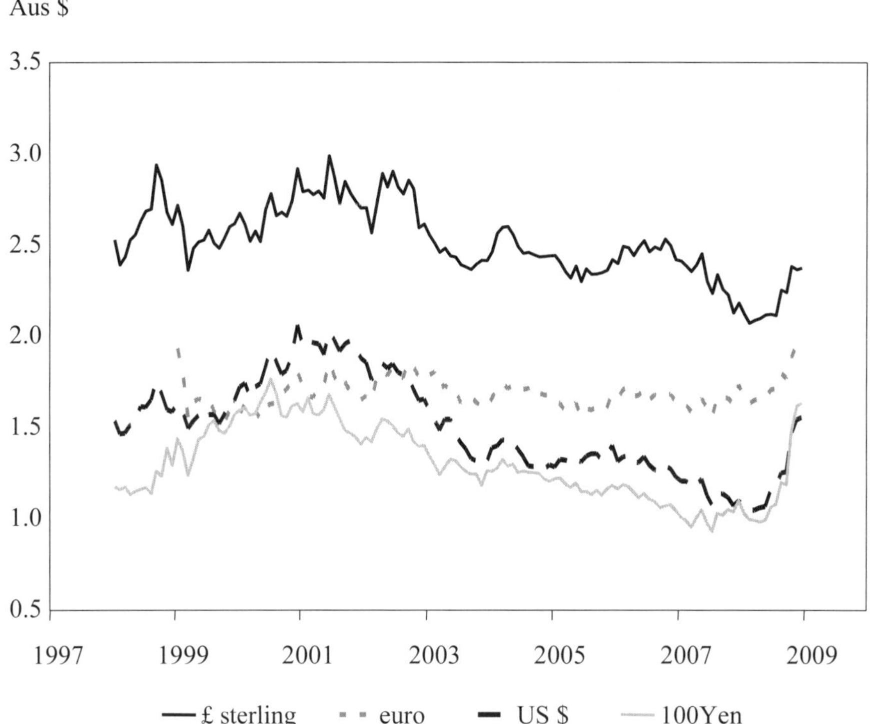

Price Inflation

The table on the next page represents general price and building materials and price inflation in Australia since 1996. Tender prices for construction have increased dramatically faster than general prices in the economy – running at approximately double the rate of CPI during the past ten years.

CONSUMER PRICE AND BUILDING COST AND PRICE INFLATION

Year	*Consumer price inflation average index(CPI)*	*Tender price index*
1996-1997	121.6	113
1997-1998	123.4	116
1998-1999	125.4	119
1999-2000	128.4	125
2000-2001	136.0	130
2001-2002	140.2	143
2002-2003	143.0	155
2003-2004	144.7	158
2004-2005	147.3	159
2005-2006	150.2	172
2006-2007	153.9	182

USEFUL ADDRESSES

Public Organizations

Australian Bureau of Statistics
ABS House
45 Benjamin Way
Belconnen
ACT 2617
Tel: (61) 2 6252 5000
Fax: (61) 2 6252 5566
Website: www.abs.gov.au

Trade And Professional Associations

Association of Consulting Engineers Australia
Level 6, 50 Clarence Street
Sydney
New South Wales 2000
Tel: (61) 2 9922 4711
Fax: (61) 2 9957 2484
E-mail: acea@acea.com.au
Website: www.acea.com.au

Association of Professional Engineers, Scientists and Managers, Australia
163 Eastern Road
South Melbourne
VIC 3205
Tel: (61) 2 9695 8800
Fax: (61) 2 9695 8902
E-mail: info@apesma.asn.au
Website: www.apesma.asn.au

Australian Building Codes Board
P.O. Box 9839
Canberra
ACT 2601
Tel: (61) 1300 134 631
Fax: (61) 2 6213 7287
E-mail: abcb.office@abcb.gov.au
Website: www.abcb.com.au

Australian Chamber of Commerce and Industry
Commerce House, Level 3
24 Brisbane Avenue
Barton
ACT 2600
Tel: (61) 2 6273 2311
Fax: (61) 2 6273 3286
E-mail: info@acci.asn.au
Website: www.acci.asn.au

Australian Institute of Building
217 Northbourne Avenue
Turner
ACT 2612
Tel: (61) 2 6247 7433
Fax: (61) 2 6248 9030
Website: www.aib.org.au

Australian Institute of Quantity Surveyors
P.O. Box 301
Deakin West
ACT 2600
Tel: (61) 2 6282 2222
Fax: (61) 2 6285 2427
E-mail: contact@aiqs.com.au
Website: www.aiqs.com.au

Australian Property Institute
6 Campion Street
Deakin
ACT 2600
Tel: (61) 2 6282 2411
Fax: (61) 2 6285 2194
E-mail: national@api.org.au
Website: www.propertyinstitute.com.au

Civil Contractors Federation
Level 1, 210 High Street
Kew
Victoria 3151
Tel: (61) 3 9851 9900
Fax: (61) 3 9851 9999
Website: www.civilcontractors.com

Green Building Council of Australia
Level 15, 179 Elizabeth Street
Sydney
NSW 2000
Tel: (61) 2 8252 8222
Fax: (61) 2 8252 8223
Website: www.gbca.org.au

Housing Industry Association
79 Constitution Avenue
Campbell
ACT 2612
Tel: (61) 2 6245 1300
Fax: (61) 2 6245 1444
E-mail: enquiry@hia.com.au
Website: hia.com.au

Master Builders Association Australia
Level 1, 16 Bentham Street
Yarralumla
P.O. Box 7170
ACT 2600
Tel: (61) 2 6202 8888
Fax: (61) 2 6202 8877
E-mail: enquiries@masterbuilders.com.au
Website: www.masterbuilders.com.au

Planning Institute of Australia
Unit 8, Level 2, Engineering House, 11 National Circuit
Barton
ACT 2600
Tel: (61) 2 6262 5933
Fax: (61) 2 6262 9970
E-mail: act@planning.org.au
Website: www.planning.org.au

Property Council of Australia
Level 1, 11 Barrack Street
Sydney
New South Wales 2000
Tel: (61) 2 9033 1900
Fax: (61) 2 9033 1991
E-mail: info@propertyoz.com.au
Website: www.propertyoz.com.au

Royal Australian Institute of Architects
Level 2, 7 National Circuit
Barton
ACT 2600
Tel: (61) 2 6121 2000
Fax: (61) 2 6121 2001
E-mail: national@raia.com.au
Website: www.architecture.com.au

BRUNEI DARUSSALAM

All data relate to 2007 unless otherwise indicated.

Population	
Population	390,000
Urban population (2001)	71.71%
Population under 15 (2005)	32.07%
Population 65 and over (2005)	2.62%
Average annual growth rate (2001 to 2007)	2.68%
Geography	
Land area	5,765 km^2
Agricultural area (1995)	3%
Capital city	Bandar Seri Begawan
Population of capital city (1997)	46,230
Economy	
Monetary unit	Brunei Dollar (B$)
Exchange rate (average fourth quarter 2007) to:	
the pound sterling	B$ 2.98
the US dollar	B$ 1.45
the euro	B$ 2.09
the yen x 100	B$ 1.27
Average annual inflation (2003 to 2007)	1.82%
Inflation rate	2.8%
Gross Domestic Product (GDP)	B$18.512 bn
GDP per capita	B$ 47,467
Average annual real change in (GDP) (1997 to 2007)	1.86%
Private consumption as a proportion of GDP	20.78%
Public consumption as a proportion of GDP	12.66%
Investment as a proportion of GDP (2005)	8.80%
Construction	
Gross value of construction output	B$ 1.383 bn
Net value of construction output	-
Gross value of construction output as a proportion of GDP	7.47%

Note:
All GDP values above are based on current prices as at 2007 except for average annual real change in GDP which is based on constant prices with year 2000 as base year

THE CONSTRUCTION INDUSTRY

Construction Output

The value of the gross output of the construction industry in 2007 based on current prices as at 2006 was B$1.383 billion, equivalent to US$0.95 billion, or 7.47% of GDP.

Brunei is heavily reliant on its oil and gas sector which accounts for more than 90% of the country's export receipts. Its contribution to GDP has however fallen from about 70% in the early 1980s to 66.56% in 2007 due to the official policy on diversification. The non-oil sector comprising the government and the non-oil private sector have performed favourably, had shown a 7.95% growth in 2007.

Although the contribution of construction to GDP is small in relation to the other sectors, it remains an important sector of the economy. Statistics on construction output at current prices are shown in the table below.

GROSS DOMESTIC PRODUCT AND CONSTRUCTION OUTPUT, 1996-2007

Year	*Gross domestic product B$ million*	*Construction output B$ million*	*% of gross domestic product*
1996	9,963	3,021	30.3%
1997	9,816	1,872	19.1%
1998	9,761	1,745	17.9%
1999	10,059	1,148	11.4%
2000	10,346	891	8.6%
2001	10,630	991	9.3%
2002	11,042	1,083	9.8%
2003	11,362	1,115	9.8%
2004	11,419	1,162	10.2%
2005	11,464	1,176	10.3%
2006	12,053	1,159	9.6%
2007	18,512	1,383	7.47%

Note: The above values are based on Constant Prices with year 2000 as base year
Source: Department of Statistics, Department of Economic Planning and Development

The level of construction activity is heavily dependent on government development projects. Under the *Ninth National Development Plan for the period between 2007 and 2012*, the overall expenditure allocated for development is B$9,500 million. The proposed main expenditure items are shown in the table below.

MAIN EXPENDITURE ITEMS UNDER NINTH NATIONAL DEVELOPMENT PLAN FOR THE PERIOD 2007-2012

Type of work	*B$ million*	*% of total*
Government and national housing	1,578.9	16.6
Public utilities	1,492.7	15.7
Info-communication technology	1,145.7	12.1
Educational facilities	822.5	8.7
Industrial and commercial development	725.8	7.6
Public buildings	672.9	7.1
Roads	568.5	6.0
Telecommunications (incl. radio, TV & postal)	357.2	3.8
Muara Besar Island Development	299.1	3.1
Public facilities and environment	182.5	1.9
Science & technology and research & development	165.1	1.7
Medical and health	149.1	1.6
Civil aviation, marine and ports	141.3	1.5
Others	1,198.7	12.6
Total	9,500	100

Source: Brunei Darussalam Long Term Development Plan

Public housing and public utilities continue to be the emphasis under this *National Development Plan* as was the case in the past plans. In addition, emphasis has been given to Info-communication technology in both the eighth and the ninth national development plans covering an overall percentage of 15.7% and 12.1% respectively of the total allocation.

Characteristics and Structure of the Industry

The Ministry of Development which was set up in 1984 is responsible for all construction activities. It consists of seven units and six departments including the Public Works Department which is further sub-divided into seven departments. The Ministry provides a range of services from human resource training to basic infrastructure development. The main departments are as follows.

- The Public Works Department (PWD) is responsible for the planning, design, implementation and construction of various government projects such as bridges, roads, water and sewerage. The Public Works Department disseminates its services through its seven departments namely Department of Administration and Finance, Department of Building Services, Department of Development, Department of Drainage and Sewerage, Department of Road, Department of Technical Services and Department of Water Services.

- The Housing Development Department is in charge with implementing the government's objective for every citizen to own a house. The department is also responsible for the management and controlling of buildings in the National Housing Scheme and Landless Citizens Scheme area.
- The Land Department is responsible for registration of privately owned land.
- The Survey Department is responsible for surveys throughout the country. In addition, this department also processes applications related to Land sub-division and consolidation, creates and maintains digital topographical database and produces customised digital maps and orthophoto maps.
- The Town and Country Planning Department is responsible for land use planning and control, covering structure, action and local development plans. This department acts as an advisory to government agencies and developers on physical planning matters and processes all earthwork and building applications within the development control areas.
- The Department of Environment, Parks and Recreation was formed in 2002. The department acts as a regulatory agency for environmental acts and regulations and is responsible for development and implementation of environmental protection policies and programme.

The units within the Ministry include the following:

- Construction Planning and Research Unit
- Research and Development Unit
- Bumiputra Guidance and Development Unit
- Lands Unit
- Housing Unit
- Financial Regulation Unit
- Istana Maintenance Unit

The Ministry of Development, through its Bumiputra Guidance and Development Unit, registers contractors and suppliers under the following classes.

CLASSES OF CONTRACTORS AND SUPPLIERS

Class	*Minimum paid up capital*	*Limit of contract value of project*
I	-	Up to B$25,000
II	-	B$25,000 - B$150,000
III	B$50,000	B$150,000 - B$500,000
IV	B$250,000	B$500,000 - B$1,500,000
V	B$500,000	B$1,500,000 - B$5,000,000
VI	B$1,000,000	Above B$5,000,000

There are seven work categories as follows which a contractor or supplier can apply for registration.

- Civil engineering works
- Building construction works
- Works on water supply, sewerage and drainage
- Maintenance (consists of eight sub-categories)
- Specialist works (consists of fifteen sub-categories)
- Supply and services (consists of eight sub-categories)
- Electrical works (consists of three sub-categories)

Plans for development approval to the Town and Country Planning Department, Land Department, Development Control Unit (Public Works Department), Municipal Board or District Offices are submitted by qualified persons. There are two categories of qualified persons: those able to submit plans for a maximum of four residential units only and those who can submit plans for all types of buildings.

Clients and Finance

The industry is very dependent on public sector projects. Government expenditure allocations are set out in five-year *National Development Plans* which aim at reducing reliance on oil and gas income and increase private sector participation in the economy. Most construction and civil engineering work from the public sector is administered by the Ministry of Development.

Of the private sector clients, the most prominent are Brunei Shell Petroleum Sdn Bhd and a few local property developers who concentrate on providing residential, retail and commercial space. These developers are mainly self-financed or receive assistance from local banks. There are a lot of small construction projects providing private housing where finance is often obtained from local commercial banks through personal loans.

Development Control and Standards

The development of land and building is basically controlled by three different government bodies in their respective control areas, namely Municipal Board, Development Control Competent Authority and Land Department.

The Development Control Competent Authority (DCCA) established and mandated under the Town and Country Planning Act 1972 regulates, plans, coordinates, controls and approves any land or building development within the declared development control areas.

Applications for private land or building developments are received, processed and approved by the Development Control Unit (DCU). Upon completion of the project, the DCU carries out a joint inspection with the other members of the approving authority, and gives recommendation to enable an occupancy permit to be issued by the approving authority.

Piawai Brunei Darussalam (PBD) standards and Guidance Documents (GD) are developed and published by the Ministry of Development, through its Construction

Planning and Research Unit, to maintain quality and consistency in materials and workmanship in the industry.

CONSTRUCTION COST DATA

Cost of Labour

The figures that follow are typical of labour costs in Bandar Seri Begawan as at the fourth quarter of 2008. The wage rate is the basis of an employee's income, while the cost of labour indicates the cost to a contractor of employing that employee. The difference between the two covers a variety of mandatory and voluntary contributions – a list of items which could be included is given in section 2.

	Wage rate (per day) B$	*Cost of labour (per day) B$*	*Number of hours worked per year*
Site operatives			
Mason/bricklayer	20 - 40	40 - 60	2,600
Carpenter	35 - 55	45 - 65	2,600
Plumber	40 - 55	55 - 70	2,600
Electrician	40 - 55	55 - 70	2,600
Structural steel erector	35 - 55	45 - 65	2,600
Semi-skilled worker	20 - 40	40 - 60	2,600
Unskilled labourer	20 - 35	35 - 60	2,600
Equipment operator	55 - 75	75 - 95	2,600
Watchman/security	35 - 55	45 - 65	2,600
	(per month)	*(per month)*	
Site supervision			
General foreman	2,200 - 3,000	4,500 - 7,500	2,600
Trades foreman	1,600 - 2,500	2,500 - 3,500	2,600
Clerk of works	1,200 - 2,400	1,800 - 3,300	2,600
Contractors' personnel			
Site manager	3,800 - 6,000	7,600 - 12,000	2,500 - 2,600
Resident engineer	3,300 - 5,000	4,500 - 7,500	2,500 - 2,600
Resident surveyor	2,000 - 4,000	3,300 - 6,000	2,500 - 2,600
Junior engineer	1,700 - 3,000	2,800 - 4,600	2,500 - 2,600
Junior surveyor	1,600 - 2,500	2,800 - 3,800	2,500 - 2,600
Planner	1,600 - 3,800	2,800 - 5,600	2,500 - 2,600

	Wage rate (per month) B$	*Cost of labour (per month) B$*	*Number of hours worked per year*
Consultants' personnel			
Senior architect	4,500 - 5,500	6,500 - 7,500	2,040
Senior engineer	4,500 - 5,500	6,500 - 7,500	2,040
Senior surveyor	4,000 - 5,000	6,000 - 7,000	2,040
Qualified architect	3,000 - 4,000	5,000 - 6,000	2,040
Qualified engineer	3,000 - 4,000	5,000 - 6,000	2,040
Qualified surveyor	3,000 - 4,000	5,000 - 6,000	2,040

Cost of Materials

The figures that follow are the costs of main construction materials, delivered to site in the Bandar Seri Begawan area, as incurred by contractors in the fourth quarter of 2008. These assume that the materials would be in quantities as required for a medium sized construction project and that the location of the works would be neither constrained nor remote.

	Unit	*Cost B$*
Cement and aggregate		
Ordinary portland cement in 50kg bags	tonne	155.00
Coarse aggregates for concrete	m^3	33.00
Fine aggregates for concrete	m^3	35.00
Ready mixed concrete (1:2:4) Grade 20	m^3	109.00
Ready mixed concrete (1:1.5:3) Grade 25	m^3	114.00
Ready mixed concrete (1:1:2) Grade 30	m^3	119.00
Steel		
Mild steel reinforcement	tonne	1,150.00
High tensile steel reinforcement	tonne	1,150.00
Structural steel sections	tonne	1,950.00
Bricks and blocks		
Common bricks (4" x 9" x 3")	each	0.15
Hollow concrete blocks (6" x 9" x 4")	each	0.45
Solid concrete blocks (4" x 9" x 3")	each	0.20
Glass blocks (8" x 8" x 3 7/8")	each	6.50
Timber and insulation		
Kapur bukit timber (Sawn)	tonne	890.00
Kapur bukit timber (Wrot)	tonne	950.00
Red Meranti timber (Sawn)	tonne	725.00
Red Meranti timber (Wrot)	tonne	785.00
Exterior quality plywood (12mm thick)	m^2	11.50
Plywood for interior joinery (6mm thick)	m^2	7.70

	Unit	Cost B$
Teak parquet flooring	m²	37.70
Chipboard sheet flooring (12mm thick)	m²	35.00
100mm thick quilt insulation	m²	6.00
Aluminium insulation foil	m²	4.50
Softwood internal quality door (single leaf) complete with frames and ironmongery	each	450.00
Gypsum board (9mm thick)	m²	4.10
Glass and ceramics		
Tinted laminated glass (8mm thick)	m²	170.00
Tinted float glass (10mm thick)	m²	80.00
Tinted float glass (6mm thick)	m²	40.00
Sealed double glazing units (50mm)	m²	400.00
Plaster and paint		
Good quality ceramic wall tiles (8" x 8")	m²	9.50
Plasterboard (12mm thick)	m²	12.00
Emulsion paint in 5 litre tins	litre	9.50
Gloss oil paint in 5 litre tins	litre	12.00
Tiles and paviors		
Clay floor tiles (8" x 4" x 0.5")	m²	N/A
Vinyl floor tiles (12" x 12" x 0.125")	m²	18.00
Precast concrete paving slabs (12" x 12" x 3")	m²	40.00
Clay roof tiles	each	3.50
Precast concrete roof tiles	each	1.50
Drainage		
WC suite complete	each	255.00
Lavatory basin complete	each	160.00
200mm diameter vitrified clay sewer pipes	m	27.00
100mm diameter ductile iron pipes	m	12.50
150mm diameter ductile iron pipes	m	17.60
Drainage composite for RC drain	m²	10.00

Unit Rates

The descriptions that follow are generally shortened versions of standard descriptions listed in full in section 4. Where an item has a two digit reference number (e.g. 05 or 33), this relates to the full description against that number in section 4. Where an item has an alphabetic suffix (e.g. 12A or 34B) this indicates that the standard description has been modified. Where a modification is major the complete modified description is included here and the standard description should be ignored; where a modification is minor (e.g. the insertion of a named hardwood) the shortened description has been modified here but, in general, the full description in section 4 prevails.

The unit rates below are for main work items on a typical construction project in the Bandar Seri Begawan area in the fourth quarter of 2008. The rates include all necessary labour, materials and equipment. Allowances to cover preliminary and general items and contractors' overheads and profit have been added to the rates.

		Unit	*Rate B$*
Excavation			
01	Mechanical excavation of foundation trenches	m^3	4.00
02	Hardcore filling making up levels	m^3	8.50
03	Earthwork support	m^2	N/A
Concrete work			
04	Plain in situ concrete in strip foundations in trenches	m^3	130.00
05	Reinforced in situ concrete in beds	m^3	133.00
06	Reinforced in situ concrete in walls	m^3	147.00
07	Reinforced in situ concrete in suspended floors or roof slabs	m^3	143.50
08	Reinforced in situ concrete in columns	m^3	149.50
09	Reinforced in situ concrete in isolated beams	m^3	146.50
10	Precast concrete slab	m^2	254.00
Formwork			
11	Softwood or metal formwork to concrete walls	m^2	14.00
12	Softwood or metal formwork to concrete columns	m^2	14.00
13	Softwood or metal formwork to horizontal soffits of slabs	m^2	14.00
Reinforcement			
14	Reinforcement in concrete walls	tonne	1,450.00
15	Reinforcement in suspended concrete slabs	tonne	1,450.00
16	Fabric reinforcement in concrete beds	m^2	6.20
Steelwork			
17	Fabricate, supply and erect steel framed structure	tonne	2,850.00
18	Framed structural steelwork in universal joist sections	tonne	2,250.00
19	Structural steelwork lattice roof trusses	tonne	3,500.00
Brickwork and blockwork			
21A	Solid (perforated) concrete blocks	m^2	33.00
22	Sand lime bricks	m^2	30.00
23	Facing bricks	m^2	42.00
Roofing			
24	Concrete interlocking roof tiles 430 x 380mm	m^2	32.00
25	Plain clay roof tiles 260 x 160mm	m^2	60.00
26	Fibre cement roof slates 600 x 300mm	m^2	N/A
27	Sawn softwood roof boarding	m^2	N/A

		Unit	Rate B$
29	3 layers glass-fibre based bitumen felt roof covering	m^2	45.00
30	Bitumen based mastic asphalt roof covering	m^2	32.00
33A	Troughed galvanised steel roof cladding (0.65mm TCT)	m^2	47.00
Woodwork and metalwork			
34	Preservative treated sawn softwood 50 x 100mm	m	5.75
35	Preservative treated sawn softwood 50 x 150mm	m	11.25
36	Single glazed casement window in Nyatoh hardwood, size 650 x 900mm	each	280.00
37	Two panel glazed door in Nyatoh hardwood, size 850 x 2000mm	each	600.00
38	Solid core half hour fire resisting hardwood internal flush doors, size 800 x 2000mm	each	380.00
39	Aluminium double glazed window, size 1200 x 1200mm	each	775.00
40	Aluminium double glazed door, size 850 x 2100mm	each	600.00
41	Hardwood skirtings	m	5.50
Plumbing			
42	UPVC half round eaves gutter	m	29.00
43	UPVC rainwater pipes	m	12.00
44	Light gauge copper cold water tubing	m	13.00
45	High pressure plastic pipes for cold water supply	m	6.50
46	Low pressure plastic pipes for cold water distribution	m	16.50
47	UPVC soil and vent pipes	m	15.00
48	White vitreous china WC suite	each	290.00
49	White vitreous china lavatory basin	each	345.00
50	Glazed fireclay shower tray	each	280.00
51	Stainless steel single bowl sink and double drainer	each	700.00
Electrical work			
52A	PVC insulated and copper sheathed cable (1.5mm^2 dual core)	m	7.15
53A	13 amp switched socket outlet	each	78.00
54	Flush mounted 20 amp, 1 way light switch	each	75.00
Finishings			
55	2 coats gypsum based plaster on brick walls	m^2	11.00
56	White glazed tiles on plaster walls	m^2	31.00
57A	Homogenous unpolished ceramic tiles to floors	m^2	29.00
58	Cement and sand screed to concrete floors	m^2	13.80
59	Thermoplastic floor tiles on screed	m^2	32.00
60	Mineral fibre tiles on concealed suspension system	m^2	27.00

		Unit	*Rate B$*
Glazing			
61	Glazing to wood	m^2	40.00
Painting			
62	Emulsion on plaster walls	m^2	5.80
63	Oil paint on timber	m^2	6.30

Approximate Estimating

The building costs per unit area given below are averages incurred by building clients for typical buildings in the Bandar Seri Begawan area as at the fourth quarter of 2008. They are based upon the total floor area of all storeys, measured between external walls and without deduction for internal walls.

Approximate estimating costs generally include mechanical and electrical installations but exclude furniture, loose or special equipment, and external works; they also exclude fees for professional services. The costs shown are for specifications and standards appropriate to Brunei and this should be borne in mind when attempting comparisons with similarly described building types in other countries. A discussion of this issue is included in section 2. Comparative data for countries covered in this publication, including construction cost data, are presented in Part Three.

Approximate estimating costs must be treated with caution; they cannot provide more than a rough guide to the probable cost of building.

	Cost m^2 *B$*	*Cost* ft^2 *B$*
Industrial buildings		
Factories for letting	500 - 600	46 - 56
Factories for owner occupation (light industrial use)	650 - 700	60 - 65
Factories for owner occupation (heavy industrial use)	800 - 850	74 - 79
Factory/office (high-tech) for letting (shell and core only)	620 - 670	58 - 62
Factory/office (high-tech) for letting (ground floor shell, first floor offices)	700 - 750	65 - 70
Factory/office (high tech) for owner occupation (controlled environment, fully finished)	1,100 - 1,200	102 - 111
High tech laboratory workshop centres (air-conditioned)	1,300 - 1,400	121 - 130
Warehouses, low bay (6 to 8m high) for letting (no heating)	625 - 675	58 - 63
Cold stores/refrigerated stores	1,200 - 1,300	111 - 121
Administrative and commercial buildings		
Civic offices, non air-conditioned	650 - 700	60 - 65
Civic offices, fully air-conditioned	800 - 850	74 - 79
Offices for letting, 5 to 10 storeys, non air-conditioned	850 - 950	79 - 88

	Cost m² B$	Cost ft² B$
Offices for letting, 5 to 10 storeys, air-conditioned	1,050 - 1,150	98 - 107
Offices for owner occupation 5 to 10 storeys, non air-conditioned	1,050 - 1,250	98 - 116
Offices for owner occupation 5 to 10 storeys, air-conditioned	1,250 - 1,450	116 - 135
Prestige/headquarters office, 5 to 10 storeys, air-conditioned	1,450 - 1,650	135 - 153
Health and education buildings		
General hospitals (500 beds)	1,500 - 1,650	139 - 153
Teaching hospitals (100 beds)	1,350 - 1,450	125 - 135
Private hospitals (100 beds)	1,300 - 1,500	121 - 139
Health centres	1,150 - 1,250	107 - 116
Nursery schools	650 - 850	60 - 79
Primary/junior schools	950 - 1,050	88 - 98
Secondary/middle schools	1,050 - 1,150	98 - 107
Management training centres	1,100 - 1,200	102 - 111
Recreation and arts buildings		
Theatres (over 500 seats) including seating and stage equipment	3,200 - 3,600	297 - 334
Theatres (less than 500 seats) including seating and stage equipment	2,800 - 3,200	260 - 297
Concert halls including seating and stage equipment	1,900 - 2,200	177 - 204
Sports halls including changing and social facilities	1,450 - 1,750	135 - 163
Swimming pools (international standard complete with changing facilities, grandstand, pool terrace excluding special equipment)	each	2,750,000
National museums including full air-conditioned and standby generator	1,900 - 2,400	177 - 223
City centre/central libraries	1,250 - 1,500	116 - 139
Branch/local libraries	1,050 - 1,250	98 - 116
Residential buildings		
Social/economic single family housing (multiple units)	550 - 650	51 - 60
Private/mass market single family housing 2 storey detached/semi detached (multiple units)	600 - 750	56 - 70
Purpose designed single family housing 2 storey detached (single unit)	800 - 1,000	74 - 93
Social/economic apartment housing, low rise (no lifts)	750 - 900	70 - 84
Social/economic apartment housing, high rise (with lifts)	800 - 1,100	74 - 102
Private sector apartment building (standard specification)	1,050 - 1,150	98 - 107
Private sector apartment buildings (luxury)	1,250 - 1,500	116 - 139

	Cost m^2 *B$*	*Cost* ft^2 *B$*
Student/nurses halls of residence	700 - 900	65 - 84
Hotel, 5 star, city centre	2,750 - 3,250	255 - 302
Hotel, 3 star, city/provincial	2,250 - 2,750	209 - 255
Motel	750 - 1,200	70 - 111

EXCHANGE RATES

The graph below plots the movement of the Brunei dollar against the sterling, the euro, the US dollar and 100 Japanese yen since 1998. The values used for the graph are average of each year. The average exchange rate in the fourth quarter of 2007 was B$2.98 to the pound sterling, B$2.09 to the euro, B$1.45 to US dollar and B$1.27 to 100 Japanese yen.

THE BRUNEI DOLLAR AGAINST STERLING, EURO, US DOLLAR AND 100 JAPANESE YEN

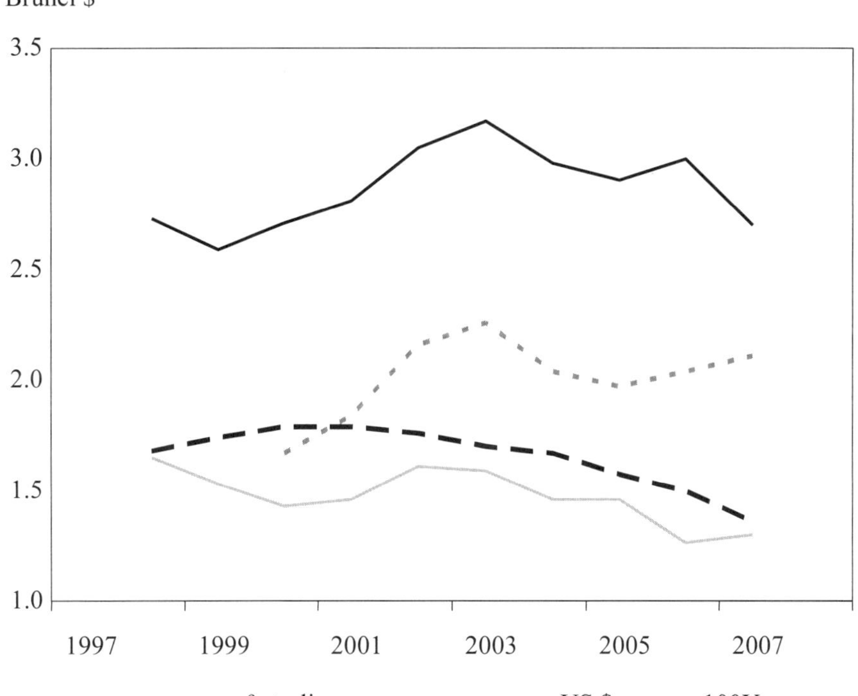

USEFUL ADDRESSES

Public Organizations

Brunei Industrial Development Authority
KM 8, Jalan Gadong BE 1118
Bandar Seri Begawan BB3910
Tel: (673) 2444100
Fax: (673) 2423300
E-mail: bruneibina@brunet.bn
Website: www.bina.gov.bn

Construction Planning and Research Unit
Ministry of Development
Old Airport, Berakas
Bandar Seri Begawan BB3510
Tel: (673) 2381033
Fax: (673) 2381541

Department of Economic Planning and Development
Prime Minister's Office
Block 2A, Jalan Ong Sum Ping
Bandar Seri Begawan BA1311
Tel: (673) 2233344
Fax: (673) 2230226
E-mail: info@jpke.gov.bn
Website: www.depd.gov.bn

Department of Electrical Services
Old Airport, Berakas
Bandar Seri Begawan BB3510
Tel: (673) 2382090
Fax: (673) 2383371
E-mail: desmail@brunet.bn
Website: www.des.gov.bn

Department of Environment, Parks and Recreation
Ministry of Development
Old Airport, Berakas
Bandar Seri Begawan BB3510
Tel: (673) 2383222
Fax: (673) 2383644
E-mail: info@env.gov.bn
Website: www.env.gov.bn

Housing Development Department
Ministry of Development
Old Airport, Berakas
Bandar Seri Begawan BB3510
Tel: (673) 2382145
Fax: (673) 2382736
E-mail: housing@brunet.bn
Website: www.housing.gov.bn

Land Department
Ministry of Development
Old Airport, Berakas
Bandar Seri Begawan BB3510
Tel: (673) 2381181
Fax: (673) 2380365
E-mail: landcomm@brunet.bn
Website: www.land.gov.bn

Ministry of Development
Old Airport, Berakas
Bandar Seri Begawan BB3510
Tel: (673) 2383222
Fax: (673) 2380298
Website: www.mod.gov.bn

Public Works Department
Ministry of Development
Old Airport, Berakas
Bandar Seri Begawan BB3510
Tel: (673) 2383911
Fax: (673) 2380595
Website: www.pwd.gov.bn

Survey Department
Ministry of Development
Old Airport, Berakas
Bandar Seri Begawan BB3510
Tel: (673) 2382171
Fax: (673) 2382900
E-mail: survey@brunet.bn
Website: www.survey.gov.bn

Town and Country Planning Department
Ministry of Development
Old Airport, Berakas
Bandar Seri Begawan BB3510
Tel: (673) 2382591
Fax: (673) 2383313
E-mail: tcpcomp@brunet.bn
Website: www.pbd.gov.bn

Trade And Professional Associations

Association of Surveyors, Engineers and Architects
Pertubuhan Ukur Jurutera dan Arkitek (PUJA)
Unit 3, Block B9, Simpang 32-66
Kawasan Anggerek Desa, Berakas
Bandar Seri Begawan BB3713
Tel: (673) 2384021
Fax: (673) 2384021
E-mail: secgeneral@puja-brunei.com
Website: www.puja-brunei.com

The Brunei Darussalam International Chamber of Commerce and Industry
Unit 402-403A, 4th Floor
Wisma Jaya
Jalan Pemancha
P O Box 2988
Bandar Seri Begawan BS8675
Tel: (673) 2228382
Fax: (673) 2228389

Brunei Institution of Geomatics (BIG)
Block C-4, First Floor, Simpang 88
Urairah Complex
Kampong Kiulap
Bandar Seri Begawan BE3978
Tel: (673) 2231618 / 2231620
Fax: (673) 2231619
Website: www.big.org.bn

CAMBODIA

All data relate to 2007 unless otherwise indicated.

Population	
Population	14.4 mn
Urban population	18%
Population under 15	35%
Population 65 and over	4%
Average annual growth rate (2004 to 2007)	1.9%
Geography	
Land area	181,035 km^2
Agricultural area	20%
Capital city	Phnom Penh
Population of capital city	2 mn
Economy	
Monetary unit	Riels
Exchange rate (average fourth quarter 2008) to:	
the pound sterling	Riels 6,130
the US dollar	Riels 4,126
the euro	Riels 5,556
the yen x 100	Riels 4,520
Average annual inflation (2003 to 2007)	4.3%
Inflation rate	5.9%
Gross Domestic Product (GDP)	Riels 35,039 bn
GDP per capita	Riels 2,440,000
Average annual real change in (GDP) (1998 to 2007)	9.4%
Private consumption as a proportion of GDP	78.2%
Public consumption as a proportion of GDP	5.7%
Investment as a proportion of GDP	20.8%
Construction	
Gross value of construction output	N/A
Net value of construction output	Riels 2,338 bn
Net value of construction output as a proportion of GDP	6.7%

THE CONSTRUCTION INDUSTRY

Construction Output

The net output of construction industry in 2007 was Riels 2,338 billion, equivalent to US$566 million, or 6.7% of GDP. The construction industry expanded rapidly since 1999 and the growth for 2007 was 17%.

Construction is booming in Cambodia on a scale never seen before. The rocketing property prices in Phnom Penh have encouraged overseas developers to invest in golf courses, housing and condominium projects.

The graph and table below show the net value of construction output for the last ten years and the construction area approved from 1998 to 2007.

CONSTRUCTION OUTPUT AT MARKET PRICE

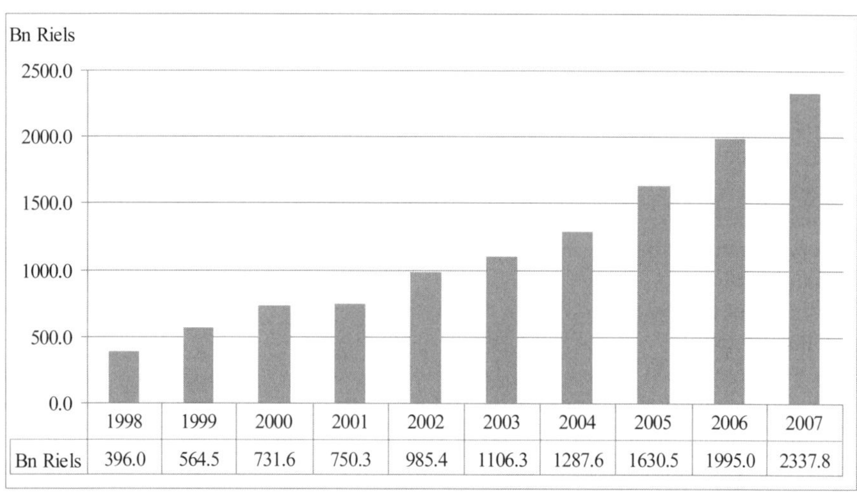

	1998	1999	2000	2001	2002	2003	2004	2005	2006	2007
Bn Riels	396.0	564.5	731.6	750.3	985.4	1106.3	1287.6	1630.5	1995.0	2337.8

Source: Asian Development Bank (ADB) – Key indicators 2008

Year	*Approved by Ministry of Land Management, Planning and Construction*		*Approved by Municipality and Province*	
	Area (m^2)	*Value (US$ '000s)*	*Area (m^2)*	*Value (US$ ' 000s)*
2001	867,920	218,844		
2002	776,916	173,138		
2003	1,534,771	267,509	335,400	41,925
2004	1,668,486	523,145	2,111,825	299,822
2005	2,605,343	608,703	1,512,056	248,526
2006	3,591,004	726,149	2,653,398	466,306
2007	9,549,838	3,000,771	1,145,502	211,124

Source: Ministry of Land Management, Urban Planning and Construction

However, due to the global recession, the industry is slowing down and the outlook for 2009 is expected to be gloomy.

Characteristics and Structure of the Industry

In the past few years, the construction industry in Cambodia expanded rapidly, moving from dominantly infrastructure works and factories developments in 2001 to commercial and residential developments.

Due to the lack of large construction firms with capability to undertake large projects, construction management method is the most commonly used form of managing construction whereby the Owner employs his own management team to manage the subcontractors and suppliers.

The construction industry in Cambodia depends heavily on imported materials mainly from China, Vietnam, Thailand and other Asian countries.

Shortage of skilled workers is still a major issue in Cambodia.

Clients and Finance

International assistance remains the main source of finance for government infrastructure projects. However, private investment has increased substantially in the last few years mainly in residential and commercial projects.

Selection of Design Consultants

There is no specific procedure and criteria for the selection of the design consultants. Developers normally select their design consultants based on their experience and fees.

Contractual Arrangements

There is no standard form of contract in Cambodia. Various standard forms of contract are used in private sector. The most common type is the FIDIC short form contract. The most common form of contract management is Construction Management whereby either the Owner employs a consultant or set up his own site team to manage the subcontractors and the suppliers.

Development Control and Standards

For a building of less than 3,000 square meters in size, the permit is issued by the Governor of the Municipality or the Province. For a building of more than 3,000 square meters in size, the permit is issued by the Ministry of Land Management, Urbanization and Construction. Construction work must commence within one year from the issuance of the permit. Upon completion, the building must be approved by the Ministry.

CONSTRUCTION COST DATA

Cost of Labour

The figures below are typical of labour costs in the Phnom Penh area as at the fourth quarter of 2008.

	Wage rate (per day) US$	*Number of hours worked per year*
Site operatives		
Bricklayer	4.00 - 6.00	2,496
Carpenter	4.00 - 6.00	2,496
Plumber	6.00 - 8.00	2,496
Electrician	6.00 - 8.00	2,496
Structural steel erector	6.00 - 8.00	2,496
Welder	6.00 - 8.00	2,496
Labourer	2.50	2,496
Equipment operator (Tower Crane)	4.00 - 5.00	2,496
	(per month)	
Site supervision		
General foreman	200.00	2,496
Trades foreman	200.00	2,496
Clerks of works	180.00	2,496
Resident Engineer	700.00	2,496
Contractors' personnel		
Site manager	600.00	2,496
Site engineer	400.00	2,496
Site quantity surveyor	400.00	2,496
Consultants' personnel		
Senior architect	700.00	2,496
Senior engineer	700.00	2,496
Senior surveyor	700.00	2,496

Cost of Materials

The figures that follow are the costs of main construction materials, delivered to site in the Phnom Penh area, as incurred by contractors in the fourth quarter of 2008. These assume that the materials would be in quantities as required for a medium sized construction project and that the location of the works would be neither constrained nor remote. All the costs in this section exclude value added tax (VAT).

	Unit	*US$*
Cement and aggregate		
Ordinary portland cement in 50kg bags	tonne	81.00
Coarse aggregates for concrete	m^3	15.00
Fine aggregates for concrete	m^3	8.50
Ready mixed concrete (mix Grade 20)	m^3	46.00
Ready mixed concrete (mix Grade 24)	m^3	50.00
Steel		
Mild steel reinforcement	tonne	970.00
High tensile steel reinforcement	tonne	970.00
Bricks and blocks		
Common bricks (160 x 35 x 70mm)	1,000 pcs	800.00
Good quality facing bricks (220 x 65 x 105mm)	1,000 pcs	800.00
Timber and insulation		
Softwood for carpentry	m^3	300.00 - 500.00
Softwood for joinery	m^3	500.00 - 600.00
Hardwood for joinery	m^3	600.00 - 800.00
Exterior quality plywood (20mm)	m^2	10.50
Plywood for interior joinery (4mm)	m^2	1.70
Plywood for interior joinery (20mm)	m^2	8.00
Chipboard sheet flooring (25mm)	m^2	20.00
Glass and ceramics		
Float glass (6mm)	m^2	18.00 - 20.00
Plaster and paint		
Good quality ceramic wall tiles (200 x 200mm)	m^2	7.00 - 10.00
Plaster in 50kg bags	tonne	70.00 - 75.00
Tiles and paviors		
Clay roof tiles (255 x 140mm)	m^2	4.00 - 11.00
Drainage		
WC suite complete (medium quality)	each	700.00 - 800.00
Lavatory basin complete (medium quality)	each	200.00 - 300.00
100mm diameter PVC drain pipes	m	2.50
150mm diameter cast iron drain pipes	m	32.50

Unit Rates

The descriptions below are generally shortened versions of standard descriptions listed in full in section 4. Where an item has a two digit reference number (e.g. 05 or 33), this relates to the full description against that number in section 4. Where an item has an alphabetic suffix (e.g. 12A or 34B) this indicates that the standard description has been modified. Where a modification is major the complete modified description is included here and the standard description should be ignored; where a modification is minor (e.g. the insertion of a named hardwood) the shortened description has been modified here but, in general, the full description in section 4 prevails.

The unit rates below are for main work items on a typical construction project in the Phnom Penh area in the fourth quarter of 2008. The rates include all necessary labour, materials and equipment. Allowance of 15% to cover preliminaries and general items and 8% to cover for Contractor's profit and overheads have been included in the unit rates. All the rates in this section exclude value added tax (VAT).

		Unit	*Rate US$*
Excavation			
01A	Mechanical excavation of foundation trenches including earthwork support	m^3	3.00
02	Hardcore filling making up levels	m^3	2.50
Concrete work			
04	Plain in situ concrete in strip foundations in trenches	m^3	50.00
05	Reinforced in situ concrete in beds	m^3	65.00
06	Reinforced in situ concrete in walls	m^3	65.00
07	Reinforced in situ concrete in suspended floors or roof slabs	m^3	65.00
08	Reinforced in situ concrete in columns	m^3	65.00
09	Reinforced in situ concrete in isolated beams	m^3	65.00
10	Precast concrete slabs	m^2	50.00
Formwork			
11	Softwood formwork to concrete walls	m^2	15.00
12	Softwood formwork to concrete columns	m^2	15.00
13	Softwood formwork to horizontal soffits of slabs	m^2	15.00
Reinforcement			
14	Reinforcement in concrete walls	tonne	700.00
15	Reinforcement in suspended concrete slabs	tonne	700.00
16	Fabric reinforcement in concrete beds	m^2	N/A
Steelwork			
17	Fabricate, supply and erect steel framed structure	tonne	1,300.00
18	Framed structural steelwork in universal joist sections	tonne	1,300.00

		Unit	Rate US$
19	Structural steelwork lattice roof trusses	tonne	1,300
Brickwork and blockwork			
21A	Solid (perforated) concrete blocks (70mm thick)	m^2	20.00
23A	Local one brick wall	m^2	6.00
Roofing			
24A	Concrete interlocking roof tiles 400 x 330mm	m^2	9.00
25A	Plain clay roof tiles 255 x 140mm	m^2	11.00
27	Sawn softwood roof boarding	m^2	1.20
Woodwork and metalwork			
34	Preservative treated sawn softwood 50 x 100mm	m	7.50
35	Preservative treated sawn softwood 50 x 150mm	m	10.80
37	Two panel glazed door in hardwood, size 850 x 2,000mm	each	260.00
41	Hardwood skirtings	m	10.00
Plumbing			
43A	PVC rainwater pipes (100mm diameter) class 8.5	m	2.50
Electrical work			
52	PVC insulated and copper sheathed cable	m	0.50
53A	10 amp unswitched socket outlet	each	6.00
54	Flush mounted 20 amp, 1 way light switch	each	6.00 - 7.00
Finishings			
55	2 coats gypsum based plaster on brick walls	m^2	4.50
56	White glazed tiles on plaster walls	m^2	24.00
58	Cement and sand screed to concrete floors	m^2	3.60
Painting			
62	Emulsion on plaster walls	m^2	4.00
63	Oil paint on timber	m^2	5.00

Approximate Estimating

The building costs per unit area given on the next page are averages incurred by building clients for typical buildings in the Phnom Penh area as at the fourth quarter of 2008. They are based upon the total floor area of all storeys, measured between external walls and without deduction for internal walls.

Approximate estimating costs generally include mechanical and electrical installations but exclude furniture, loose or special equipment, and external works; they also exclude fees for professional services. The costs shown are for specifications and standards appropriate to Cambodia and this should be borne in mind when attempting comparisons with similarly described building types in other countries. A discussion of this issue is included in section 2. Comparative data for countries covered in this publication, including construction cost data, are presented in Part Three.

Approximate estimating costs must be treated with caution; they cannot provide more than a rough guide to the probable cost of building. All the rates in this section exclude value added tax (VAT).

	Cost *m^2 US$*	*Cost* *ft^2 US$*
Industrial		
Light duty flatted factories, 150 lb loading	300	28
Single storey conventional factory of structural steelwork	450	42
Office/commercial		
Average standard offices, high rise	600	56
Prestige offices, high rise	700	65
Domestic		
Detached houses and bungalows	400	37
Average standard apartments, high rise	500	46
Luxury apartments, high rise	800	74
Hotels		
3 star budget hotel inclusive of fixtures and fittings	1,100	102
5 star luxury hotels inclusive of fixtures and fittings	1,500	139
Others		
Car parks, above ground	350	33
Retail/department stores (without finishes)	700	65

Value Added Tax (VAT)

The standard rate of value added tax (VAT) is currently 10%.

EXCHANGE RATES

The graph overleaf plots the movement of the Riels against the sterling, the euro, the US dollar and 100 Japanese yen since 1998. The figures used for the graph are quarterly and the method of calculating these is described and general guidance on the interpretation of the graph provided in section 2. The average exchange rate at the fourth quarter of 2008 was Riels 6,130 to pound sterling, Riels 5,556 to euro, Riels 4,126 to US dollar and Riels 4,520 to 100 Japanese yen.

THE RIELS AGAINST STERLING, EURO, US DOLLAR AND 100 JAPANESE YEN

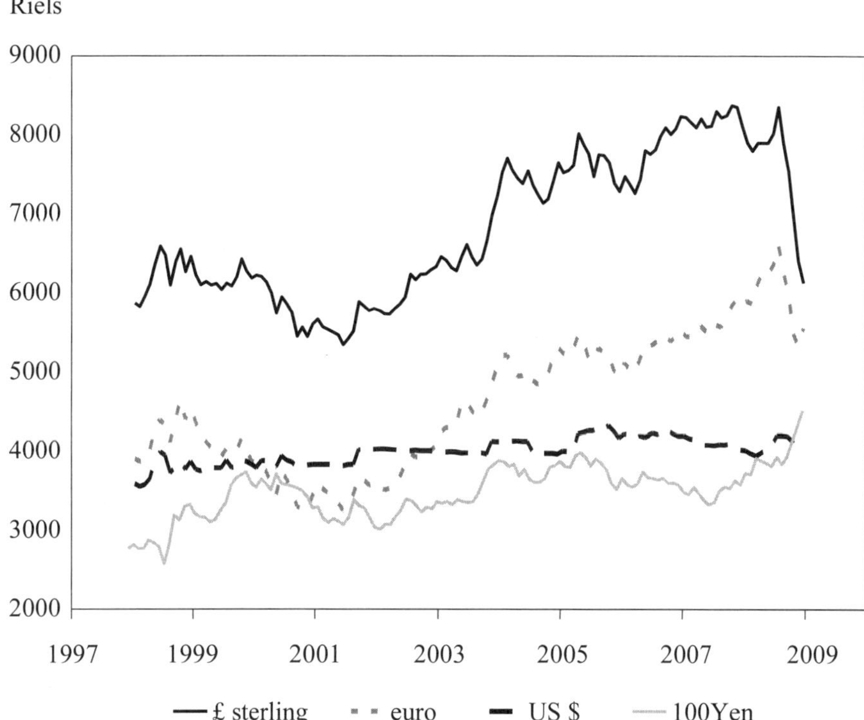

USEFUL ADDRESSES

Council for the Development of Cambodia (CDC) – Cambodian Investment Board
Government Palace, Sisowath Quay
Wat Phnom, Phnom Penh
Tel: (855) 23 981154, 981156, 981183
Fax: (855) 23 428426 / 23 427597, 428953
E-mail: CDC.cib@online.com.kh
Website: www.cambodiainvestment.gov.kh

Ministry of Land Management, Urban Planning and Construction
No. 771 773, Monivong Boulevard
Phnom Penh
Tel: (855) 23 215660
Fax: (855) 23 217035 / 215277
E-mail: gdlmup-mlmupc@camnet.com.kh (General Dept of Land Mgmt)
E-mail: gdc-mlmupc@camnet.com.kh (General Dept of Construction)
Website: www.mlmupc.gov.kh

CHINA

All data relate to 2007 unless otherwise indicated.

Population	
Population	1,321 mn
Urban population	45%
Population under 15	18%
Population 65 and over	9.4%
Average annual growth rate (2004 to 2007)	0.55%
Geography	
Land area	9,425,290 km^2
Agricultural area (2005)	60%
Capital city	Beijing
Population of capital city	16.3 mn
Economy	
Monetary unit	Renminbi (Rmb)
Exchange rate (average fourth quarter 2008) to:	
the pound sterling	Rmb 10.70
the US dollar	Rmb 6.83
the euro	Rmb 9.00
the yen x 100	Rmb 7.15
Average annual inflation (1998 to 2007)	1.3%
Inflation rate	4.8%
Gross Domestic Product (GDP)	Rmb 25,148 bn
GDP per capita	Rmb 18,934
Average annual real change in (GDP) (1998 to 2007)	10.0%
Private consumption as a proportion of GDP	35.4%
Public consumption as a proportion of GDP	13.3%
Investment as a proportion of GDP	42.3%
Construction	
Gross value of construction output	Rmb 5,104 bn
Net value of construction output	Rmb 1,401 bn
Net value of construction output as a proportion of GDP	5.6%

THE CONSTRUCTION INDUSTRY

Construction Output

The property market in China was at an unprecedented peak in 2007. The hot property market led to a boom in construction activities. Unlike 10 years ago, this boom was no longer concentrated only in the coastal cities and provinces but had, following the implementation of the Western Development Strategy by the Central Government, spread to the inland and western areas too. Construction works were initiated not only by government and foreign investments but also by the extensive developments from the newly established local private companies. The net value of construction output in 2007 was Rmb1,401 billion, equivalent to US$189 billion, representing 5.6% of GDP. In line with the remarkable growth of China's economy in recent years, the construction industry in China is now one of the largest in the world.

Many political observers have pointed out that a large proportion of senior ministers in the Chinese Government have an engineering education. This is in contrast to that in western developed countries where political leaders tend to come from a legal or social science background. It is therefore not surprising that the Chinese Government has always had a particular love for infrastructure developments. Such policies are usually taken much to their heart by many of China's leaders. An example is how Past Premier Li Peng pushed forward the Three Gorges Dam project in the face of fierce criticism from both outside and inside the country and the enormous resettlement problem of whole towns and villages affected. China's new infrastructure is the envy of many developing countries. Many articles have been written comparing China with another major rising economy, India, and China's infrastructure is always quoted as one of its main advantage. However the breakdown in electricity supply and railway service in the central and southern parts of China at the beginning of 2008 due to a series of unusually harsh snow storms has put into question its sufficiency for modern China. It appears that investment in infrastructure works will not be reduced for quite some time.

On the building side, the passing of the long awaited Property Rights Law in 2007 (the law was first initiated in 1994 and has gone through 8 readings since then) aroused the long suppressed Chinese Dream of owning one's own house. The rush to buy newly constructed homes has made property development so hot and property prices rising so fast that the government has had to implement a series of administrative measures to control its activities. Foreign investments in this sector of the economy were basically discouraged. A few of the more radical actions taken include reserving certain parcels of land for purchase by local development companies only; extra administrative requirements on exchange conversion on money remitted into China; and a near banning of bank financing for purchase of land.

On another aspect, international environmental pressure groups have been raising concerns over the effects on the environment due to China's massive building process. To address such criticisms, the government has promised to boost its efforts in tackling with environmental problems by the introduction of a new Energy Law which will call for the development of energy-efficient production processes, energy-efficient consumption approaches and sustainable development for new construction projects. The new Energy Law would increase the cost of construction but the effect on demand for building works would be very minimal. On the other hand, it is expected that it will elevate the design and construction capability of design institutes and construction companies and will generate new types of industries and projects, like wind farms and biomass plants, etc. for the construction industry.

China's construction industry has been growing at an average annual rate of nearly 19% for the past 10 years. The 2008 financial tsunami may slow its growth in the short term, but with the continual investment in the infrastructure by the government, the large scale rebuilding projects in Sichuan after the 12 May 2008 earthquake and the spreading of private developments into second tier cities and inland provinces due to insatiable demand for new modern homes and good commercial spaces, the medium to long term growth still looks very robust.

The distribution of construction activities in China in 2007 are shown in the table below.

REGIONAL DISTRIBUTION OF CONSTRUCTION OUTPUT COMPARED TO POPULATION, 2007

Region	*% of population*	*% of construction output*	*Rank by population*	*Rank by construction output*
Beijing	1.2	5.0	26	5
Tianjin	0.8	2.4	27	15
Hebei	5.3	3.2	6	12
Shanxi	2.6	2.1	19	18
Inner Mongolia	1.8	1.3	23	23
Liaoning	3.3	4.1	14	10
Jilin	2.1	1.4	21	22
Heilongjiang	2.9	1.7	15	19
Shanghai	1.4	4.9	25	6
Jiangsu	5.8	13.7	5	1
Zhejiang	3.8	13.7	10	2
Anhui	4.6	3.0	8	14
Fujian	2.7	3.0	18	13
Jiangxi	3.3	1.5	13	20
Shandong	7.1	6.4	2	3
Henan	7.1	4.2	3	7
Hubei	4.3	4.1	9	8
Hunan	4.8	3.6	7	11
Guangdong	7.2	5.9	1	4

Region	*% of population*	*% of construction output*	*Rank by population*	*Rank by construction output*
Guangxi	3.6	1.2	11	24
Hainan	0.6	0.2	28	30
Chongqing	2.1	2.2	20	17
Sichuan	6.2	4.1	4	9
Guizhou	2.8	0.7	16	27
Yunnan	3.4	1.5	12	21
Tibet	0.2	0.1	31	31
Shaanxi	2.8	2.3	17	16
Gansu	2.0	0.9	22	26
Qinghai	0.4	0.2	30	29
Ningxia	0.5	0.3	29	28
Xinjiang	1.6	0.9	24	25

Note: Due to sampling and survey errors, the total of region populations are not necessarily equal to 100%

Source: China Statistical Yearbook, 2008

Construction work is classified according to the following categories:

(1) Building works including building and housing
(2) Civil engineering works, including highways, bridges, dams, harbours, power stations, airports, etc.
(3) Equipment installation
(4) Decoration and fitting out works
(5) Other construction works

The Gross Construction Output for the various categories in 2007 is shown in the table below.

GROSS CONSTRUCTION OUTPUT OF CONSTRUCTION ENTERPRISES IN CATEGORIES IN 2007

Type of work	*Rmb 1,000,000*
Building works	3,066,095
Civil engineering works	1,359,032
Equipment installation (Equipment supplied by others)	417,179
Decoration and fitting out works	184,674
Other construction works	77,391
Total	5,104,371

Source: China Statistical Yearbook, 2008

Characteristics and Structure of the Industry

The Chinese has always been masters in construction. From the Great Wall of China in the distant past to the modern Yangtze River Bridge built during the early days of modern China, the Chinese has shown their expertise in complicated building technology. Even during the Cultural Revolution days, China exported construction to African and East European countries, undertaking construction projects for their comrades in need. Lu Ban (507 BC to 444 BC), a legendary carpenter that lived in the Warring States Period is honoured by builders all over China, including Hong Kong. It is therefore not surprising that the contractors in China have taken only 20 years to chase up with foreign contractors. Nowadays, Chinese contractors are performing as well as any international contractor in the structure and building works. A slight lag might still exist in the building services installations and high standard decoration works. However the gap is closing fast. Over 90% of the works in such famous iconic developments as the CCTV Building, the Olympic Main Stadium and the National Opera House, etc. are done by the Chinese contractors.

The Ministry of Construction (MOC) supervises and regulates all construction activities in China. Its administration covers town planning, surveying, design, tendering, construction, quality inspection and material standards, etc. Its control in certain parts of the industry is so detailed and specific that many international developers would find it hard to understand.

China adopts a Main Contractor approach. The Main Contractor is made responsible for all matters happening on the construction site and the overall quality of the works, including work done by his subcontractors. Contractors are divided into different categories: Building, Highway, Railway, Port and Waterway, Hydraulic, Electricity, Mining, Refining, Chemical, Civil, Information System and Electrical & Mechanical. Building contractors are divided into 4 grades: Special Grade, Grade 1, Grade 2 and Grade 3. The permitted scopes of work that can be undertaken by the various grades are as follows:

PERMITTED SCOPES OF WORK BY DIFFERENT GRADES OF BUILDING MAIN CONTRACTORS

Special Grade	*Grade 1*	*Grade 2*	*Grade 3*
Unlimited	Maximum 240m height Maximum 40 storeys Maximum 200,000m^2 floor area	Maximum 120m height Maximum 28 storeys Maximum 120,000m^2 floor area	Maximum 70m height Maximum 14 storeys Maximum 60,000m^2 floor area

Contractors in China tend to be very large and employ a high percentage of direct labour. Specialist works are subcontracted (with the permission of the developer) to qualified subcontractors. The number of levels of subcontracting is relatively shallow when compared to the markets in Hong Kong or other countries. Most contractors are State-owned enterprises (SOEs) or branches of SOEs. They are enormous conglomerates but many are burdened by the welfare and retirement responsibility for their workers. Some SOEs are mini societies with their own schools and even hospitals. To be more competitive, the SOEs establish subsidiary companies to make them leaner to compete in the market. For example, the First Construction Unit of China State Construction has Engineering Bureaus 1-8 as main contractors in the local construction industry. These subsidiary companies compete against each other for business and even against the parent company, China State Construction Company, for work.

Apart from working locally and on a national scale, many Chinese contractors are increasingly getting involved in international contracts mostly in the field of civil engineering works in developing countries. The table below shows the major building contractors in China as ranked in the *Engineering News Record's Top 225 International Contractors.*

TOP 10 CHINESE CONTRACTORS RANKED BY REVENUE FROM PROJECTS OUTSIDE HOME COUNTRY, 2007-2008

	ENR rank by revenue	
Major contractors	*2007*	*2008*
China Communications Construction Group	14	18
China State Construction Engineering Corporation	18	21
China National Machinery Industrial Corporation	55	48
Sinohydro Corporation	51	50
China Railway Group Ltd.	67	71
CITIC Construction	98	72
China Petroleum Engineering Construction (Group) Corporation	70	76
China Metallurgical Group Corporation	95	81
Dongfang Electric Corporation	138	86
Shanghai Construction (Group) General Co.	73	90

Source: Engineering News Record Top 225 International Contractors, 2007 and 2008

Foreign contractors working in China used to operate under a project branch registration which was approved on a project by project basis under the *Interim Measures for the Administration of Qualifications of Foreign Enterprises Contracting Projects in China ("Decree 32").* Since 2003, following the issue of the *Regulation on Management of Foreign-Invested Construction Enterprises ("Decree 113"),* this arrangement has been abolished and any foreign contractor that wish to enter China's construction industry must either joint venture with a qualified Chinese contractor or obtain a qualification certificate from the MOC. The threshold for becoming a Special Grade Contractor (the only Grade worth being for many international contractors) is the same as that for local companies but it would be very difficult for foreign companies to meet as the MOC recognises only their investment and track record in China. As at the beginning of 2008, there are approximately 120 Special Grade Building Main Contractors with only one being a Foreign Invested Enterprise. Most foreign contractors therefore work under Construction Joint Ventures with local construction companies.

Competitive tendering has become the accepted practice for selection of contractors including government projects. It should be noted, however, that tendering in China is very different from that in other countries. All tenders, even for privately funded developments, are subject to the *Tender Law of the People's Republic of China.* The Law is intended to regulate tendering and bidding activities, maintain and improve fair and positive competition, enhance efficiency in resource allocation and prevent corruption in the tendering process. However, the rigid procedures stipulated by the Law basically deprive decision making by the Clients themselves and is therefore highly unpopular with private developers. The tendering process is handled by a qualified tendering agent who prepares the documents and sets the evaluation criteria in consultation with the Client. The tender award is later made upon recommendation by an evaluation committee that consists of at least 5 members of which two-thirds must be "specialists" which are randomly selected from a list of names approved by the local Construction Committee. The process is more appropriate for public funded projects and might initially have been intended for such. However, later interpretation of the term "projects of public interest and public safety" as stated in Article 3 of the Law has extended it to cover virtually all types of building developments. Tendering agency is a relatively new business. Firms are categorised into Grade A, B and C by the MOC, with Grade A being the highest and unlimited in the project size it can handle. As at the beginning of 2008, there are a total of approximately 790 Grade A tendering agents in China.

The Land System

There is no freehold land in China. Under the *1982 Constitution of the People's Republic of China*, all land is owned either by the State or by Agricultural Collectives. Basically land in the cities is owned by the State and land in the rural and suburban areas are owned by collectives. Despite such arrangement, all land is actually controlled by the State as the State has absolute discretion to requisite land from the Agricultural Collectives for public purposes, subject to payment of compensation and settlement fees. The *Land Administration Law* and its *Implementing Regulations* stipulates that only state owned land can be leased for commercial development purposes and therefore collective owned land may only be used for property developments after it is converted into state owned land through the stipulated land acquisition procedures.

Stated owned land use rights can be obtained by entering into a land grant contract with the government and the paying of a land grant premium. To increase government transparency and ensure fair trade, the Ministry of Land and Resources issued the *Rule of Granting of State-owned Land by Biding, Auction or Quotation* in 2002, requiring granting of land leases for commercial purposes must be done through land auctions or bidding at the State or provincial Land Exchanges. The land so leased can be developed, mortgaged or transferred.

The term of the leases vary according to the specified usage for the land and are normally as follows:

Land lease for industry purpose	:	50 years
Land lease for hotel, commercial and entertainment purpose	:	40 years
Land lease for science, technology, education, cultural, health purpose	:	50 years
Land lease for residential purpose	:	70 years

The *Rule of Granting of State-owned Land by the Agreement* issued in 2003 allows application for conversion of land use purpose (e.g. from industrial to commercial) for land which was acquired by agreement. The approval will be subjected to compliance with town planning rules and the payment of a conversion premium for the value thus added. The success of such applications are uncertain and can be quite time consuming but conversion premiums charged would normally be lower than the land prices demanded at auctions or tenders.

Speculative buying and selling of land is not encouraged. Land leased through auctions or tenders normally have to be developed within 4 years. Land not developed within this period face the imposing of penalties and, in extreme cases, confiscation.

Clients and Finance

Due to its closely controlled economy and not freely convertible renminbi, China was not affected much by the Asian Financial Crisis in 1998. Whilst other countries in Asia struggled to recover, China's economy started its strong growth and has been growing at an average rate of 11.5% in GDP since 1998. As mentioned above, President Hu Jintao's implementation of the Western Development Strategy boosted infrastructure investment and channelled private investment from the coastal cities to the inland provinces. The State Planning Commission (SPC) is responsible for preparing long term investment plans and approves all the major projects initiated by the various ministries and municipal governments. In general, however, decision-making and policy has become less centralised and the municipal governments are acting with increasing autonomy. This has made the initiation and implementation of new projects much faster than before.

The restructuring of the banking system in China in preparation for the opening of its finance market under the terms of the World Trade Organisation (WTO) Agreement has increased the availability of credit to both the business sector and ordinary individuals. Property mortgages, which used to be scarce previously, are now a major business for all Chinese banks. This has directly fuelled the increase in property prices. The emerging stock market has also provided an alternative cheap source of funds for many companies. In 2007, many local developers used initial public offerings (IPOs) to raise the necessary money to settle outstanding land grant payments. Such developers have been growing at an astonishing rate propelled by their vast land bank and have been building massive developments all over the country. The ease to obtain finance fuelled a property bubble which quickly became a concern for the Central Government.

Apart from spending by government and local developers, the property market also attracted a lot of interest from overseas investors. All major Hong Kong property conglomerates increased their investment in China at unprecedented rates. A large number of private equity funds have also been very active in China, snatching up anything that is feasible in the market. Most of these funds participate as joint venture partners with the original local developers, providing them with the necessary finance to complete the project.

Investors interested in property development must comply with the *Administrative Regulations of the People's Republic of China on Urban Real Estate Development and Operation*, effective 1998 which stipulates that only qualified Real Estate Development Enterprises (RDE) are eligible to conduct real estate development and operation. An RDE can be a permanent enterprise or a Special Purpose Vehicle (SPV) established for the development of a particular project and will cease to exist at the completion of the project. RDEs will need a real estate qualification which is divided into four grades: Grades 1, 2, 3 and 4.

The Central Government has been very concerned on the run away prices of properties and the increasing gap between the haves and have nots. To comply with President Hu Jintao's commitment to build a harmonious society, the government has been encouraging development of cheaper and smaller residential units that are more affordable by the masses. In 2006, a mandatory requirement was issued that all large housing developments must have 70% of its units of $90m^2$ gross floor area or below. Seeing opportunity in this trend, some companies with construction background have entered the low cost housing market. The returns are not high as the sale prices are dictated by the government who, on the other hand, virtually guarantees purchase of the units built.

Selection of Consultants

The designs of all construction works must be submitted for approval by a qualified Local Design Institute (LDI). These are multidisciplinary organisations, often several thousand strong, which provide the designs for local projects. As at the beginning of 2008, there are about 5,600 LDIs in China. Most of these institutes seek work independently whilst others are attached to various ministries, large enterprises or local government.

Official fee scales exist but competitive fee-bidding is also becoming an accepted practice now. Typically, fees will be in the region of 2.5% of the construction cost.

Any foreign architect or engineer working on a project in China must work with an LDI because only the latter is authorized to submit drawings to the local authority for approval. In addition, the foreign architect or engineer would also need the LDI to advise them on the Chinese planning and building codes and what technology and materials are locally available. The most common arrangement would be to employ a foreign architect and engineer to carry out the schematic design and the preliminary design for the project. The LDI may be employed by the architect/engineer or employed directly by the client to advise on the planning and building codes and make the necessary submissions. The LDI will then take over the design and produce construction drawings for the contractor to execute his works. Construction drawings in China are more detailed than that in Hong Kong or many other countries.

LDIs are allocated to one of the four classifications, namely

- Comprehensive Class, of which they are qualified to carry out design works for all 21 specified construction sectors, including building construction
- Designated Sector Class, of which they are only qualified to carry out design works for designated construction sector(s), e.g. building construction
- Designated Sub-Sector Class, of which they are only qualified to carry out design works for designated sub-sector(s), e.g. civil defence works under building construction sector
- Specialist Class, of which they are only qualified to carry out design works for specialised works under corresponding sectors, e.g. fitting out works, curtain wall, etc.

Each institute will be further graded according to the types and size of the projects that they are qualified to handle. The table below shows the grading for the building construction sector.

PROJECTS PERMITTED FOR VARIOUS GRADES OF DESIGN INSTITUTES IN BUILDING CONSTRUCTION SECTOR

Public Buildings	*Residential buildings*	*Industrial buildings and warehouses*	*Other*
Class A			
All types	All types	All types	All types
Class B			
Building not exceeding 20,000m^2 in size	Estate not exceeding 300,000m^2 in size	Warehouse and factory of all sizes	Fitting out works for 3-star hotel and standards below
Building Height not exceeding 50m	Building not exceeding 20 storeys	Building height not exceeding 50m	Building height not exceeding 50m
		Single storey building with maximum span less than 30m and working load of hoisting beam not exceeding 30 tons	

Public Buildings	*Residential buildings*	*Industrial buildings and warehouses*	*Other*
		6-storey building with maximum span less than 12m	
Class C			
Building not exceeding 5,000m^2 in size		Small-sized warehouse and simple-equipped factory	Fitting out works for 1-star hotel and standards below
Building height not exceeding 24m	Building not exceeding 12 storeys	Building Height not exceeding 24m Single storey building with maximum span less than 24m and working load of hoisting beam not exceeding 10 tons	Building height not exceeding 24m
		3-storey building with maximum span less than 6m and no live load at roof top	
Class D			
Building not exceeding 2,000m^2 in size	Not exceeding 2,000m^2 per block		Standard chimney not exceeding 20m high
	Reinforced concrete and blockwork structure		Water tank, with capacity not exceeding 50m^3

Public Buildings	*Residential buildings*	*Industrial buildings and warehouses*	*Other*
Building height not exceeding 12m	Not exceeding 4-storey high	Single storey building with maximum span of 15m and working load of hoisting beam not exceeding 5 tons or	Water pool, with capacity not exceeding 300m^3
		2-storey building with maximum span of 7.5m and no live load at roof top	Material storage tank, with diameter not exceeding 6m

The classification accords to the number of qualified staff employed by the institute, their experience and other related criteria. As at the beginning of 2008, there are approximately 1,300 Class A LDIs.

LDIs do not normally carry out supervision duties of the construction works. Under the communist system, LDIs and contractors are on the same standing and they virtually split the work required of a construction project: the LDI does the design and the contractor does the construction. This arrangement worked fine when all projects were publicly funded and everybody was working for the State. When the market opened in the 1980s, the MOC realised the lack of supervision on the Contractor's works. This gave rise to the construction supervisors' profession. The MOC stipulates that all new construction works and all major alteration works must be supervised by a qualified construction supervising firm who should be entrusted with the role of supervising the contractor and inspection of the quality of the works performed. The construction supervisors report to the Quality Inspection Department of the MOC. Construction supervising firms are classified into 3 grades: Grades A, B and C respectively, with Grade A being the highest. As at the beginning of 2008, there are approximately 1,220 Grade A construction supervising firms in China.

Construction cost control can be carried out by the construction supervisors or by construction cost engineers. Construction cost engineers are the equivalent of quantity surveyors in the British system. The local cost engineering firms tend to be localised and carry out specific tasks for submission purposes under local regulations (e.g. a preliminary estimate, a bills of quantities or an assessment of the final account) although a few large ones have adopted the UK style comprehensive approach. A number of international quantity surveying firms, mostly branches from their base in Hong Kong, are also active in the Chinese market. Hong Kong quantity surveying firms have been working in China since the early 1980s and it was them who introduced the competitive tendering and cost control concepts to

China. All Hong Kong quantity surveying firms offer the full comprehensive service as they do in Hong Kong. Projects carried out by international quantity surveying firms include not only those by foreign investors but also projects by local developers and semi-government corporations. International quantity surveying firms used to operate in China either in joint venture with local firms or under licences granted on a project by project basis. In 2007, the government issued *The Provision for Administration of Foreign-Funded Engineering Service Enterprises ("Decree 155")*, allowing the establishment of wholly foreign owned quantity surveying firms to practise in China provided they meet the same requirements stipulated for local firms. The employment of a quantity surveyor is not mandatory, but for many foreign investors, the employment of an international quantity surveyor would offer reassurance on tendering and contractual arrangements which can be complex or even appear irregular.

Contractual Arrangements

One of the most important laws relevant to the construction industry in China must be *The Construction Law* of the People's Republic of China which became effective from 1st March 1998. The Construction Law stipulates how construction activities should be carried out. It also specifically requires the employer and contractor of a construction project to enter into a written contract which clearly defines the rights and obligations of the two parties. However, the form of the contract to be employed is not stated. The most popular local standard form of contract is the *GF_1999-0201 Standard Form of Chinese Construction Contract*, issued by the Industry and Commerce Administration Bureau and the Ministry of Construction. Localised versions issued by the provincial construction departments are also available. These contracts should be used with care as the apportionment of risks and responsibilities between the client and the contractor differs quite widely with other popular standard forms like the JCT, HKIA or FIDIC forms originated from other countries. Of all the foreign standard forms introduced into China, the *Conditions of Contract for Work of Civil Engineering Construction* issued by the *Federation Internationale des Ingenieurs-Conseils* (FIDIC) has had the widest acceptance. The FIDIC also has an official Chinese translated version which gives it convenience of use as all contracts in China have to be translated into Chinese for registration purposes.

It should be noted that China has a business tax that is applicable to all transactions for services and works. A value added tax is applicable to sales of materials and equipment. The business tax for general services is 5% whilst that for construction works is around 3%. Contractors pay value added tax for materials and equipment bought plus an additional construction business tax on the total payment received from the client. For certain specifically listed equipment purchased directly by the client, the contractor will only pay for the value added tax. The contractual arrangements will have an impact on the total amount of tax payable and should therefore be carefully planned and arranged.

In the event of a contractual dispute, it is probable, depending on the provisions of the contract, that some form of arbitration or conciliation will be employed. Arbitration in China is governed by the *Arbitration Law of the People's*

Republic of China. The China International Economic and Trade Arbitration Commission (CIETAC) is the main arbitration commission to hear disputes with a foreign element. However if it is desired to hold the arbitration outside mainland China, popular choices include the Hong Kong International Arbitration Centre, Stockholm Chamber of Commerce and the Singapore International Arbitration Centre. Although arbitration is common in China, there is still a strong bias in favour of settling disputes without recourse to arbitration or litigation. Conciliation has long played a prominent role in Chinese society and culture and has always been actively employed on disputes arising from Chinese – Foreign business contracts.

Development Control and Standards

The master planning and elevation treatment of developments are controlled by the relevant planning commissions of the various cities and provinces. Detail design of the buildings must comply with the design codes and standards issued by the relevant Ministry Commission under the State Council. The followings are some commonly used codes and standards in building developments:

- Code for design of building foundation — GB50007-2002
- Load code for the design of building structures — GB50009-2001
- Code for design of concrete structures — GB50010-2002
- Code for seismic design of buildings — GB50011-2001
- Code for design of building water supply and sewerage — GB50015-2003
- Code for fire protection design of buildings — GB50016-2006
- Code for design of steel structures — GB50017-2003
- Code for design of heating, ventilation and air-conditioning — GB50019-2003
- Standard for lighting design of buildings — GB50034-2004
- Code for fire protection design of tall buildings — GB50045-95 (partly revised in 1999)
- Code for design of electric power supply systems — GB50052-95
- Code for design of dwelling houses — GB50096-99
- Code for design of high-rise structures — GB50135-2006
- Code for planning design of urban residential area — GB50180-93 (partly revised in 2002)
- Code for fire prevention design of interior decoration of buildings — GB50222-95 (partly revised in 2001)
- Standard for classification of urban land and for planning of constructional land — GBJ137-90
- Standard for daylighting design of buildings — GB/T50033-2001
- Standard for design of intelligent building — GB/T50314-2000

The *Construction Law of the People's Republic of China* addresses also construction controls and construction quality. The Construction Law requires that, prior to commencement of construction, construction permits be obtained from the administrative department in charge of construction for the local government at county level or above. The construction permit is granted only after planning and development approvals are obtained and evidence of funds for the project are provided, demonstrating that the design has been developed and drawings have been prepared, quality and safety procedures have been considered, etc. A construction permit will normally be issued within 15 days if an application meets the conditions. Construction must, however, start within three months of issue otherwise the permit will become void.

To ensure liability for the quality of the construction works remains intact with the designated main contractor, the Construction Law prohibits dissection of the works into different parcels to be carried out by different contractors, i.e. it is not allowed to build the podium of a project by one contractor and the towers on top by a different contractor.

CONSTRUCTION COST DATA

Cost of Labour

The figures below are typical labour costs in Shanghai as at the fourth quarter of 2008. The wage rate indicated represent the amount paid to the employee, however, for the cost of employing that employee, additions would have to be made to these rates to cover a variety of mandatory and voluntary contributions – a list of items which could be included is given in section 2.

	Wage rate (per month) Rmb	*Number of hours worked per year*
Site operatives		
Mason/bricklayer	1,000 - 1,500	2,080
Carpenter	1,000 - 1,500	2,080
Plumber	1,000 - 1,500	2,080
Electrician	1,000 - 1,500	2,080
Structural steel erector	1,000 - 1,500	2,080
HVAC installer	1,500 - 1,800	2,080
Semi-skilled worker	800 - 1,000	2,080
Unskilled labourer	700 - 900	2,080
Equipment operator	1,000 - 1,500	2,080
Watchman/security	500 - 700	2,080
Site supervision		
General foreman	3,450 - 4,600	2,080
Trades foreman	2,300 - 3,450	2,080
Clerk of works	3,450 - 4,600	2,080

	Wage rate (per month) Rmb	*Number of hours worked per year*
Contractors' personnel		
Site manager	5,750 - 6,900	2,080
Resident engineer	3,450 - 4,600	2,080
Resident surveyor	2,300 - 3,450	2,080
Junior engineer	1,750 - 2,900	2,080
Junior surveyor	1,750 - 2,900	2,080
Planner	1,750 - 2,900	2,080
Consultants' personnel		
Senior architect	8,000 - 10,350	2,080
Senior engineer	8,000 - 10,350	2,080
Senior surveyor	8,000 - 10,350	2,080
Qualified architect	5,200 - 6,350	2,080
Qualified engineer	5,200 - 6,350	2,080
Qualified surveyor	5,200 - 6,350	2,080

Labour costs all across the country are on the rise. Inflation in 2008 has been at 2% - 3% p.a. To keep pace with inflation, the Shanghai government raised the minimum rate from Rmb 840/month to Rmb 960/month to be effective in April 2008. This represents an increase of 14.3% over 7 months.

Cost of Materials

China presently produces most of the materials and equipment required for development projects. The figures below are the costs of main construction materials, delivered to site in Shanghai, as incurred by contractors in the fourth quarter of 2008. These assume that the materials would be in quantities as required for a medium sized construction project and that the location of the works would be neither constrained nor remote. All the costs in this section exclude business tax.

	Unit	*Cost Rmb*
Cement and aggregate		
Ordinary portland cement in 50kg bags	tonne	300
Coarse aggregates for concrete	tonne	52
Fine aggregates for concrete	tonne	53
Steel		
Mild steel reinforcement	tonne	4,550
High tensile steel reinforcement	tonne	4,850
Bricks and blocks		
Common bricks (240 x 115 x 53mm)	1,000	Not Used
Good quality facing bricks (240 x 115 x 53mm)	1,000	Not Used
Hollow concrete blocks (390 x 115 x 190mm)	1,000	1,880

	Unit	Cost Rmb
Timber and insulation		
Softwood sections for carpentry	m^3	1,500
Softwood for joinery	m^3	1,950
Hardwood for joinery	m^3	3,500
Exterior quality plywood (18mm thick)	m^2	52
Plywood for interior joinery (12mm thick)	m^2	35
Hardwood strip flooring (15mm thick)	m^2	50
100mm thick quilt insulation	m^2	35
Hardwood internal door complete with frames and ironmongery	each	800
Glass and ceramics		
Float glass (4mm)	m^2	35
Sealed double glazing units	m^2	220
Plaster and paint		
Good quality ceramic wall tiles (152 x 152mm)	m^2	35
Plaster in 50kg bags	tonne	250
Plasterboard (12mm thick)	m^2	26
Emulsion paint in 5 litre tins	kg	14
Gloss oil paint in 5 litre tins	kg	15
Tiles and paviors		
Clay floor tiles (200 x 200 x 8mm)	m^2	30
Non-slip vinyl floor tiles (305 x 305 x 1.5mm)	m^2	40
Precast concrete paving slabs (490 x 490 x 40mm)	m^2	40
Clay roof tiles (200 x 500mm)	1,000	3,200
Precast concrete roof tiles (390 x 390 x 40mm)	1,000	3,800
Drainage		
WC suite complete (set)	each	500
Lavatory basin complete (set)	each	400
100mm diameter clay drain pipes (2500mm long)	m	35
150mm diameter cast iron drain pipes (1830mm long)	m	160

Unit Rates

The descriptions on the next page are generally shortened versions of standard descriptions listed in full in section 4. Where an item has a two digit reference number (e.g. 05 or 33), this relates to the full description against that number in section 4. Where an item has an alphabetic suffix (e.g. 12A or 34B) this indicates that the standard description has been modified. Where a modification is major the complete modified description is included here and the standard description should be ignored; where a modification is minor (e.g. the insertion of a named hardwood) the shortened description has been modified here but, in general, the full description in section 4 prevails.

The unit rates below are for main work items on a typical construction project in Shanghai in the fourth quarter of 2008. The rates include all necessary labour, materials, equipment and an allowance to cover preliminary and general items. Five per cent should be added to the rates to cover contractors' overheads and profit. All the rates in this section exclude business tax.

		Unit	*Rate Rmb*
Excavation			
01	Mechanical excavation of foundation trenches	m^3	20
02	Hardcore filling making up levels (150mm)	m^2	20
Concrete work			
04A	Plain in situ concrete in strip foundations in trenches (C30)*	m^3	380
05A	Reinforced in situ concrete in beds (C40)*	m^3	420
06A	Reinforced in situ concrete in walls (200mm thick) (C40)*	m^3	420
07A	Reinforced in situ concrete in suspended floors or roof slabs (C40)*	m^3	420
08A	Reinforced in situ concrete in columns (C40)*	m^3	420
09A	Reinforced in situ concrete in isolated beams (C40)*	m^3	420
10	Precast concrete slab	m^2	250
* Note : Concrete strength grades are based on PRC standard			
Formwork			
11	Softwood formwork to concrete walls	m^2	40
12	Softwood or metal formwork to concrete columns	m^2	40
13	Softwood or metal formwork to horizontal soffits of slabs	m^2	40
Reinforcement			
14	Reinforcement in concrete walls	tonne	5,200
15	Reinforcement in suspended concrete slabs	tonne	5,200
Steelwork			
17	Fabricate, supply and erect steel framed structure	tonne	12,000
Brickwork and blockwork			
23A	Red brick wall (half brick thick)	m^2	Not used
23B	Red brick wall (one brick thick)	m^2	Not used
Roofing			
24A	Concrete interlocking roof tiles 490 x 490mm	m^2	50
29A	Felt roof covering	m^2	35

		Unit	*Rate Rmb*
Woodwork and metalwork			
34	Preservative treated sawn softwood 50 x 100mm	m	17
35	Preservative treated sawn softwood 50 x 150mm	m	20
37	Two panel glazed door in hardwood, size 850 x 2000mm	each	1,000
38	Solid core half hour fire resisting hardwood internal flush door, size 800 x 2000mm	each	900
41	Hardwood skirtings	m	25
Plumbing			
44A	Light gauge copper cold water tubing (20mm diameter)	m	30
Sanitary ware			
48	White vitreous china WC suite	each	750
49	White vitreous china lavatory basin	each	650
Electrical work			
52A	PVC insulated and copper sheathed cable ($4mm^2$)	m	15
53	13 amp unswitched socket outlet	each	55
54	Flush mounted 20 amp, 1 way light switch	each	45
Finishings			
55	2 coats gypsum based plaster on brick walls	m^2	28
56A	White glazed tiles on plaster walls (P.C Rmb $80/m^2$)	m^2	130
57A	Red clay quarry tiles on concrete floors (Ditto)	m^2	85
59	Thermoplastic floor tiles on screed	m^2	100
60	Mineral fibre tiles on concealed suspension system	m^2	160
Glazing			
61	Glazing to wood	m^2	65
Painting			
62	Emulsion on plaster walls	m^2	25
63	Oil paint on timber	m^2	30

Approximate Estimating

The building costs per unit area that follow are averages incurred by building clients for typical buildings in Shanghai as at the fourth quarter of 2008. They are based upon the total floor area of all storeys, measured between external walls and without deduction for internal walls.

Approximate estimating costs generally include mechanical and electrical installations but exclude furniture, loose or special equipment, and external works; they also exclude fees for professional services. The costs shown are for specifications and standards appropriate to China and this should be borne in mind when attempting comparisons with similarly described building types in other countries. A discussion of this issue is included in section 2. Comparative data for countries covered in this publication, including construction cost data, are presented in Part Three.

Approximate estimating costs must be treated with caution; they cannot provide more than a rough guide to the probable cost of building.

	Cost m^2 *Rmb*	*Cost* ft^2 *Rmb*
Industrial buildings		
Factories for letting	N/A	N/A
Factories for owner occupation (light industrial use)	4,300	399
Factories for owner occupation (heavy industrial use)	4,800	446
Factory/office (high-tech) for letting (shell and core only)	N/A	N/A
Factory/office (high-tech) for letting (ground floor shell, first floor offices)	N/A	N/A
Factory/office (high tech) for owner occupation (controlled environment, fully finished)	7,000	650
High tech laboratory workshop centres (air-conditioned)	7,500	697
Warehouses, low bay (6 to 8m high) for letting (no heating)	N/A	N/A
Warehouses, low bay for owner occupation	4,900	455
Warehouses, high bay for owner occupation	5,500	511
Cold stores/refrigerated stores	9,000	836
Administrative and commercial buildings		
Civic offices, fully air-conditioned	9,000	836
Offices for letting, 5 to 10 storeys, air-conditioned	5,700	530
Offices for letting, high rise, air-conditioned	6,200	576
Offices for owner occupation high rise, air-conditioned	7,200	669
Prestige/headquarters office, 5 to 10 storeys, air-conditioned	7,000	650
Prestige/headquarters office, high rise, air-conditioned	7,500	697
Health and education buildings		
General hospitals (1000 beds)	8,000	743
Private hospitals (500 beds)	9,200	855
Health centres	6,300	585
Nursery schools	4,000	372
Primary/junior schools	3,500	325
Secondary/middle schools	3,600	334
University (arts) buildings	8,500	790
University (science) buildings	9,500	883
Management training centres	9,600	892

	Cost m² Rmb	*Cost ft² Rmb*
Recreation and arts buildings		
Theatres (over 500 seats) including seating and stage equipment	11,000	1,022
Theatres (less than 500 seats) including seating and stage equipment	11,500	1,068
Concert halls including seating and stage equipment	11,000	1,022
Swimming pools (international standard) including changing and social facilities (outdoor)	each	4,000,000
Swimming pools (schools standard) including changing facilities (outdoor)	each	1,800,000
National museums including full air-conditioning and standby generator	14,000	1,301
Local museums including air-conditioning	12,000	1,115
City centre/central libraries	8,000	743
Branch/local libraries	6,500	604
Residential buildings		
Private/mass market single family housing 2 storey detached/semi detached (multiple units)	7,500	697
Purpose designed single family housing 2 storey detached (single unit)	9,000	836
Social/economic apartment housing, high rise (with lifts)	2,000	186
Private sector apartment building (standard specification)	3,500	325
Private sector apartment buildings (luxury)	4,600	427
Student/nurses halls of residence	2,100	195
Homes for the elderly (shared accommodation)	2,400	223
Homes for the elderly (self contained with shared communal facilities)	2,800	260
Hotel, 5 star, city centre	10,400	966
Hotel, 3 star, city/provincial	6,700	622

Regional Variations

The approximate estimating costs shown above are based on projects in Shanghai. For other parts of China, adjust these costs by the following factors:

Beijing	±0%
Guangzhou	-5%
Chongqing	-7%
Wuhan	-8%
Shenyang	-10%
Tianjin	-7%

Tax

The basic tax obligations of the construction industry are stipulated in the state finance and tax regulations.

Building materials and equipment imported to China must pay import duty. The rates chargeable for import duty vary from material to material and currently the average rate is 23% plus Value Added Tax (VAT) of 17%. It is likely that this will be subject to further change as China implements the terms of the World Trade Organisation.

VAT is levied on the sales revenue of manufacturing companies, imported and exported goods and repair and processing services. Business tax is levied on the sales revenue of service firms, immovable property sales and intangible goods. The amount of business tax is dependent upon the industry, for construction it is currently 3%.

In certain regions, Land Appreciation Tax may be levied which applies to the gain arising from the transfer of real estate and is imposed on all entities or individuals, whether foreign or domestic.

Overall the rules governing tax are complex and subject to change and it is recommended that a tax consultant's advice be sought on a case by case basis.

EXCHANGE RATES AND INFLATION

The combined effect of exchange rates and inflation on prices within a country and price comparisons between countries is discussed in section 2.

Exchange Rates

The United States and Europe has recently been exerting great pressure on China to appreciate the value of its currency, the Renminbi. China has been responding by a controlled escalation. The graph overleaf plots the movement of the Chinese Renminbi against the sterling, the euro, the US dollar and 100 Japanese yen since 1998. The values used for the graph are quarterly and the method of calculating these is described and general guidance on the interpretation of the graph provided in section 2. The exchange rate in the fourth quarter of 2008 was Rmb 10.70 to pound sterling, Rmb 9.00 to euro, Rmb 6.83 to US dollar and Rmb 7.15 to 100 Japanese yen.

THE CHINESE RENMINBI AGAINST STERLING, EURO, US DOLLAR AND 100 JAPANESE YEN

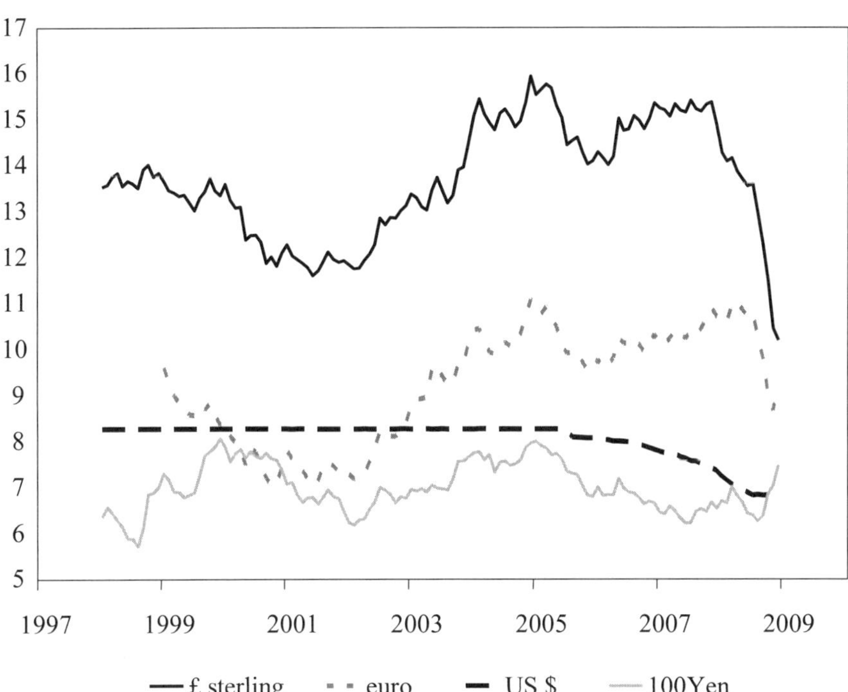

Price Inflation

The Consumer Price Index in February 2008 recorded an annual increase of 8.7%, the highest since 1997. Part of the reason was due to an extraordinary surge of 23.3% in the Foods category, caused by the worst winter in 50 years that damaged power supply lines and caused transport disruptions, making it difficult or impossible to replenish supplies. The Chinese have a dreaded memory of inflation that dated back to the pre-communist days. Prevention of inflation has therefore always been high on the agenda of the Central Government and strong administrative measures to keep it in check are quickly implemented. However, the inflation rate peaked in April and then started to drop due to the effects of the global financial tsunami caused by the sub-prime crises in America. The table on the next page presents the consumer price and construction cost inflation in China since 1998.

CONSUMER PRICE AND CONSTRUCTION COST INFLATION

Year	Consumer price inflation average index	Consumer price inflation average change %	Construction cost index average index	Construction cost index average change %
1998	100.0	-0.8	100.0	0.5
1999	98.6	-1.4	100.3	0.3
2000	99.0	0.4	102.7	2.4
2001	99.7	0.7	104.1	1.4
2002	98.9	-0.8	105.1	1.0
2003	100.1	1.2	109.5	4.2
2004	104.0	3.9	118.5	8.2
2005	105.8	1.7	120.6	1.8
2006	107.4	1.5	122.2	1.3
2007	112.6	4.8	128.4	5.1
2008	117.8*	4.6*	149.2*	16.2*

** Figures for Third Quarter of the year*

Source: Consumer Price Index and Construction & Installation Price Index, National Bureau of Statistics

USEFUL ADDRESSES

Public Organizations

Beijing Urban Engineering Design & Research Institute Co Ltd
5 Fuchengmen Bei Da Street
Xicheng District
Beijing 100 037
Tel: (86) 010 8833 6666
Fax: (86) 010 6830 0793
E-mail: service@buedri.com
Website: www.buedri.com

Bureau of Foreign Trade and Economic Cooperation of Guangzhou Municipality
11/F-14/F, 158 Dongfeng Road West
Guangzhou 510 170
Tel: (86) 020 8109 7472
Fax: (86) 020 3892 0747/3892 0724
E-mail: info@investguangzhou.gov.cn
Website: www.gzboftec.gov.cn

Bureau of Urban Planning of Guangzhou Municipality
80 Jixiang Road
Guangzhou 510 030
Tel: (86) 020 8319 1057
E-mail: webupo@upo.gov.cn
Website: www.upo.gov.cn

China Academy of Urban Planning & Design
5 Chegongzhuang Xi Road
Beijing 100 044
Tel: (86) 010 5832 2222
Fax: (86) 010 5832 2000
E-mail: caup@caupd.com
Website: www.caupd.com.cn

China Building Material Industry Association
11 Sanlihe Road
Haidian District
Beijing 100 831
Tel: (86) 010 8837 6372
E-mail: hyx@cbminfo.com
Website: www.bm.cei.gov.cn

China National Democratic Construction Association
208 Jixiang Lane
Chaoyangmen Wai Da Street
Beijing 100 020
Tel: (86) 010 8569 8008
Fax: (86) 010 8569 8007
E-mail: webmaster@cndca.org.cn
Website: www.cndca.org.cn

Chinese Academy of Sciences
52 Sanlihe Road
Beijing 100 864
Tel: (86) 010 6859 7289
Fax: (86) 010 6851 2458
E-mail: bulletin@mail.casipm.ac.cn
Website: www.cas.cn

Chinese Architecture Design & Research Group
19 Chegongzhuang Street
Beijing 100 044
Tel: (86) 010 6830 2001
Fax: (86) 010 6834 8832
E-mail: yb@cadg.cn
Website: www.cadreg.cn

Construction and Communications Commission of Shanghai Municipal Government
100 Dagu Road
Shanghai 200 003
Tel: (86) 021 6443 1576
E-mail: xxyjzx@public.shucm.sh.cn
Website: www.shucm.sh.cn

Construction Commission of Guangzhou Municipality
1 Fuqian Road
Guangzhou 510 032
Guangzhou
Tel: (86) 020 8312 5910
Fax: (86) 020 8312 5810
E-mail: gzcc@gzcc.gov.cn
Website: www.gzcc.gov.cn

General Administration of Quality Supervision, Inspection and Quarantine
9 Madian East Road
Haidian District
Beijing 100 088
E-mail: webmaster@aqsiq.gov.cn
Website: www.aqsiq.gov.cn

Guangzhou Environmental Protection Bureau
20 Zengsha Street, Huilong Road
Guangzhou 510 115
Tel: (86) 020 8337 3017
Fax: (86) 020 8332 9508
E-mail: gzeic@gzepb.gov.cn
Website: www.gzepb.gov.cn

Guangzhou Prices Bureau
Block A2 Jinguiyuan
1382 Jiefang Road North
Guangzhou 510400
Tel: (86) 020 8619 6800
Fax: (86) 020 8619 6820
E-mail: admin@gzwjj.gov.cn
Website: www.gzwjj.gov.cn

Ministry of Construction
9 Sanlihe Road
Haidian District
Beijing 100 835
Tel: (86) 010 5893 4114
E-mail: cin@mail.cin.gov.cn
Website: www.cin.gov.cn

Ministry of Foreign Trade & Economic Cooperation
2 Dong Chang'an Street
Beijing 100 731
Tel: (86) 010 6519 8114
Fax: (86) 010 6519 8173
E-mail: moftec@moftec.gov.cn
Website: www.cofortune.com.cn/moftec_cn

Ministry of Land and Resources
64 Fu Nei Street
Xicheng District
Beijing 100 812
Tel: (86) 010 6655 8407/08/20
Fax: (86) 010 6612 7247
Website: www.mlr.gov.cn

National Bureau of Statistics of China
57 Yuetan Nanjie Street South
Sanlihe, Xicheng District
Beijing100 826
Fax: (86) 010 6878 2000
E-mail: info@stats.gov.cn
Website: www.stats.gov.cn

National Development and Reform Commission
38 Yuetan Street South
Xicheng District
Beijing 100 824
E-mail: ndrc@ndrc.gov.cn
Website: www.sdpc.gov.cn

Policy Research Centre, Ministry of Construction (China Urban and Rural Construction Economy Research Institute)*
Sanlihe Road
Beijing 100 835
Tel: (86) 010 5893 3439

Shanghai Construction Engineering Tender Administration*
3/F 683 Xiaomuqiao Road
Shanghai 200 032
Tel: (86) 021 5461 4788
Fax: (86) 021 6404 4550

Shanghai Development & Reform Commission Cum Shanghai Municipal Bureau of Price Control
200 Renmin Da Road
Shanghai 200 003
Tel: (86) 021 6321 2810
Fax: (86) 021 6358 6443
E-mail: bgs@shdpc.gov.cn
Website: www.shdpc.gov.cn

Shanghai Foreign Economic Relation and Trade Commission
16/F, New Town Mansion
55 Lou Shan Guan Road
Shanghai 200 336
Tel: (86) 021 6275 2200
Fax: (86) 021 6275 8166
E-mail: chairman@smert.gov.cn
Website: www.smert.gov.cn

Trade And Professional Associations

Architectural Society of China
9 Sanlihe Road
Beijing 100 835
Tel: (86) 010 8808 2224
Fax: (86) 010 8808 2223
E-mail: asc@chinaasc.org
Website: www.chinaasc.org

China Civil Engineering Society
9 Sanlihe Road
Beijing 100 835
Tel: (86) 010 5893 3071
E-mail: international@cces.net.cn
Website: www.cces.net.cn

China Construction Industry Association*
Level 7, Block 7, Jiulong Commercial Centre
48 Zhongguancun Street South
Beijing 100 081
Tel: (86) 010 6213 7390
Fax: (86) 010 6213 7191
E-mail: zgjzyxh@163.com
Website: www.zgjzy.org

China Engineering Cost Association
9 Sanlihe Road
Haidian District
Beijing 100 835
Tel: (86) 010 5893 4013
Fax: (86) 010 5893 4648
E-mail: ceca@ceca.org.cn
Website: www.ceca.org.cn

Chinese Mechanical Engineering Society
46 Sanlihe Road
Xicheng District
Beijing 100 823
Tel: (86) 010 6859 5316
Fax: (86) 010 6853 3613
E-mail: headquarters@cmes.org
Website: www.cmes.org

* Those organisations marked with an asterisk have no official English name, the name indicated is an approximation based on the translation of the Chinese name.

HONG KONG

All data relate to 2007 unless otherwise indicated.

Population	
Population	6.95 mn
Urban population (Census 2006)	94%
Population under 15	13.0%
Population 65 and over	12.7%
Average annual growth rate (2004 to 2007)	0.75%
Geography	
Land area	1,079 km^2
Agricultural area	6.1%
Capital city	-
Economy	
Monetary unit	Hong Kong Dollar (HK$)
Exchange rate (average fourth quarter 2008) to:	
the pound sterling	HK$ 12.19
the US dollar	HK$ 7.75
the euro	HK$ 10.20
the yen x 100	HK$ 8.09
Average annual inflation (1998 to 2007)	-0.7%
Inflation rate (2007)	2%
Gross Domestic Product (GDP) #	HK$ 1,616 bn
GDP per capita #	HK$ 233,358
Average annual real change in (GDP) (1998 to 2007) #	2.5%
Private consumption as a proportion of GDP #	59.8%
Public consumption as a proportion of GDP #	8.0%
Investment as a proportion of GDP #	20.3%
Construction	
Gross value of construction output	HK$ 93 bn
Net value of construction output	HK$ 37.9 bn
Net value of construction output as a proportion of GDP	2.3%

 # Provisional figures

THE CONSTRUCTION INDUSTRY

Construction Output

The construction industry suffered a severe setback after the Asian Financial Crisis in 1998. The gross value of construction work dropped to HK$93 billion in 2007 as compared to HK$133 billion at its peak in 1998. Conditions started to improve by the later part of 2005 and continued until 2008 when the global financial tsunami hit the economy again.

In the 2007-2008 Policy Address, the Chief Executive announced the undertaking of 10 major infrastructure projects to boost Hong Kong's regional competitiveness and to promote further integration with the adjoining cities of the Pearl River Delta and beyond. These 10 major infrastructure projects total to an estimated cost of HK$250 billion and comprises: (1) the South Island Rail Line, (2) the Shatin to Central Rail Link, (3) The Tuen Mun Western Bypass and Tuen Mun to Chek Lap Kok Link, (4) the Guangzhou–Shenzhen–Hong Kong Intercity Express Rail Link, (5) the Hong Kong–Zhuhai–Macao Bridge, (6) the Hong Kong–Shenzhen Airport Rail Link, (7) the Hong Kong–Shenzhen joint development of the Lok Ma Chau Loop, (8) the West Kowloon Cultural District, (9) the Kai Tak Development which includes the new Cruise Terminal and (10) the establishment of New Development Areas.

In addition to the 10 major infrastructure projects, all Universities will be expanding their facilities to cope with extra student intake by 2012 as the education system in Hong Kong commenced its change to a 3-3-4 academic structure in 2006.

On the public housing side, the Hong Kong Housing Authority is targeting to produce 77,000 rental flats from 2007/08 to 2011/12, an average of 15,500 flats per year, in order to comply with their target of maintaining the average waiting time for public rental flats to 3 years.

Whilst spending on public works has been increased, private sector work is experiencing a slow recovery. The private property market was one of the hardest hit by the Asian Financial Crisis. Investment confidence was seriously dampened. According to the Hong Kong Monetary Authority, the number of residential mortgage loans with negative equity peaked at 106,000 cases in mid-2003. The number was gradually reduced and had been more or less eliminated by end 2007. Revived confidence has pushed up demand and prices, however new developments have been severely restricted by the difficulty to secure land. During the recession, the Hong Kong Government adopted an Application List for Land Sales system in lieu of the old regular public auction system to curb the fall in land value of new sites. It has since continued the use of the system to date although there are calls from developers to revert to the old system. The new system reduces the availability of new land plots for sale and directly affects the land supply, one of the most scarce and valuable commodities in Hong Kong. Conversion of use for privately owned sites has also been more difficult in face of growing awareness of environmental concerns and historical heritage by the public. In order to resolve concerns regarding "wall effects" caused by high-density buildings which affect ventilation and lead to heat islands, the Government has promised to review the outline zoning plans of each district with an aim to slightly lowering the overall plot ratio to reduce development density. According to figures released by the Transport and Housing Bureau, the number of residential units completed in 2008 was only 8,800 units as compared with 10,500 units in 2007, 16,600 units in 2006 and 17,300 units in 2005.

The difficulty in securing land for development has prompted many Hong Kong developers to cross the border and venture into the hot property market in China. From 2006 to 2008, Hong Kong developers have been buying land in China of sizes and at prices unheard of before. Their purchases are not restricted only to the main cities of Beijing, Shanghai, Guangzhou and Shenzhen, but also in many second tier cities like Chengdu, Chongqing, Dalian, Shenyang, Suzhou, Wuxi, Hangzhou, Zhengzhou and Xian, etc.

In general terms, overall expenditure on building and construction in Hong Kong is forecasted to increase in 2009, mainly due to an increase in Government spending.

The table below shows the expenditure on different types of construction works in 2007.

DISTRIBUTION OF GROSS VALUE OF CONSTRUCTION WORKS (2007)

Types of work	*%*
Residential	36
Commercial	25
Industrial and storage	3
Service building	12
Transport	14
Other utilities and plant	3
Environment	4
Sports and recreation	3

Source: Works Digest, Government of the Hong Kong S.A.R.

Characteristics and Structure of the Industry

As at 2008, there are around 260 approved contractors on the Hong Kong Government lists of approved contractors for public works. Contractors are classified into three groups as follows.

- Group A
  Bidding contracts of value not exceeding HK$20 million.
- Group B
  Bidding contracts of value not exceeding HK$50 million.
- Group C
  Bidding contracts of value exceeding HK$50 million.

Contractors are classified into either Group A, B or C. They are further classified under their respective expertise into any or all of five possible categories: Buildings, Port Works, Roads and Drainage, Site Formation and Waterworks. In 2008, there were 58 contractors approved for building work contracts of value exceeding HK$50 million (Group C), 47 contractors approved for contracts up to HK$50 million (Group B) and 45 contractors approved for contracts up to HK$20 million (Group A).

The listings are mandatory when tendering for government funded projects only. There are no restrictions on private developers. However, in practice, private developers will usually make reference to the *Government List of Approved Contractors* when considering a particular contractor for their works.

For most projects, a main contractor is chosen and made responsible for constructing the project and employing approved subcontractors for major sections of works such as curtain walling or windows and building services. These specialist works may also be tendered separately and nominated to the main contractor as nominated subcontractors.

Recently, Government has been using Design-and-Build contracts for selected projects to enhance administration and budgetary control. Notable projects awarded under such arrangement include the Civil Aviation Department Building, the Customs and Excise Department Building and the highly publicized Tamar Development which covers the Central Government Complex, the Legislative Council Complex, the Chief Executive's Office and public open space. However this arrangement is not widespread amongst the private sector.

The main contracting system consists of a multi-layered subcontracting system whereas the main contractor employs subcontractors who in turn employ a host of smaller subcontractors usually providing workers only. The system gives the whole industry a high flexibility and assists to keep costs down. However, it also gives rise to vague lines of responsibility and difficulties in tracking the workers on site. The majority of construction workers are engaged on a daily basis with only a very small percentage being employed by the month. Unionization is not strong and there have been numerous cases of the head of a subcontractor abandoning their employees without paying them their due salaries. These incidents lead to demonstrations on sites which are usually politicized. To ensure that main contractors administer proper governance of their subcontractors and to minimize the number of incidents, Government passed a legislation making it mandatory for the main contractors to be liable for a maximum of 2 months of wages in arrears for workers working on their sites, irrespective of whether they were directly employed by the main contractors themselves or not. The main contractor may try to recover the sum from the direct employers of the workers through civil claims proceedings.

Another drawback of the multi-layered subcontracting system is the difficulty to ensure whether the workers themselves were skilled in their respective trades or not. To address the problem, the Government established a Construction Workers Registration System under the Construction Workers Registration Authority to keep a central record of the skill levels of all construction workers in Hong Kong. The Construction Industry Council is delegated by the Authority to run the Registrar which commenced operation in December 2005. A levy of 0.03% on the value of all construction works exceeding HK$1 million in contract value is charged to pay for the working of the registration system.

The Construction Industry Council also runs training centres and provides basic courses for craftsmen and technicians. On completion of these courses, trainees are apprenticed to building contractors for a period of two or three years. The Council is funded by a special levy on newly completed construction works. Currently the rate paid by contractors is 0.4% of the value of work all construction works with contract values exceeding HK$1 million.

The construction industry depends heavily on imported materials. Government specifications still rely heavily on British Standards to define the pre-requisite quality levels. However, the use of "equivalents" subject to testing and approvals are becoming common. Hong Kong is a free port and has few import restrictions. Import licences are generally not required and duties are not applicable except for some specific items. Sales or purchase taxes are not applicable to construction materials or equipment. Because of the dependence of Hong Kong on imports, the prices of materials are heavily influenced by exchange rates. A significant amount of basic building materials are imported from China nowadays. The pegging of the Hong Kong dollar to the US dollar has caused the Hong Kong dollar to depreciate against the Renminbi and has been pushing up the price of imports from China, causing construction prices to rise.

Clients and Finance

In 2007, the private sector accounted for about 67% and the public sector about 33% of all new works in Hong Kong.

Public sector works mainly come from the Transport and Housing Bureau which supervises the Transport Department, the Civil Aviation Department, the Highways Department, the Marine Department and the Housing Department. The Secretary for Transport and Housing also serves as the Chairman of the Hong Kong Housing Authority. The Secretary for Transport and Housing is assisted by the Permanent Secretary for Transport and Housing (Housing), who also assumes the office of the Director of Housing. The Hong Kong Housing Authority is a statutory body established in April 1973 under the Housing Ordinance. It was previously responsible for coordinating public housing but since April 1988, it has been responsible for nearly all public housing in Hong Kong including policy formulation. The Housing Authority is required to be self-financing. It operates a comprehensive construction division encompassing architectural, engineering, building services, quantity surveying, planning, construction supervision and maintenance divisions.

Other sources of public works come from the numerous quasi-government organizations. The University Grants Committee controls funding to the eight higher education institutions including the City University of Hong Kong (CityU), the Hong Kong Baptist University (HKBU), the Lingnan University (LU), the Chinese University of Hong Kong (CUHK), the Hong Kong Institute of Education (HKIEd), the Hong Kong Polytechnic University (PolyU), the Hong Kong University of Science and Technology (HKUST) and the University of Hong Kong (HKU). The Hospital Authority manages all 41 public hospitals and institutions. The Urban Renewal Authority is entrusted with acquisition of deteriorating properties for whole block redevelopments and tends to carry out most of its developments in joint venture with private developers.

Private building works are mainly financed by property developers and by large enterprises such as Cheung Kong, Hutchison Whampoa, Sun Hung Kai, Henderson, Wharf, Swire, Hong Kong Land, New World, Hysan, Kerry, China Resources, K Wah, Sino, USI, China Overseas, PCCW, Hong Kong Bank, etc. The property developers in Hong Kong tend to be large and sophisticated. All of them have their own in-house project management departments and some, like Sun Hung Kai, Henderson and China Resources, even have their own contracting arm and/or sales division.

Another major source of funds for development is The Hong Kong Jockey Club. Of the very large sums of money which accrue to the Club from racing and lotteries, the majority is of course returned to the punters. However, the Club donates heavily to charity organizations and regularly finances community development projects. For example, the University of Science and Technology, Kowloon Park, the Hong Kong Academy for Performing Arts, the Ocean Park, the Hong Kong Park, the Hong Kong Stadium and the Kau Sai Chau Public Golf Course were either partially or wholly funded by this body.

Another new source of work for contractors and consultants in Hong Kong is the large amount of construction works built in nearby Macau. Most of the new casinos, hotels and luxury apartments built in this small enclave from 2004 to 2008 were carried out by Hong Kong contractors and workers. However, following tightening of permit issues by China for mainland visitors and the credit crunch caused by the global financial tsunami in 2008, works in Macau reduced drastically in 2009.

Selection of Design Consultants

Hong Kong has a large number of well-established firms of architects, both of overseas and local origin. There are many foreign architects, mainly from the UK, Australia and the USA, working in Hong Kong. Small interior design and minor work practices proliferate on the back of the never-ending fitting out of offices, shops, flats, etc. All public work, such as hospitals, schools, museums and council chambers fall under the ambit of the Architectural Services Department who may carry out the design themselves or outsource the work to private architects, engineers and quantity surveyors. The Housing Authority also has a similar policy of outsourcing.

The main professional bodies in the construction industry are the Hong Kong Institute of Architects with approximately 2,500 members and the Hong Kong Institute of Engineers with approximately 12,000 members. On many larger schemes, a project manager may be appointed to replace the architect as the project coordinator. Most of the large engineering firms also provide project management/coordination services.

The profession of the quantity surveyor is recognized in Hong Kong but there is no compulsory registration requirement except for undertaking government projects. Anyone can set up as a quantity surveyor or cost consultant practice and undertake quantity surveying work for private developments. The Hong Kong Institute of Surveyors (HKIS) was established in 1984 to take over the role of the Royal Institution of Chartered Surveyors [RICS] for education and professional qualification assessment in the territory. The HKIS and the RICS have a reciprocity agreement whereby members of the RICS are eligible for HKIS membership after working for 1 year in Hong Kong and vice versa. The RICS re-opened their Hong Kong Branch in 2003 mainly to serve their members in Hong Kong.

There are no hard and fast guidelines for selection of consultants and the government changes their selection criteria every now and then. The Government requires their consultants, as well as contractors, to operate a Quality Assurance System to ISO 9000. Tenders are called for nearly all consultancy appointments and are handled on a two-envelope system with different weightings given for the technical and the financial submissions.

Architects are appointed directly by the client and may or may not be required to include the structural engineer and the building services engineer under their appointment.

Quantity surveyors and cost consultants are normally appointed separately by the client and reports directly to him. Price is always a substantial criterion in both the public and private sector and usually a lump sum fee is used in the consultancy contracts.

All the main professional bodies publish non-mandatory fee scales and they are sometimes used for public work as well as for private sector work. However, it is not known how often they are used or what discounts are negotiated. Fees generally are being driven down by competition.

Contractual Arrangements

The majority of contracts are awarded on a lump sum basis based on competitive tenders received from a list of firms that have been selected by the clients and his consultants. The Government has a mandatory tendering process to ensure fairness and openness. This is followed by the quasi-government organizations and even by many private developers. Generally tenders are based on measured bills of quantities though other methods may sometimes be used.

The most common form of contract used in Hong Kong for private clients is the *Agreement and Schedule of Conditions of Building Contract for use in the Hong Kong SAR (the Hong Kong Standard Form of Contract), 2005 Edition,* issued jointly by the Hong Kong Institute of Architects, the Hong Kong Institute of Surveyors and the Hong Kong Institute of Construction Managers. There are two versions available; one is the 'with quantities' form which assumes the use of measured bills of quantities that are normally prepared by the quantity surveyor; the other is the 'without quantities' version.

It is normal for these contracts to be modified by a series of special conditions, introduced by the client or his advisors which serve to supplement the general conditions of contract to suit particular requirements of individual projects.

The Hong Kong Government has three separate forms of contract for use in the construction industry:

- General Conditions of Contract for Building Works (1999 Edition);
- General Conditions of Contract for Civil Engineering Works (1999 Edition) and
- General Conditions of Contract for Electrical and Mechanical Engineering Works (1999 Edition)

Contracts are generally let on a firm price basis, the client being keen to eliminate the cost uncertainty associated with fluctuations. Public clients, such as the Architectural Services Department and the Housing Authority, adopt a contract price fluctuation system (CPFS) for their projects unless there are practical problems for not doing so. Under the CPFS provisions stipulated by the Development Bureau, fluctuations are normally given for both labour and materials, calculated by means of a formula using data issued by the Government Census and Statistics Department. Other alternative contractual arrangements used include management contracting, design-and-build, cost plus and construction management. However, most of these alternatives do not have standard contract forms for use in Hong Kong and if adopted, standard forms from foreign countries (usually from UK) would have to be tailored to fit.

Liability and Insurance

Any departure by either party from the strict terms of the agreement is a technical breach of contract and therefore actionable at law with the remedy of damages. However, most contracts will have settlement procedures (e.g. dispute resolution, mediation and arbitration) built into the agreement. The courts will, in most cases, seek to ensure that such procedures have been tried or were inapplicable to the circumstances which have arisen. For a contract executed under hand, claims for breach of contract may only be made within six years from the date of the breach. Where a contract is executed under seal, that period is extended to twelve years.

The contractor is also liable in common law to exercise proper skill and care in his operations. Such liability would apply between two parties even if they have no direct contractual connection. It would be sufficient to demonstrate that one party owes a duty of care to the other party.

The consultants have a contractual relationship only with their Employer (the client) via their contract of service. Breach of their contract of service which affects the contractor may only be remedied by the contractor suing the Employer and the Employer in turn taking action against the consultants. Most clients require their consultants to carry professional indemnity insurance to ensure appropriate protection. Consultants and contractors may also be subject to criminal sanctions or claims for breach of statutory duties if the parties fail to comply with the local legislative framework. A typical example would be a breach of *Health & Safety Statutes.*

All nominated subcontractors are required to be subject to broadly the same terms as the main contractor and to indemnify the main contractor against the same liabilities as the main contractor himself might bear. In the *Standard Hong Kong Form of Contract*, there is no contractual relationship between the Employer and the nominated subcontractor. In response to this, it is common for the Employer to enter into a collateral agreement with the nominated subcontractor in order to create a contractual link.

Insurances covering the construction works themselves, people/employees engaged in the works, and third party/public liability are addressed in the *Standard Forms of Contract.*

Development Control and Standards

There is no freehold property in Hong Kong and the Government owns all land and allocates its use in accordance with outline zoning plans. Planning laws are not too onerous but are currently being tightened. Planning and control of development is the responsibility of the Town Planning Board and the Planning Department. The use of a site is designated by means of outline *Zoning Plans* and a plot ratio and height limit are set and then leases are auctioned. The shortage of land in Hong Kong means that typically more than 70% of the cost of a city centre development is in land cost. Land auctions are an important source of revenue for the Exchequer. Applications to modify the zoning of a particular site may be made to the Town Planning Board. If approved, the case will be passed to the Lands Department who will determine the conversion premium. The conversion premium is a payment made to the Government calculated on the enhanced market value of a site based on its new designated use or increased development area. Often the premium can be quite significant.

Planning applications are made to the Buildings Department and decisions are normally given within three months of submission. Typically there will be some negotiation with the Planners and a resubmission may be necessary. All projects must comply with the government regulations of which the main one is the *Buildings Ordinance (Cap 123 of the Laws of Hong Kong).* Building control is exercised through the appropriate department of the Buildings Department which approves development and building plans ensuring that they comply with the relevant regulations. During construction, the Architect is directly responsible to the Buildings Department and "spot" site inspections are carried out regularly.

Geotechnical conditions in Hong Kong are a major factor in any development. Piling or other major foundation work together with slope stabilization is the "norm". Calculations and construction details for these items are submitted to the Buildings Department for approval. An advisory service is also provided by the Geotechnical Engineering Office.

There is also a move to raise environmental sustainability in new buildings. In line with the Action Blue Sky Campaign, Hong Kong has pledged to achieve a reduction in energy intensity of at least 25% by 2030 (with 2005 as the base year). To this end, the Government will be looking into mandatory implementation of Building Energy Codes by means of legislation. It will also set examples by conducting Carbon Audits and implementing emissions reduction provisions to their own new major projects, such as the Tamar Complex.

CONSTRUCTION COST DATA

Cost of Labour

The figures below are typical of labour costs in Hong Kong as at the fourth quarter of 2008. The wage rate is the basis of an employee's income, while the cost of labour indicates the cost to a contractor of employing that employee. The difference between the two covers a variety of mandatory and voluntary contributions – a list of items which could be included is given in section 2.

	Wage rate (per day) HK$	*Cost of labour (per day) HK$*	*Number of hours worked per year*
Site operatives			
Bricklayer	825	910	2,400
Mason	780	860	2,400
Carpenter	965	1,060	2,400
Plumber	795	875	2,400
Electrician	690	760	2,400
Structural steel erector	885	975	2,400
Semi-skilled worker	600	660	2,400
Unskilled labourer	565	620	2,400
Equipment operator	745	820	2,400
	(per month)	*(per month)*	
Watchman/security	8,000	9,600	2,400
Site supervision			
General foreman	40,000	48,000	2,400
Trades foreman	25,000	30,000	2,400
Clerk of works	25,000	30,000	2,400
Contractors' personnel			
Site manager	50,000	60,000	2,400
Resident engineer	40,000	48,000	2,400
Resident surveyor	30,000	36,000	2,400
Junior engineer	17,000	20,400	2,400
Junior surveyor	17,000	20,400	2,400

	Wage rate (per month) HK$	*Cost of labour (per month) HK$*	*Number of hours worked per year*
Planner	25,000	30,000	2,400
Consultants' personnel			
Senior architect	60,000	72,000	2,100
Senior engineer	45,000	54,000	2,100
Senior surveyor	45,000	54,000	2,100
Qualified architect	40,000	48,000	2,100
Qualified engineer	35,000	42,000	2,100
Qualified surveyor	35,000	42,000	2,100

Cost of Materials

The figures that follow are the costs of main construction materials, delivered to site in the Hong Kong area, as incurred by contractors in the fourth quarter of 2008. These assume that the materials would be in quantities as required for a medium sized construction project and that the location of the works would be neither constrained nor remote.

	Unit	*Cost HK$*
Cement and aggregate		
Ordinary portland cement in 50kg bags	tonne	560
Coarse aggregates for concrete	m^3	62
Fine aggregates for concrete	m^3	80
Ready mixed concrete (Grade 40)	m^3	700
Ready mixed concrete (Grade 10)	m^3	550
Steel		
Mild steel reinforcement	tonne	6,000
High tensile steel reinforcement	tonne	6,000
Structural steel sections	tonne	8,500
Bricks and blocks		
Common bricks (225 x 105 x 70mm)	1,000	1,200
Good quality facing bricks (225 x 105 x 70mm)	1,000	1,500
Hollow concrete blocks (300 x 150 x 75mm)	1,000	2,000
Solid concrete blocks (300 x 150 x 75mm)	1,000	2,400
Precast concrete cladding units with exposed aggregate finish	m^2	700
Timber and insulation		
Softwood sections for carpentry	m^3	3,100
Softwood for joinery	m^3	3,300
Hardwood for joinery	m^3	3,700

	Unit	*Cost HK$*
Exterior quality plywood (19mm)	m²	67
Plywood for interior joinery (19mm)	m²	60
Softwood strip flooring (25mm)	m²	240
Chipboard sheet flooring (19mm)	m²	55
100mm thick quilt insulation	m²	55
100mm thick rigid slab insulation	m²	85
Softwood internal door complete with frames and ironmongery	each	1,600
Glass and ceramics		
Float glass (6mm)	m²	99
Sealed double glazing units (4 x 4m)	m²	600
Plaster and paint		
Good quality ceramic wall tiles (150 x 75mm)	m²	65
Plasterboard (12mm thick)	m²	140
Emulsion paint in 5 litre tins	litre	39
Gloss oil paint in 5 litre tins	litre	50
Tiles and paviors		
Clay floor tiles (250 x 250 x 20mm)	m²	55
Vinyl floor tiles (300 x 300 x 2.3mm)	m²	70
Precast concrete paving slabs (250 x 250 x 25mm)	m²	120
Clay roof tiles	1,000	4,000
Precast concrete roof tiles (300 x 300 x 25mm)	1,000	6,300
Drainage		
WC suite complete	each	2,200
Lavatory basin complete	each	2,000
150mm diameter cast iron drain pipes	m	200

Unit Rates

The descriptions overleaf are generally shortened versions of standard descriptions listed in full in section 4. Where an item has a two digit reference number (e.g. 05 or 33), this relates to the full description against that number in section 4. Where an item has an alphabetic suffix (e.g. 12A or 34B) this indicates that the standard description has been modified. Where a modification is major the complete modified description is included here and the standard description should be ignored; where a modification is minor (e.g. the insertion of a named hardwood) the shortened description has been modified here but, in general, the full description in section 4 prevails.

The unit rates below are for main work items on a typical construction project in Hong Kong in the fourth quarter of 2008. The rates include all necessary labour, materials and equipment. Allowances to cover preliminary and general items and contractor's overheads and profit have been added to the rates.

		Unit	*Rate HK$*
Excavation			
01	Mechanical excavation of foundation trenches	m^3	135
02	Hardcore filling making up levels	m^3	100
03	Earthwork support	m^2	55
Concrete work			
04	Plain in situ concrete in strip foundations in trenches	m^3	750
05	Reinforced in situ concrete in beds	m^3	900
06	Reinforced in situ concrete in walls	m^3	900
07	Reinforced in situ concrete in suspended floors or roof slabs	m^3	900
08	Reinforced in situ concrete in columns	m^3	900
09	Reinforced in situ concrete in isolated beams	m^3	900
10	Precast concrete slab	m^2	950
Formwork			
11	Softwood formwork to concrete walls	m^2	150
12	Softwood formwork to concrete columns	m^2	150
13	Softwood formwork to horizontal soffits of slabs	m^2	150
Reinforcement			
14	Reinforcement in concrete walls	tonne	9,000
15	Reinforcement in suspended concrete slabs	tonne	9,000
16	Fabric reinforcement in concrete beds	m^2	80
Steelwork			
17	Fabricate, supply and erect steel framed structure	tonne	24,000
18	Framed structural steelwork in universal joist sections	tonne	24,000
19	Structural steelwork lattice roof trusses	tonne	26,000
Brickwork and blockwork			
20	Precast lightweight aggregate hollow concrete block walls	m^2	140
21	Solid (perforated) concrete blocks	m^2	160
22	Sand lime bricks	m^2	180
23	Facing bricks	m^2	230

		Unit	*Rate HK$*
Roofing			
24	Concrete interlocking roof tiles 430 x 380mm	m^2	400
25	Plain clay roof tiles 260 x 160mm	m^2	350
27	Sawn softwood roof boarding	m^2	250
28	Particle board roof coverings	m^2	230
29	3 layers glass-fibre based bitumen felt roof covering	m^2	180
30	Bitumen based mastic asphalt roof covering	m^2	150
31	Glass-fibre mat roof insulation 160mm thick	m^2	150
32	Rigid sheet load bearing roof insulation 75mm thick	m^2	130
33	Troughed galvanized steel roof cladding	m^2	400
Woodwork and metalwork			
34	Preservative treated sawn softwood 50 x 100mm	m	90
35	Preservative treated sawn softwood 50 x 150mm	m	110
36	Single glazed casement window in hardwood, 650 x 900mm	each	1,600
37	Two panel glazed door in hardwood, size 850 x 2000mm	each	4,500
38	Solid core half hour fire resisting hardwood internal flush door, size 800 x 2000mm	each	3,500
39	Aluminium double glazed window, size 1200 x 1200mm	each	3,400
40	Aluminium double glazed door, size 850 x 2100mm	each	6,500
41	Hardwood skirtings	m	50
Plumbing			
42	UPVC half round eaves gutter	m	200
43	UPVC rainwater pipes	m	250
44	Light gauge copper cold water tubing	m	160
45	High pressure plastic pipes for cold water supply	m	100
46	Low pressure plastic pipes for cold water distribution	m	120
47	UPVC soil and vent pipes	m	240
48	White vitreous china WC suite	each	2,800
49	White vitreous china lavatory basin	each	2,600
50	Glazed fireclay shower tray	each	2,000
51	Stainless steel single bowl sink and double drainer	each	2,500
Electrical work			
52	PVC insulated and copper sheathed cable	m	30
53	13 amp unswitched socket outlet	each	330
54	Flush mounted 20 amp, 1 way light switch	each	580

		Unit	Rate HK$
Finishings			
55	2 coats gypsum based plaster on brick walls	m^2	120
56	White glazed tiles on plaster walls	m^2	160
57	Red clay quarry tiles on concrete floors	m^2	250
58	Cement and sand screed to concrete floors	m^2	80
59	Thermoplastic floor tiles on screed	m^2	150
60	Mineral fibre tiles on concealed suspension system	m^2	300
Glazing			
61	Glazing to wood	m^2	230
Painting			
62	Emulsion on plaster walls	m^2	40
63	Oil paint on timber	m^2	80

Approximate Estimating

The building costs per unit area given below are averages incurred by building clients for typical buildings in the Hong Kong area as at the fourth quarter of 2008. They are based upon the total floor area of all storeys, measured between external walls and without deduction for internal walls.

Approximate estimating costs generally include mechanical and electrical installations but exclude furniture, loose or special equipment, and external works; they also exclude fees for professional services. The costs shown are for specifications and standards appropriate to Hong Kong and this should be borne in mind when attempting comparisons with similarly described building types in other countries. A discussion of this issue is included in section 2. Comparative data for countries covered in this publication, including construction cost data, are presented in Part Three.

Approximate estimating costs must be treated with caution; they cannot provide more than a rough guide to the probable cost of building.

	Cost m^2 HK$	Cost ft^2 HK$
Industrial buildings		
Factories for letting	7,900	734
Factories for owner occupation (light industrial use)	8,500	790
Factories for owner occupation (heavy industrial use)	9,400	873
Factory/office (high-tech) for letting (shell and core only)	9,900	920
Factory/office (high-tech) for letting (ground floor shell, first floor offices)	12,000	1,115
Factory/office (high tech) for owner occupation (controlled environment, fully finished)	18,200	1,691
High tech laboratory workshop centres (air-conditioned)	22,000	2,044

	Cost m^2 HK$	Cost ft^2 HK$
Warehouses, low bay (6 to 8m high) for letting (no heating)	9,200	855
Warehouses, low bay for owner occupation	10,500	975
Warehouses, high bay for owner occupation	11,500	1,068
Cold stores/refrigerated stores	16,500	1,533
Administrative and commercial buildings		
Civic offices, fully air-conditioned	22,000	2,044
Offices for letting, 5 to 10 storeys, air-conditioned	14,300	1,329
Offices for letting, high rise, air-conditioned	15,500	1,440
Offices for owner occupation high rise, air-conditioned	18,200	1,691
Prestige/headquarters office, 5 to 10 storeys, air-conditioned	18,700	1,737
Prestige/headquarters office, high rise, air-conditioned	19,800	1,839
Health and education buildings		
General hospitals (1000 beds)	22,000	2,044
Private hospitals (500 beds)	25,300	2,350
Health centres	15,400	1,431
Nursery schools	13,200	1,226
Primary/junior schools	9,600	892
Secondary/middle schools	9,900	920
University (arts) buildings	18,700	1,737
University (science) buildings	19,800	1,839
Management training centres	16,500	1,533
Recreation and arts buildings		
Theatres (over 500 seats) including seating and stage equipment	27,500	2,555
Theatres (less than 500 seats) including seating and stage equipment	28,600	2,657
Concert halls including seating and stage equipment	27,500	2,555
Swimming pools (international standard) including changing and social facilities (outdoor)	each	16,500,000
Swimming pools (schools standard) including changing facilities (outdoor)	each	5,500,000
National museums including full air-conditioning and standby generator	33,000	3,066
Local museums including air-conditioning	27,500	2,555
City centre/central libraries	22,000	2,044
Branch/local libraries	18,700	1,737

	Cost m² HK$	*Cost ft² HK$*
Residential buildings		
Private/mass market single family housing 2 storey detached/semi detached (multiple units)	18,200	1,691
Purpose designed single family housing 2 storey detached (single unit)	23,000	2,137
Social/economic apartment housing, high rise (with lifts)	5,800	539
Private sector apartment building (standard specification)	11,400	1,059
Private sector apartment buildings (luxury)	14,900	1,384
Student/nurses halls of residence	12,000	1,115
Homes for the elderly (shared accommodation)	14,300	1,329
Homes for the elderly (self contained with shared communal facilities)	15,200	1,412
Hotel, 5 star, city centre	23,400	2,174
Hotel, 3 star, city/provincial	18,400	1,709

EXCHANGE RATES AND INFLATION

The combined effect of exchange rates and inflation on prices within a country and price comparisons between countries is discussed in section 2.

Exchange Rates

The graph on the next page plots the movement of the Hong Kong dollar against the sterling, the euro, the US dollar and 100 Japanese yen since 1998. The values used for the graph are quarterly and the method of calculating these is described and general guidance on the interpretation of the graph provided in section 2. The average exchange rate in the fourth quarter of 2008 was HK$12.19 to pound sterling, HK$10.20 to euro, HK$7.75 to US dollar and HK$ 8.09 to 100 Japanese yen.

THE HONG KONG DOLLAR AGAINST STERLING, EURO, US DOLLAR AND 100 JAPANESE YEN

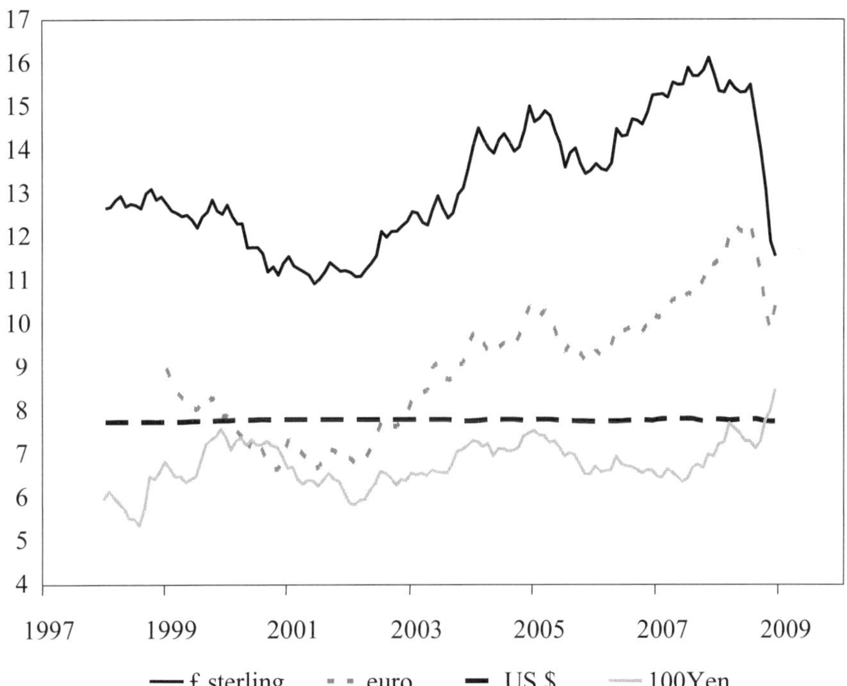

Price Inflation

The table below presents consumer price, building cost and tender price inflation in Hong Kong since 1998.

CONSUMER PRICE, BUILDING PRICE AND TENDER PRICE INFLATION

Year	*Consumer price inflation*		*Building cost index*		*Tender price index*	
	average index	*average Change %*	*average index*	*average change %*	*average index*	*average change %*
1998	100.0	2.9	100.0	7.4	100.0	9.1
1999	96.1	-3.9	101.4	1.4	95.6	-4.4
2000	92.5	-3.7	103.2	1.8	83.0	-13.2
2001	91.0	-1.6	103.5	0.3	75.9	-8.6
2002	88.2	-3.1	103.2	-0.3	67.1	-11.6
2003	85.9	-2.6	102.2	-1.0	66.9	-0.3
2004	85.6	-0.3	101.0	-1.2	65.8	-1.6
2005	86.4	0.9	98.9	-2.1	66.8	1.5
2006	88.1	2.0	98.9	-	70.1	4.9
2007	89.9	2.0	99.9	1.0	84.2	20.1
2008	93.8	4.3	Ceased publication		113.9@	35.3@

@ Figures for Second Quarter of the year

Source : Census and Statistics Department, Hong Kong SAR
Architectural Services Department, Hong Kong SAR

The year-on-year rate of increase in the tender price index has been increasing since 2005. Following the reduction in work since the Asian Financial Crisis in 1998, both the number of construction firms and site workers has shrunk. According to the Census and Statistics Department, there are approximately 299 main contractors and 50,185 workers on site at 2007, down from 371 and 70,941 respectively in 1999. Supply has therefore become more inelastic than before. With the increase in demand brought in by the Government spending and the work in Macau and China, tender prices had been rising at a steep pace from 2007 to 2008.

Since the later part of 2008, tender prices have dropped due to sharp declines in the cost of steel reinforcement, copper and other materials. On the other hand, tenderers have shown higher caution towards tenders due to the uncertain financial outlook. The result has been varying tender results from project to project.

Based on the above, construction prices are expected to be volatile in the near future before resuming an upward trend after the effects of thc global financial tsunami subsides.

USEFUL ADDRESSES

Public Organizations

Airport Authority
HKIA Tower, 1 Sky Plaza Road
Hong Kong International Airport
Lantau
Hong Kong
Tel: (852) 2181 8888
Website: www.hongkongairport.com

Architectural Services Department
36/F, Queensway
Government Offices
66 Queensway
Hong Kong
Tel: (852) 2867 3628
Fax: (852) 2869 0289
E-mail: archsd@archsd.gov.hk
Website: www.archsd.gov.hk

Buildings Department
12/F Pioneer Centre
750 Nathan Road
Mongkok
Kowloon
Tel: (852) 2626 1616
Fax: (852) 2537 4992
E-mail: enquiry@bd.gov.hk
Website: www.bd.gov.hk

Census and Statistics Department
16/F-22/F and 25/F,Wanchai Tower
12 Harbour Road
Wan Chai
Hong Kong
Tel: (852) 2582 4807
Fax: (852) 2802 4000
E-mail: gen-enquiry@censtatd.gov.hk
Website: www.censtatd.gov.hk

Civil Engineering and Development Department
101 Princess Margaret Road
Kowloon
Hong Kong
Tel: (852) 2762 5111
Fax: (852) 2714 0140
E-mail: enquiry@cedd.gov.hk
Website: www.cedd.gov.hk

Construction Industry Council
10/F, Murray Building
Garden Road, Central
Hong Kong
Tel: (852) 2848 6251
Fax: (852) 2869 6095
E-mail: enquiry@hkcic.org
Website: www.hkcic.org

Electrical and Mechanical Services Department
3 Kai Shing Street
Kowloon
Hong Kong
Tel: (852) 2333 3762
Fax: (852) 2890 7493
E-mail: info@emsd.gov.hk
Website: www.emsd.gov.hk

Environmental Protection Department
46/F Revenue Tower
5 Gloucester Road
Wan Chai
Hong Kong
Tel: (852) 2838 3111
Fax: (852) 2838 3111
E-mail: enquiry@epd.gov.hk
Website: www.epd.gov.hk

Highways Department
5/F, Ho Man Tin Government Offices
88 Chung Hau Street
Ho Man Tin
Kowloon
Tel: (852) 2926 4111
Fax: (852) 2714 5216
E-mail: enquiry@hyd.gov.hk
Website: www.hyd.gov.hk

Hong Kong Housing Authority
Hong Kong Housing Authority Headquarters
33 Fat Kwong Street
Ho Man Tin
Kowloon
Hong Kong
Tel: (852) 2712 2712
Fax: (852) 2624 5685
E-mail: hkha@housingauthority.gov.hk
Website: www.housingauthority.gov.hk

Lands Department
20/F, North Point Government Offices
333 Java Road
North Point
Hong Kong
Tel: (852) 2231 3294
Fax: (852) 2868 4707
E-mail: landsd@landsd.gov.hk
Website: www.landsd.gov.hk

Planning and Lands Branch
18/F Murray Building
Garden Road, Central
Hong Kong
Tel: (852) 2848 2718
Fax: (852) 2845 3489
E-mail: plbenq@devb.gov.hk
Website: www.devb-plb.gov.hk

Planning Department
17/F, North Point Government Offices
333 Java Road
North Point
Hong Kong
Tel: (852) 2231 5000
Fax: (852) 2877 0389
E-mail: enquire@pland.gov.hk
Website: www.pland.gov.hk

Urban Renewal Authority
10/F Low Block, Grand Millennium Plaza
181 Queen's Road Central
Hong Kong
Tel: (852) 2588 2333
Fax: (852) 2827 0176
E-mail: inquiry@mail1.ura.org.hk
Website: www.ura.org.hk

Water Supplies Department
48/F, Immigration Tower
7 Gloucester Road
Wanchai
Hong Kong
Tel: (852) 2824 5000
Fax: (852) 2824 0578
E-mail: wsdinfo@wsd.gov.hk
Website: www.wsd.gov.hk

Works Branch
10/F, Murray Building
Garden Road, Central
Hong Kong
Tel: (852) 2848 2111
Fax: (852) 2523 5327
E-mail: wbenq@devb.gov.hk
Website: www.devb-wb.gov.hk

Trade And Professional Associations

The Hong Kong Construction Association, Limited
3/F, 180-82 Hennessy Road
Wanchai
Hong Kong
Tel: (852) 2572 4414
Fax: (852) 2572 7104
E-mail: admin@hkca.com.hk
Website: www.hkca.com.hk

Hong Kong Construction Industry Employees General Union
2/F Wah Hing Commercial Centre
383 Shanghai Street
Yaumatei, Kowloon
Hong Kong
Tel: (852) 2388 6887
Fax: (852) 2385 5002
E-mail: master@hkciegu.org.hk
Website: www.hkciegu.org.hk

Hong Kong General Chamber of Commerce
22/F United Centre
95 Queensway
Admiralty
Hong Kong
Tel: (852) 2529 9229
Fax: (852) 2527 9843
E-mail: chamber@chamber.org.hk
Website: www.chamber.org.hk

The Hong Kong Institute of Architects
19/F, 1 Hysan Avenue
Causeway Bay
Hong Kong
Tel: (852) 2511 6323
Fax: (852) 2519 6011
E-mail: info@hkia.net
Website: www.hkia.net

Hong Kong Institute of Construction Managers, Limited
Room 801-2, 8/F, On Lok Yuen Building
25 Des Voeux Road Central
Hong Kong
Tel: (852) 2523 2081
Fax: (852) 2845 4749
E-mail: info@hkicm.org.hk
Website: hkicm.org.hk

The Hong Kong Institute of Engineers
9/F, Island Beverley
No 1 Great George Street
Causeway Bay
Hong Kong
Tel: (852) 2895 4446
Fax: (852) 2577 7791
E-mail: hkie-sec@hkie.org.hk
Website: www.hkie.org.hk

The Hong Kong Institute of Surveyors
Suite 801, Jardine House
1 Connaught Place
Central
Hong Kong
Tel: (852) 2526 3679
Fax: (852) 2868 4612
E-mail: info@hkis.org.hk
Website: www.hkis.org.hk

Real Estate Developers Association of Hong Kong
1403, World Wide House
19 Des Voeux Road Central
Hong Kong
Tel: (852) 2826 0111
Fax: (852) 2845 2521

INDIA

All data relate to 2008 unless otherwise indicated.

Population	
Population	1,149 mn
Urban population	29%
Population under 15	31.5%
Population 65 and over	5.2%
Average annual growth rate	1.57%
Geography	
Total area	3,287,590 km^2
Agricultural area	61%
Capital city	New Delhi
Population of Delhi including New Delhi	17 mn
Economy	
Monetary unit	Indian Rupees (Rs)
Exchange rate (average fourth quarter 2008) to:	
the pound sterling	Rs 76.61
the US dollar	Rs 48.31
the euro	Rs 64.34
the yen x 100	Rs 50.35
Average annual inflation (2000 to 2008)	6.0%
Inflation rate (1st Quarter 2009)	5.2%
Gross Domestic Product (GDP-PPP)	US$ 3.1 trillion
GDP per capita	US$ 2,600
Average annual real change in (GDP) (2007 to 2008)	8.7%
Private consumption as a proportion of GDP (2007 to 2008)	57.6%
Public consumption as a proportion of GDP (2007 to 2008)	9.5%
Investment as a proportion of GDP (2006 to 2007)	33.8%
Construction	
Output at Basic Prices (2005)	Rs 7,484 bn
Output at Basic Prices (2004)	Rs 6,010 bn

THE CONSTRUCTION INDUSTRY

Construction Output

The present size of Construction Industry in terms of annual monetary values is estimated at Rs 310,000 crores (includes Public & Private Investments), equivalent to approximately USD 64.58 billions (1USD=INR 48).

The country's development has been shaped through a series of strategic Five Year Plans, the first of which was drawn up in 1951. The Five Year Plans cover all sectors of economic and social performance and provide the directions and framework for future development and investment. Increasing emphasis has been placed since the first Five Year Plan (1951-56) on the development of infrastructure, irrigation, energy, transportation and communications, health, housing, social welfare, rural development and other activities.

The official Five Year Plans do not include the two private sector investment sectors: private household and corporate. While the scale of private corporate investment is most difficult to estimate, it is clear that there has been a lot of activity in this type of work, particularly in the commercial, tourism and residential sectors.

The construction industry is dependent on investments on the infrastructure, industrial and real estate sectors. The Planning Commission has envisaged an outlay of approximately US$ 300 billion during the 11th Five Year Plan for infrastructure development in the country. These investments would be achieved through a combination of Public and Public–Private Partnerships.

The total investment would ultimately translate into an effective construction investment of about Rs 10,000 billion in the next 4 to 5 years.

The principal sectors with major construction activities are:

- Housing – particularly private housing
- Corporate – commercial, retail and leisure
- Utilities – water supply, sewage and irrigation, energy (thermo and hydro)
- Transport – roads, railways, ports and airports
- Industrial – new industrial parks

While all sectors have been expanding, considerable investment remains in energy and infrastructure. The country's still rising population and continuing population shift from rural to urban areas and natural disasters have further added to the already existing shortfall in housing. According to the report of the Technical Group on Estimation of Housing Shortage, the estimated shortage was 24.71 million as at 2007 and the shortage during the 11th year plan (2007-2012) including backlog is estimated at 26.53 million.

The construction industry continues to show a high growth like previous years due to the high demand in housing and infrastructure. It is anticipated that the rate of construction in the private sector will have a downward trend due to the credit crisis as part of current global economic recession.

The following table indicates the construction investment by sector.

CONSTRUCTION INVESTMENT (RS BILLION)

Sector	*2004-2005*	*2005-2006*	*2006-2007*	*2007-2008*
A. PUBLIC				
1. Residential	54.53	63.87	69.85	80.07
2. Commercial	37.50	44.16	50.52	56.09
3. Industrial	652.88	768.09	876.87	975.25
4. Infrastructure	979.32	1,152.14	1,315.31	1,462.87
Total Public	1,724.23	2,028.26	2,312.55	2,574.28
B. PRIVATE				
1. Residential	72.97	86.13	100.15	109.93
2. Commercial	90.00	105.84	119.48	133.91
3. Industrial	265.12	311.91	347.13	392.75
4. Infrastructure	397.68	467.86	520.69	589.13
Total Private	825.77	971.74	1087.45	1,225.72
Grand Total	2,550.00	3,000.00	3,400.00	3,800.00

Source: Construction Industry Development Council

Characteristics and Structure of the Industry

As per year 2005 statistics the construction industry in India employs about 31 million people comprising 0.8 million engineers, 1.3 million supervisors and qualified personnel, and 28.8 million workers.

Construction companies are classified under the following categories which are based on contract values:

- Group I – Companies that can bid for contracts over Rs 300 million
- Group II – Companies that can bid for contracts over Rs 100 million
- Group III – Companies that can bid for contracts less than Rs 100 million

There are over 28,000 construction companies in India, the majority of the Group I category who are involved mainly in housing construction and repair and maintenance work. About 200 companies are involved in major contracts or in turnkey and engineering, procurement and construction (EPC) type contracts.

The top 10 major contractors in India and their characteristics are given in the table next page.

MAJOR CONTRACTORS AND THEIR CHARACTERISTICS

Major contractors	*Turnover US$ million*	*Main work/types*
Larsen & Toubro Ltd (ECC Group)	3,835.43 (2007-08)	Heavy industrial construction, institutional buildings, special structures
Gammon India Ltd	488.52 (2007-08)	Hydraulic structures, tunnelling, natural draft cooling towers, heavy industrial construction, bridges and flyover
Hindustan Construction Company Ltd	646.67 (2007-08)	Hydraulic structures, bridges, flyover, irrigation structures, heavy industrial construction
Jaiprakash Industries Ltd	890.42 (2007-08)	Hydraulic structures, hydro electric power plants, heavy industrial construction
Unitech Ltd	891.67(2007-08)	Roads, bridges, heavy industrial construction, housing and institutional building, real estates
Kvaerner Cementation India Ltd	191.31 (2007-08)	Hydraulic structures, heavy industrial construction
National Building Construction Corp'n Ltd	315.42 (2006-07)	Hydraulic structures, roads and highways, hydro power plants and cooling towers, directional drilling
Bridges & Roof Co. Ltd	115.96 (2006-07)	Hydraulic structures, roads and highways, hydro power plants and cooling towers, directional drilling
Punj Lloyd Ltd	421.35 (2007-08)	Hydraulic structures, roads and highways, hydro power plants and cooling towers, directional drilling, pipelines, heavy construction

Source: Construction Industry Development Council

Punj Lloyd Ltd is ranked 55 and Larson & Toubro Ltd is ranked 67 in the *Engineering News Record's 2008 Top 225 International Contractors listing*.

One feature of the major Indian contractors is their capability beyond general contracting in infrastructure, heavy industrial and transport projects. This has enabled them to work in several countries such as Iraq, Libya, North Yemen and United Arab Emirates. Between 1975 and 1980, the work undertaken was worth about US$5 billion. Due to political changes, overseas contracts declined to about US$106 million in 1996-97, from a peak of US$443 million in 1986-87.

The Indian workforce is categorized as follows:

- Segment I – University qualified managerial and supervisory staff
- Segment II – Workmen with on-site work experience but little or no formal education

In the past, workmen were trained by master craftsmen over a period of time. Recognizing the need for more formal structured training, the government has established several industrial training institutes over the last 20 years and in conjunction with various academic bodies, launched trade training programmes through distant learning.

Design work is mainly undertaken by architects and engineers. There are about 29,085 architects in India who are currently registered with the Council of Architecture as at April 2007. Of this number, 5% are in the public sector, 49% in the private practice and 46% self-employed. The title of architect is protected under the *Architect Act 1972* and to qualify, registration with the Council of Architecture is mandatory.

The number of practising civil engineers is approximately 250,000. There is no uniform system of registration and one can become a civil engineer upon graduation from the university.

The number of quantity surveyors is increasing and presently more than 3,000 are engaged in various roles. The Royal Institution of Chartered Surveyors has initiated the process of regulating the profession and there are a number of chartered surveyors practising at present in both public and private organizations.

Clients and Finance

The major government clients (central, state or union territories) which comprise the Indian Railways, Central and State Public Works Department, Indian Army, Indian Oil Ltd, Oil and Natural Gas Commission Ltd, Steel Authority of India Ltd and Gas Authority of India Ltd.

Until very recently, the government was the predominant client for all major construction projects. With current strategic initiatives being implemented, thc private sector market share is increasing year-on-year.

Finance for developers are provided through any registered finance institutions. Loans are commonly in the order of 40% to 60% of the construction cost, for two to two and a half years, and at rates of interest of about 12% to 20%, depending on market conditions.

For individuals, loans are available for up to 80% of the value of the property with average repayment periods of 8 to 9 years up to a limit of 20 years at interest rates varying between 8% and 14%.

Selection of Design Consultants

It is mandatory for public sector clients to select on the basis of fee competition through a public tender.

In the private sector, there is more flexibility when selecting design consultants. A pre-qualification process to identify a list of suitable consultants can be drawn up through a public advertisement or the selection is based on client referrals where the experience, track record and capability of the firm are considered.

Contractual Arrangements

The usual procedure for inviting tenders is through public notices seeking an expression of interest from contractors for public projects and for shortlisted tenders for private projects.

The traditional method of procurement where the Contractor is appointed by the client is still adopted for the majority of projects. For larger contracts or contracts requiring multidisciplinary expertise, design and build is used and has become increasingly important. The decision to award is made by the client on the basis of price, previous experience, capability, referrals and design proposals.

The following are the commonly adopted procurement routes in India.

- BQ Contract/Remeasurement Contract
- Lump Sum Contract
- Design & Build/Turnkey
- Management Contract

And commonly used contract forms are:

- FIDIC
- Indian Institute of Architects form
- Central Public Works Department forms
- Municipal bodies form

The industry does not have a standard form of contract, although it is estimated that there are some 30 types of contract forms in use by various client organizations.

The Construction Industry Development Council, which comprises the Planning Commission and several leading construction organizations, publishes a standard bidding document.

Bills of quantities are prepared using the standard method of measurement known as IS-1200 which is available from the Bureau of Indian Standards.

Advance payments to the Contractor are a norm and usually comprise the following:

- Mobilisation advance, payable as a lump sum upon commencement of the project
- Mobilisation advance for construction plant and equipment, payable as a lump sum upon commencement of the project or when the plant or equipment is brought to site
- Advance for main construction materials, payable after delivery to site

Repayment of the advance payment is through adjustments made at each interim payment claim.

Liability and Insurance

Contractors are required to furnish bank guarantees as Performance Security and Retention Money Guarantee in the order of 5% to 10% of contract sum respectively. The insurance policies for the projects are generally in three sections such as Contractors All Risk (CAR), Third party liability and Workmen's Compensation insurances.

Development Control and Standards

Master Plans, Zonal Plans and Zonal Regulations dictate the land use and type of development that can be carried out.

Permits are required before commencement of construction work on site, and upon completion before handover to the client. Some of the authorities involved include the Municipal, Urban Art Commission, Fire Department, Aviation Department, Ministry of Environmental and Forest, and Services Department.

CONSTRUCTION COST DATA

Cost of Labour

The figures below are typical of labour costs in the Bangalore area as at the third quarter of 2008.

	Wage rate (per day) Rs	*Cost of labour (per day) Rs*	*Number of hours worked per year*
Site operatives			
Bricklayer	250	320	2,520 - 2,560
Shuttering carpenter	250	320	2,520 - 2,560
Plumber	250	320	2,520 - 2,560
Electrician	250	320	2,520 - 2,560
Structural steel erector	250	320	2,520 - 2,560

	Wage rate (per day) Rs	*Cost of labour (per day) Rs*	*Number of hours worked per year*
Welder	250	320	2,520 - 2,560
Labourer	100	130	2,520 - 2,560
Equipment operator	150	195	2,520 - 2,560
Hoist operator	200	260	2,520 - 2,560
JCB operator	200	260	2,520 - 2,560
	(per month)	*(per month)*	
Site supervision			
General foreman	10,000	15,000	2,400
Trades foreman	10,000	15,000	2,400
Clerk of works	9,000	13,500	2,400
Resident engineer	60,000	90,000	2,400
Contractors' personnel			
Site manager	60,000	90,000	2,400
Site engineer	30,000	45,000	2,400
Site quantity surveyor	35,000	52,500	2,400
Consultants' personnel			
Senior architect	60,000	90,000	2,400
Senior engineer	40,000	60,000	2,400
Senior surveyor	45,000	67,500	2,400
Qualified architect	30,000	45,000	2,400
Qualified engineer	25,000	37,500	2,400
Qualified surveyor	25,000	37,500	2,400

Cost of Materials

The figures that follow are the costs of main construction materials, delivered to site in the Bangalore area, as incurred by contractors in the fourth quarter of 2008. These assume that the materials would be in quantities as required for a medium sized construction project and that the location of the works would be neither constrained nor remote.

All the costs in this section exclude duties and value added tax.

	Unit	*Cost Rs*
Cement and aggregate		
Ordinary portland cement in 50kg bags	bag	222
Coarse aggregates for concrete	tonne	775
Sharp sand for concrete	tonne	750
Ready mixed concrete (Grade 30)	m^3	3,500
Ready mixed concrete (Grade 20)	m^3	3,250

	Unit	Cost Rs
Steel		
Mild steel reinforcement	tonne	36,000
High tensile steel reinforcement	tonne	38,000
Structural steel sections	tonne	41,000
Bricks and blocks		
Common bricks (215 x 102.5 x 65mm)	1,000	2,500
Good quality facing bricks (215 x 102.5 x 65mm)	1,000	3,000
Hollow concrete blocks (390 x 190 x 100mm)	100	2,800
Timber and insulation		
Hardwood for joinery	m^3	16,000
Exterior quality plywood (12mm)	m^2	450
50mm thick quilt insulation (16 kg/m^3)	m^2	365
Hardwood internal door complete with frame and ironmongery	m^2	3,000
Glass and ceramics		
Float glass (10mm)	m^2	1,200
Plaster and paint		
Good quality ceramic wall tiles (6mm)	m^2	350
Plasterboard (13mm thick) – gypsum	m^2	175
Emulsion paint in 5 litre tins	litre	170
Gloss oil paint in 5 litre tins	litre	1,000
Tiles and paviors		
Clay floor tiles (100 x 200 x 8mm)	m^2	220
Vinyl floor tiles (300 x 300 x 2mm)	m^2	300
Clay roof tiles	1,000	7,000
Precast concrete roof tiles	1,000	8,000
Drainage		
WC suite complete	each	5,000
Lavatory basin complete	each	950
100mm diameter clay drain pipes	m	250
150mm diameter cast iron drain pipes (medium grade)	m	550

Unit Rates

The descriptions on the next page are generally shortened versions of standard descriptions listed in full in section 4. Where an item has a two digit reference number (e.g. 05 or 33), this relates to the full description against that number in section 4. Where an item has an alphabetic suffix (e.g. 12A or 34B) this indicates that the standard description has been modified. Where a modification is major the complete modified description is included here and the standard description should be ignored. Where a modification is minor (e.g. the insertion of a named hardwood) the shortened description has been modified here but, in general, the full description in section 4 prevails.

		Unit	*Rate Rs*
Excavation			
01	Mechanical excavation of foundation trenches	m^3	300
02	Hardcore filling making up levels	m^3	90
Concrete work			
04	Plain in situ concrete in strip foundations in trenches – Grade 20	m^3	4,250
05	Reinforced in situ concrete in beds – Grade 20	m^3	4,200
06	Reinforced in situ concrete in walls – Grade 20	m^3	4,240
07	Reinforced in situ concrete in suspended floors or roof slabs – Grade 30	m^3	4,500
08	Reinforced in situ concrete in columns – Grade 30	m^3	4,500
09	Reinforced in situ concrete in isolated beams – Grade 30	m^3	4,500
Formwork			
11A	Waterproof plywood formwork to concrete walls	m^2	350
12A	Waterproof steel formwork to concrete isolated columns	m^2	350
12B	Waterproof plywood formwork to concrete columns	m^2	350
13A	Waterproof plywood formwork to horizontal soffits of slabs	m^2	375
Reinforcement			
14	Reinforcement in concrete walls	tonne	45,000
15A	Reinforcement in suspended concrete slabs	tonne	45,000
16	Fabric reinforcement in concrete beds	m^2	750
Steelwork			
17	Fabricate, supply and erect steel framed structure	tonne	70,000
18	Framed structural steelwork in universal joist sections	tonne	70,000
19A	Structural steelwork lattice roof trusses	tonne	70,000
Roofing			
24	Concrete interlocking roof tiles 430 x 380mm	m^2	300
25	Plain clay roof tiles 260 x 160mm	m^2	180
29A	3 layers polyester based bitumen felt roof covering	m^2	80
33	Troughed galvanized steel roof cladding	m^2	200

		Unit	Rate Rs
Woodwork and metalwork			
34	Preservative treated sawn softwood 50 x 100mm	m	120
35	Preservative treated sawn softwood 50 x 150mm	m	120
37A	Two panel glazed door in kapur hardwood size 850x2000mm	each	15,000
38A	Solid core half hour fire resisting hardwood	each	25,000
40A	Aluminium double glazed door, size 1200 x 2100mm	each	9,000
41A	Hardwood skirtings	m	175
Plumbing			
42A	UPVC half round caves gutter (100mm diameter)	m	275
43A	UPVC rainwater pipes (100mm diameter)	m	230
44	Light gauge copper cold water tubing	m	80
45A	High pressure plastic pipes for cold water supply (100mm diameter)	m	280
47	UPVC soil and vent pipes	m	230
48	White vitreous china WC suite	each	2,800
49A	White vitreous wash hand basin	each	750
51	Stainless steel single bowl sink and double drainer	each	11,000
Electrical work			
52A	PVC insulated and PVC sheathed (4 mm^2 copper wire)	m	50
53	13 amp unswitched socket outlet	each	125
54A	Flush mounted 20 amp, MCB control	each	325
Finishings			
55A	2 coats cement and sand (1:4) plaster on brick walls	m^2	240
56	White glazed tiles on plaster walls	m^2	600
57	Red clay quarry tiles on concrete floors	m^2	350
58	Cement and sand screed to concrete floors	m^2	250
59	Thermoplastic floor tiles on screed	m^2	480
60	Mineral fibre tiles on concealed suspension	m^2	600
Painting			
62	Emulsion on plaster walls	m^2	120
63	Oil paint on timber	m^2	60

Approximate Estimating

The building costs per unit area given on the next page are averages incurred by building clients for typical buildings in the India as at the fourth quarter of 2008. They are based upon the total floor area of all storeys, measured between external walls and without deduction for internal walls.

Approximate estimating costs generally include mechanical and electrical installations but exclude furniture, loose or special equipment, and external works; they also exclude fees for professional services. The costs shown are for specifications and standards appropriate to India and this should be borne in mind when attempting comparisons with similarly described building types in other countries. A discussion of this issue is included in section 2. Comparative data for countries covered in this publication, including construction cost data, are presented in Part Three.

Approximate estimating costs must be treated with caution; they cannot provide more than a rough guide to the probable cost of building. All the rates in this section exclude sales tax.

	Cost m^2 Rs	*Cost ft^2 Rs*
Industrial buildings		
Factories for letting – reinforced concrete	10,000	929
Factories for owner occupation (light industrial use) – reinforced concrete	13,450	1250
Factories for owner occupation (heavy industrial use) – reinforced concrete	15,600	1449
Factory/office (high tech) for letting (shell and core only) – reinforced concrete	6,500	604
Factory/office (high tech) for letting (shell and core only)	4,900	455
Factory/office (high tech) for owner occupation (controlled environment, fully finished)	37,700	3502
High tech laboratory (air-conditioned)	37,700	3502
Warehouses, low bay (6 to 8m high) for letting	8,650	804
Warehouses, low bay for owner occupation	8,650	804
Warehouses, high bay for owner occupation (including heating)	11,850	1,101
Administrative and commercial buildings (Warm Shell)		
Offices for letting, 5 to 10 storeys, non air-conditioned	16,150	1,500
Offices for letting, 5 to 10 storeys, air-conditioned	19,900	1,849
Offices for letting, high rise, air-conditioned	21,000	1,951
Prestige/headquarters office, 5 to 10 storeys, air-conditioned	22,600	2,100
Prestige/headquarters office, high rise, air-conditioned	23,700	2,202
Health and education buildings (Warm Shell)		
General hospitals (100 beds)	16,200	1,505
Private hospitals	18,300	1,700
Health centres	16,200	1,505
Primary/junior schools	6,500	604
Secondary/middle schools	7,000	650
University	8,600	799

	Cost *m^2 Rs*	*Cost* *ft^2 Rs*
Recreation and arts buildings (Warm Shell)		
Theatres (less than 500 seats)	9,700	901
Sports halls including changing and social facilities	5,400	502
Swimming pools (international standard) (Olympic size)	6,500	604
Swimming pools (schools standard) including changing facilities	7,000	650
City centre/central libraries	6,500	604
Branch/local libraries	6,500	604
Residential buildings		
Private/mass market single family housing 2 storey detached/semi detached (multiple units)	14,000	1,301
Purpose designed single family housing 2 storey detached (single unit)	16,150	1,500
Social/economic apartment housing, high rise with lifts & 5-10 storey	14,000	1,301
Private sector apartment building (standard specification)	16,800	1,561
Private sector apartment buildings (luxury)	20,875	1,939
Student/nurses halls of residence – without lifts	11,000	1,022
Homes for the elderly (shared accommodation) – without lifts	11,000	1,022
Hotel, 5 star, city centre	48,450	4,501
Hotel, 3 star, city/provincial	37,700	3,502

Regional Variations

The approximate estimating costs are based on average rates in the Bangalore. Adjust these costs by the following factors for regional variations:

Delhi	:	+2%
Mumbai	:	+4%
Chennai (formerly Madras)	:	-1%
Calcutta	:	-2%

Value Added Tax and Service Tax

The standard rates of Value Added Tax are currently 0%, 4% and 12.5%, chargeable on the material component when computed on the regular scheme. Where VAT is calculated on the composite scheme the rate should not exceed 8.75% of the total contract sum. Service Tax is 12.36% of the labour component and where this is not identified can be applied at 4.08% of the total contract sum.

EXCHANGE RATES AND INFLATION

The combined effect of exchange rates and inflation on prices within a country and price comparisons between countries is discussed in section 2.

Exchange Rates

The graph below plots the movement of Indian rupee against the sterling, the euro, the US dollar and 100 Japanese yen since 1998. The values used for the graph are quarterly and the method of calculating these is described and general guidance on the interpretation of the graph provided in section 2. The average exchange rate in the fourth quarter of 2008 was Rs 76.61 to pound sterling, Rs 64.34 to euro, Rs 48.30 to US dollar and Rs 50.35 to 100 Japanese yen.

THE INDIAN RUPEE AGAINST STERLING, EURO, US DOLLAR AND 100 JAPANESE YEN

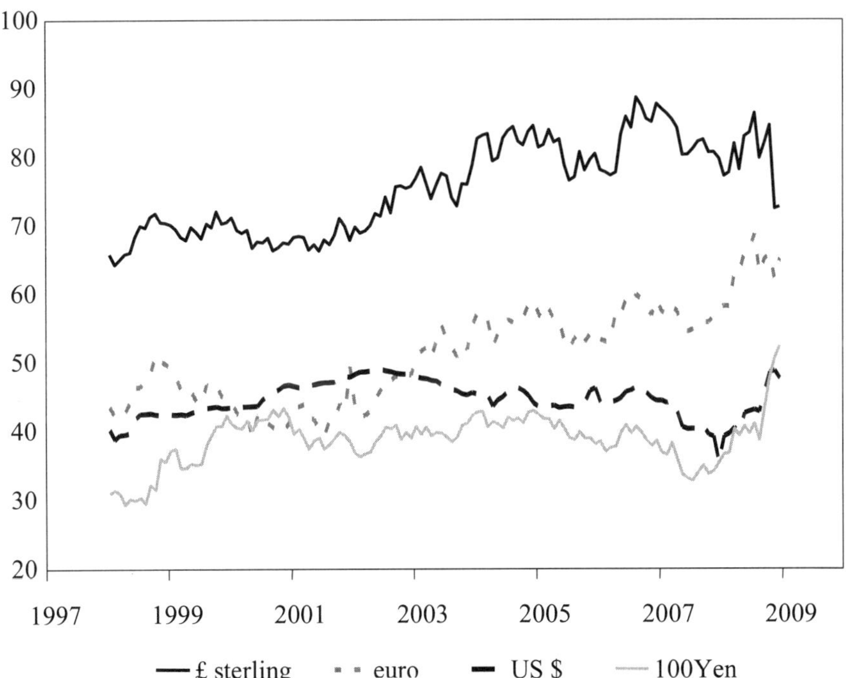

USEFUL ADDRESSES

Builders' Association of India
101 Shivam House
Karampura Commercial Complex
New Delhi 110015
Tel: (91) 11 5435856
Fax: (91) 11 5451423
E-mail: rpcc101@giasdl01.vsnl.net.in

and

G-I/G 20, Commerce Centre 7th Floor
Dadajee Road
Tardeo
Mumbai 400034
Tel / Fax: (91) 22 4940707

Building Materials & Technology Promotion Council
(Ministry of Urban Affairs and Employment)
'G' Wing, Nirman Bhavan
New Delhi 110001
Tel: (91) 11 3016174
Fax: (91) 11 3010145

Bureau of Indian Standards
Manak Bhavan
9 Bahadur Shah Zafar Marg
New Delhi 110002
Tel: (91) 11 3310131
Fax: (91) 11 3314062

Cement Manufacturers' Association
Vishnu Kiran Chambers
1st Floor
2142-47 Gurudwara Road Karol Bagh
New Delhi 110005
Tel: (91) 11 5710347
Fax: (91) 11 5715286

Civil Engineering and Construction Review
Circulating and Marketing Trend-Set Engineers Pvt Ltd
B 10 Som Dutt Chambers
II9 Bhikaji Cama Place
New Delhi 110066
Tel: (91) 11 6875465
Fax: (91) 11 6884049

Confederation of Indian Industry
23-26 Institutional Area
Lodi Road
New Delhi 110003
Tel: (91) 11 4629994
Fax: (91) 11 4633168
E-mail: cii-co@sdalt.ernet.in

Construction Industry Development Council
801-813 Hemkunt Chambers
89 Nehru Place
New Delhi 110019
Tel: (91) 11 6489991 / 2
Fax: (91) 11 6234770

Consulting Engineers Association of India
East Court, Zone 4, Core 4B
2nd Floor, India Habitat Center
New Delhi 110011
Tel: (91) 11 3013040
Fax: (91) 11 3016857

Council of Architecture
India Habitat Centre
Lodi Road
New Delhi 110003
Tel: (91) 11 4648415
Fax: (91) 11 4647746

Housing and Urban Development Corporation
HUDCO House
Lodi Road
New Delhi 110003
Tel: (91) 11 3018495
Fax: (91) 11 4693022

Housing Development Finance Corporation Limited
Ramon House
169 Backbay Reclamation
Mumbai 400020
Tel: (91) 22 2831920, 2820282
Fax: (91) 22 2046758, 2852701

Ministry of Electronics
Electronics Niketan
6 CGO Complex
New Delhi 110003
Tel: (91) 11 4364041
Fax: (91) 11 4363134
E-mail: sghosh@doe.ernet.in

Ministry of Energy
Sham Shakti Bhavan
New Delhi 110011
Tel: (91) 11 385946

Ministry of Environment and Forests
Paryavaran Bhavan
CGO Complex Phase II
Lodhi Road
New Delhi 110003
Tel: (91) 11 4360721
Fax: (91) 11 4360678

Ministry of Petroleum and Natural Gas
Shastri Bhavan
Dr. Rajendra Prasad Road
New Delhi 110001
Tel: (91) 11 3383501
Fax: (91) 11 3383585

Ministry of Railways
Ril Bhavan
Rafi Marg
New Delhi 110001
Tel: (91) 11 3384010
Fax: (91) 11 3381453
E-mail: crb@del2.vsnl.net.in

Ministry of Statistics
Sardar Patel Bhavan
Parliament Street
New Delhi 110001
Tel: (91) 11 3732150
Fax: (91) 11 3344689

Ministry of Steel
Udyog Bhavan
Maulana Azad Marg
New Delhi 110011
Tel: (91) 11 3015489
Fax: (91) 11 3013236

Ministry of Surface Transport
Parivahant Bhavan
Parliament Street
New Delhi 110001
Tel: (91) 11 3714938
Fax: (91) 11 3716656

Overseas Construction Council of India
H118, Himalaya House
11th Floor
23 Kasturba Gandhi Marg
New Delhi 110001
Tel: (91) 11 3327550, 3722425
Fax: (91) 11 3312936

Planning Commission
Yojana Bhavan
Sansad Marg
New Delhi 110001
Tel: (91) 11 3714388

Public Works Department

Government of National Capital Territory of Delhi
Curzon Road Barracks
Kasturba Gandhi Marg
New Delhi 110001
Tel: (91) 11 3018556
Fax: (91) 11 3019097

Steel Authority of India Ltd
Express Building 9/10
B.S.Z. Marg
New Delhi 110002
Tel: (91) 11 3323911
Fax: (91) 11 3317337

PT. DAVIS LANGDON & SEAH INDONESIA

PT. Davis Langdon & Seah Indonesia is dedicated to maintaining its position as the market leader and consultant of choice in its chosen fields.

TYPICAL PROJECT STAGES, INTEGRATED SERVICES AND THEIR EFFECT

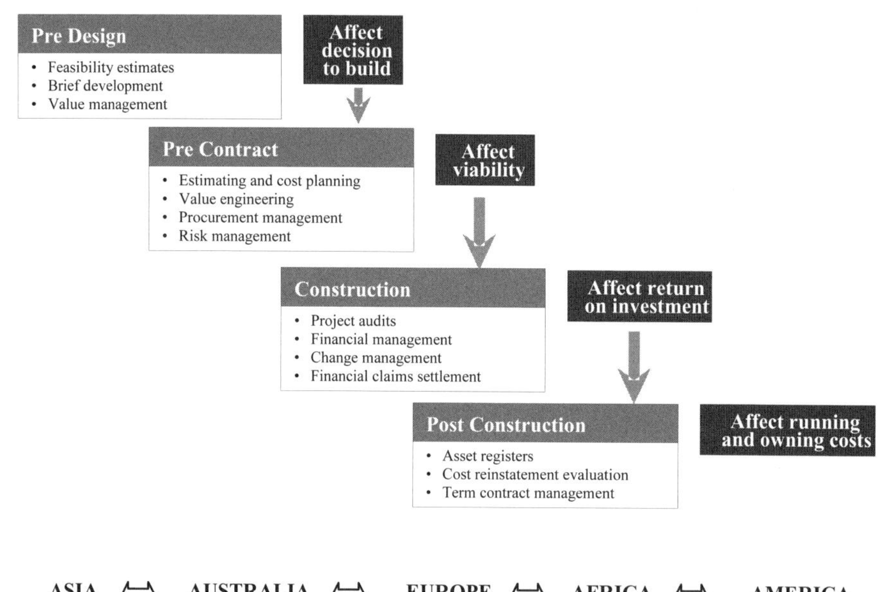

ASIA ⟺ AUSTRALIA ⟺ EUROPE ⟺ AFRICA ⟺ AMERICA

GLOBAL REACH ⇨ LOCAL DELIVERY

BALI OFFICE
Kuta Poleng Blok B/3A
Jalan Setiabudi
Kuta, Bali 80361
Tel : (62-361) 766 260
Fax: (62-361) 750 312

JAKARTA OFFICE
18th Floor, Ratu Plaza Office Tower
Jalan Jenderal Sudirman 9
Jakarta 10270
Tel : (62-21) 739 7550
Fax: (62-21) 739 7846

SURABAYA OFFICE
Room 601-A, Bank Mandiri Building
Jalan Basuki Rachmat 129-137
Surabaya 60271
Tel : (62-31) 546 5857
Fax: (62-31) 531 6579

DAVIS LANGDON & SEAH INTERNATIONAL
www.davislangdon.com

INDONESIA

All data relate to 2007 unless otherwise indicated.

Population	
Population	225.6 mn
Urban population (2007)	44%
Population under 15 (2007)	27.5%
Population 65 and over (2007)	5.07%
Average annual growth rate (2004 to 2007)	1.33%
Geography	
Land area	1,890,754 km^2
Agricultural area	24%
Capital city	Jakarta
Population of capital city (2007)*	9.13 mn
Economy	
Monetary unit	Indonesia Rupiah (Rp)
Exchange rate (average fourth quarter 2008) to:	
the pound sterling	Rp 17,214
the US dollar	Rp 10,977
the euro	Rp 14,446
the yen x 100	Rp 11,456
Average annual inflation (1998 to 2007)	15.26%
Inflation rate	6.48%
Gross Domestic Product (GDP)	Rp 3,957,404 bn
GDP per capita	Rp 15,482,800
Average annual real change in (GDP) (1998 to 2007)	17%
Private consumption as a proportion of GDP	63.55%
Public consumption as a proportion of GDP	8.3%
Investment as a proportion of GDP	24.9%
Construction	
Gross value of construction output	Rp 305,215 bn
Net value of construction output	Rp 67,318 bn
Net value of construction output as a proportion of GDP	1.7%

* *Preliminary figure*

THE CONSTRUCTION INDUSTRY

Construction Output

The gross value of construction output in 2007 was Rp305,215 billion, equivalent to US$33.09 million, or 7.71% of GDP. The net value of construction output in 2007 was Rp67,318 billion, equivalent to US$7.3 million or 1.7% of GDP.

The table below shows the value of construction by type of work completed by contractors who are members and non-members of the Indonesian Contractors Association (ICA).

VALUE OF CONSTRUCTION COMPLETED BY TYPE OF WORK, 2005

Type of work	*% of total*
Building	51
Residential	11
Non-residential	31
Specialist work related mainly to building	9
Civil Engineering	49
Water supply and gas supply network	1
Electricity and network	1
Construction or improvement of roads/bridges	28
Irrigation and drainage	6
Airports, harbours, bus stations, etc.	3
Electrical power plant and telecommunications	4
Others	6
Total	100

Source: Government statistics

Building work continues to be the predominant type of activity in 2005 accounting for 51% of total value as compared to 47% in 2003. The main component was non-residential building which made up more than half of the total building work. In contrast, there was a decline in civil engineering work which fell from 53% of the total value in 2003 to 49% in 2005. Much of the difference was attributed to the construction or improvement of roads and bridges.

CONSTRUCTION OUTPUT AND POPULATION BY REGIONS, 2005

Province	*Proportion of population*	*Proportion of construction output*	*Rank by population*	*Rank by construction output*
Nanggroe Aceh Darussalam	1.87	2.38	14	11
North Sumatera	5.67	5.71	4	5
West Sumatera	2.09	1.95	11	13
Riau	2.21	3.33	10	7
Jambi	1.21	1.45	20	18
South Sumatera	3.11	2.75	9	9
Bengkulu	0.72	0.32	26	32
Lampung	3.24	1.80	8	14
Bangka Belitung	0.49	0.51	29	27
Riau Island *	0.58	1.13	27	20
DKI Jakarta	4.06	21.03	6	1
West Java	17.89	19.05	1	2
Central Java	14.56	5.91	3	4
D.I. Yogyakarta	1.54	1.76	17	15
East Java	16.67	6.59	2	3
Banten	4.14	2.98	5	8
Bali	1.56	1.07	16	21
West Nusa Tenggara	1.90	0.47	13	28
East Nusa Tenggara	1.96	1.24	12	19
West Kalimantan	1.84	1.72	15	16
Central Kalimantan	0.90	0.45	23	29
South Kalimantan	1.51	2.07	18	12
East Kalimantan	1.32	4.95	19	6
North Sulawesi	0.98	0.33	22	31
Central Sulawesi	1.06	0.82	21	24
South Sulawesi	3.42	2.72	7	10
South East Sulawesi	0.89	0.78	24	25
West Sulawesi	Part of Central Sulawesi	0.25	Part of Central Sulawesi	33
Gorontalo *	0.43	0.86	30	23
Maluku	0.58	0.63	28	26
North Maluku	0.42	0.44	31	30
Papua	0.88	1.61	25	17
West Irian Jaya*	0.31	0.93	32	22
Total	100	100	-	-

Source: Statistics Indonesia of The Republic of Indonesia

** New province after Year 2000*

Most of the construction activity (21%) is centred in the capital city, Jakarta, which is home to 4% of the national population. West Java, Central Java and East Java, on the other hand, have construction outputs which are significantly lower than their respective share of the national population.

Characteristics and Structure of the Industry

The construction industry started to recover gradually from the 1997 Asian Financial Crisis in 2000 and construction activities increased significantly in 2003. The average annual increase of the construction output increased from 12% in year 2001-2003 to 50% in year 2003-2004.

Contractors are classified under three major groups and as of 2005, the total number registered with the National Contractors Association of Indonesia or *Gabungan Pelaksana Konstruksi Nasional Indonesia* (GAPENSI) are as follows.

NUMBER OF REGISTERED CONTRACTORS BY QUALIFICATION CLASS, 2005

Qualification	*Eligibility*	*Number of registered contractors*	*% of total*
B1 to B2 – Large	Bidding for work above Rp 3 billion	2,025	4
M – Medium	Bidding for work Rp 1 to Rp 3 billion	3,534	7
K1 to K3 – Small	Bidding for work below Rp 1 billion	46,803	89
Total		52,362	100

Source: Gabungan Pelaksana Konstruksi Nasional Indonesia

The major contractors in Indonesia are as follows:

- PT. Adhi Karya (Persero)
- PT. Balfour Beaty Sakti Indonesia
- PT. Decorient Indonesia
- PT. Hutama Karya (Persero)
- PT. Pembangunan Perumahaan (Persero)
- PT. Murinda Iron Steel
- PT. Total Bangun Persada

Clients and Finance

The client for public works is generally the government (central and regional), although private companies are now able to become involved to a significant extent in funding, constructing and operating public works such as toll roads and electricity generation plants.

The government obtains finance from various external sources as well as the national development budget, the Asian Development Bank, the World Bank and other multilateral and bilateral aid agencies. Private sector financing is obtained through local and foreign financial institutions including state and private banks, pension funds and private investors.

Selection of Design Consultants

For government projects, there must be a single stage open tender for consultancy work with pre-qualification for specialist projects.

Private sector procurement is not regulated by law and the procedure is therefore more flexible. With selective tendering, normal design is mostly outsourced.

Fees are generally negotiated. There are no recommended or mandatory published fee scales.

Contractual Arrangements

Construction in the government sector is restricted to government owned contractors (BUMN) except for large specialist projects. For the private sector, on the other hand, BUMN contractors, private contractors and foreign joint operation contractors, are allowed to participate.

The process for selection of design consultants applies also to the process for selection of contractors and subcontractors.

In the public sector, the tender is usually called on a lump sum basis by way of traditional 'construction-only' contracting. Private sector procurement is also largely traditional but also involves a variety of other methods. Traditional bills of quantities are normal. Other procurement methods such as:

- Turnkey
- Contractor financed
- Design and build

are also used but are generally restricted to relatively small number of projects.

Private contracts are usually based on internationally recognized forms adapted to suit the Indonesian conditions and statutes (typically FIDIC). Bank guarantees are normally required and cash retention is usually preferred to bonds. Advance payments are usually made. Variations are authorized through contract instructions and change orders with consequent adjustment in the contract sum.

Government contracts are based on official custom written departmental contracts.

Liability and Insurance

Insurance is compulsory for all parties. A Contractors' All Risks Policy covering the works and third party liability is normally taken out by the owner or the contractor in joint names for the full contract value and is valid until taking over of the project. By law, contractors must insure their workers with the preference being the government own scheme. Insurance companies are common in Indonesia but risks are normally reinsured offshore. Insurance claims are usually settled amicably.

Development Control and Standards

The Directorate of Regional and City Planning provides general development guidance and the regional government ensures implementation by the land user. If all other requirements are fulfilled, full or partial building permission takes about three months. The process is more stringent in Jakarta than elsewhere.

At the planning stage, architectural, structural and building services reviews are carried out by the appropriate authority to check that the design is in compliance with laws, rules and standards.

Seismic structural codes must be applied throughout Indonesia.

Before buildings can obtain an *Occupation Permit*, approval from the Fire Authority must be obtained.

The national standard for building materials/products is *Standard Industrial Indonesia (SII) – Indonesian Standard for Industry*. Foreign standards such as *ASTM, BS, DIN, PSB* and *JIS* are also applied extensively.

CONSTRUCTION COST DATA

Cost of Labour

The figures below are typical of labour costs in the Jakarta area as at the fourth quarter of 2008. The wage rate is the basis of an employee's income, while the cost of labour indicates the cost to a contractor of employing that employee. The difference between the two covers a variety of mandatory and voluntary contributions – a list of items which could be included is given in section 2.

	Wage rate (per hour) Rp	*Cost of labour (per day) Rp*	*Number of hours worked per year*
Site operatives			
Mason/bricklayer	6,500	70,000	2,208
Carpenter	7,100	75,000	2,208
Plumber	7,100	75,000	2,208
Electrician	7,100	75,000	2,208
Structural steel erector	5,250	65,000	2,208
HVAC installer	7,100	75,000	2,208
Semi-skilled worker	6,000	65,000	2,208

	Wage rate (per hour) Rp	*Cost of labour (per day) Rp*	*Number of hours worked per year*
Unskilled labourer	5,250	60,500	2,208
Equipment operator	10,000	110,000	2,208
Watchman/security	4,500	50,000	2,208
Site supervision			
General foreman	6,500	70,000	2,208
Trades foreman	6,500	70,000	2,208

Cost of Materials

The figures that follow are the costs of main construction materials, delivered to site in the Jakarta area, as incurred by contractors in the fourth quarter of 2008. These assume that the materials would be in quantities as required for a medium sized construction project and that the location of the works would be neither constrained nor remote.

All the costs in this section exclude value added tax (VAT).

	Unit	*Cost Rp*
Cement and aggregate		
Ordinary portland cement in 40kg bags	tonne	700,000
Coarse aggregates for concrete	m^3	140,000
Fine aggregates for concrete	m^3	140,000
Ready mixed concrete (K-350) slump 10	m^3	460,000
Ready mixed concrete (K-225) slump 12	m^3	400,000
Steel		
Mild steel reinforcement	tonne	9,250,000
High tensile steel reinforcement	tonne	9,250,000
Structural steel sections	tonne	15,500,000
Bricks and blocks		
Common bricks (220 x 100 x 50mm)	1,000	360,000
Light weight concrete blocks (590 x 190 x 100mm)	1,000	6,700,000
Precast concrete cladding units with exposed aggregate finish	m^2	500,000
Timber and insulation		
Softwood sections for carpentry	m^3	850,000
Softwood for joinery (Kamper)	m^3	5,000,000
Hardwood for joinery (Teak)	m^3	12,000,000
Plywood for interior joinery (18mm) 1200 x 2400mm	pc	75,000
100mm thick quilt insulation rockwool, density 80kg/m^2	m^2	130,000

	Unit	*Cost Rp*
100mm thick rigid slab insulation	m^2	52,000
Softwood internal door complete with frames and ironmongery	each	1,800,000
Glass and ceramics		
Float glass (8mm)	m^2	85,000
Good quality ceramic wall tiles (200 x 100mm)	m^2	60,000
Plaster and paint		
Plasterboard (12mm thick) 1200 x 2400mm	pc	48,000
Emulsion paint in 25 kg tins	kg	21,150
Gloss oil paint in 20 kg tins	kg	44,000
Tiles and paviors		
Clay floor tiles (150 x 150 x 10mm)	m^2	50,000
Vinyl floor tiles (300 x 300 x 2mm)	m^2	95,000
Precast concrete paving block (100 x 200 x 80mm)	m^2	32,250
Clay roof tiles	1,000	3,700,000
Precast concrete roof tiles (size 425 x 330mm)	1,000	3,850,000
Drainage		
WC suite complete	each	1,050,000
Lavatory basin complete	each	900,000
100mm diameter UPVC drain pipes	m	45,500

Unit Rates

The descriptions on the next page are generally shortened versions of standard descriptions listed in full in section 4. Where an item has a two digit reference number (e.g. 05 or 33), this relates to the full description against that number in section 4. Where an item has an alphabetic suffix (e.g. 12A or 34B) this indicates that the standard description has been modified. Where a modification is major the complete modified description is included here and the standard description should be ignored; where a modification is minor (e.g. the insertion of a named hardwood) the shortened description has been modified here but, in general, the full description in section 4 prevails.

The unit rates on the next page are for main work items on a typical construction project in the Jakarta area in the fourth quarter of 2008. The rates include all necessary labour, materials and equipment. Allowances have been included to cover contractors' overheads and profit and preliminary and general items.

All the rates in this section exclude value added tax (VAT).

		Unit	*Rate Rp*
Excavation			
01	Mechanical excavation of foundation trenches	m^3	35,000
02	Hardcore filling making up levels	m^3	55,000
Concrete work			
04	Plain in situ concrete in strip foundations in trenches	m^3	510,000
05	Reinforced in situ concrete in beds	m^3	535,000
06	Reinforced in situ concrete in walls	m^3	535,000
07	Reinforced in situ concrete in suspended floors or roof slabs	m^3	530,000
08	Reinforced in situ concrete in columns	m^3	535,000
09	Reinforced in situ concrete in isolated beams	m^3	550,000
10	Precast concrete slabs	m^3	556,000
Formwork			
11	Softwood formwork to concrete walls	m^2	75,000
12	Softwood formwork to concrete columns	m^2	80,000
13	Softwood formwork to horizontal soffits of slabs	m^2	88,000
Reinforcement			
14	Reinforcement in concrete walls	tonne	11,000,000
15	Reinforcement in suspended concrete slabs	tonne	11,000,000
16	Fabric reinforcement in concrete beds	m^2	51,000
Steelwork			
17	Fabricate, supply and erect steel framed structure	tonne	17,000,000
18	Framed structural steelwork in universal joist sections	tonne	18,600,000
19	Structural steelwork lattice roof trusses	tonne	18,500,000
Brickwork and blockwork			
21A	Solid (perforated) concrete blocks	m^2	52,000
Roofing			
24A	Concrete interlocking roof tiles 350 x 225mm	m^2	52,500
25	Plain clay roof tiles 260 x 160mm	m^2	43,000
33	Troughed galvanized steel roof cladding	m^2	170,910
Woodwork and metalwork			
34	Preservative treated sawn softwood 50 x 100mm	m	25,000
35	Preservative treated sawn softwood 50 x 150mm	m	36,750
36A	Single glazed casement window in Kamper hardwood, size 650 x 900mm	each	420,000
37A	Two panel glazed door in Kamper hardwood, size 850 x 900mm	each	950,000
38	Solid core half hour fire resisting hardwood internal flush door, size 800 x 2000mm	each	1,750,000

		Unit	*Rate Rp*
39	Aluminium double glazed window, size 1200 x 1200mm	each	2,500,000
40	Aluminium double glazed door, size 850 x 2100mm	each	3,150,000
41	Hardwood skirtings	m	45,000
Plumbing			
42	UPVC half round eaves gutter	m	95,000
43	UPVC rainwater pipes	m	65,000
44	Light gauge copper cold water tubing	m	95,000
46	Low pressure plastic pipes for cold water distribution	m	15,000
47	UPVC soil and vent pipes	m	43,000
48	White vitreous china WC suite	each	1,250,000
49	White vitreous china lavatory basin	each	850,000
50	Glazed fireclay shower tray	each	900,000
51	Stainless steel single bowl sink and double drainer	each	1,200,000
Electrical work			
52	PVC insulated and copper sheathed cable	m	4,500
53	13 amp unswitched socket outlet	each	75,000
54	Flush mounted 20 amp, 1 way light switch	each	47,000
Finishings			
56	White glazed tiles on plaster walls	m^2	60,000
57	Red clay quarry tiles on concrete floors	m^2	48,500
58	Cement and sand screed to concrete floors	m^2	22,500
59	Thermoplastic floor tiles on screed	m^2	120,000
60A	Mineral fibre tiles on concealed suspension system, 600 x 600mm	m^2	160,000
Glazing			
61A	5mm Glazing to wood	m^2	55,000
Painting			
62	Emulsion on plaster walls	m^2	17,000
63	Oil paint on timber	m^2	32,000

Approximate Estimating

The building costs per unit area that follow are expressed in US$ and are averages incurred by building clients for typical buildings in the Jakarta area as at the fourth quarter of 2008. They are based upon the total floor area of all storeys, measured between external walls and without deduction for internal walls.

Approximate estimating costs generally include mechanical and electrical installations but exclude furniture, loose or special equipment, and external works; they also exclude fees for professional services. The costs shown are for specifications and standards appropriate to Indonesia and this should be borne in mind when attempting comparisons with similarly described building types in other countries. A discussion of this issue is included in section 2. Comparative data for countries covered in this publication, including construction cost data, are presented in Part Three.

Approximate estimating costs must be treated with caution; they cannot provide more than a rough guide to the probable cost of building.

All the rates in this section exclude value added tax (VAT).

	Cost *m² US$*	*Cost* *ft² US$*
Industrial buildings		
Factories for letting	270	25
Factories for owner occupation (light industrial use)	325	30
Factories for owner occupation (heavy industrial use)	415	39
Factory/office (high-tech) for letting (shell and core only)	300	28
Factory/office (high-tech) for letting (ground floor shell, first floor offices)	390	36
Factory/office (high tech) for owner occupation (controlled environment, fully finished)	552	51
High tech laboratory workshop centres (air-conditioned)	555	52
Warehouses, low bay (6 to 8m high) for letting (no air-conditioned)	265	25
Warehouses, low bay for owner occupation (including air-conditioned)	390	36
Warehouses, high bay for owner occupation (10mm) (excluding air-conditioned)	465	43
Cold stores/refrigerated stores	650	60
Administrative and commercial buildings		
Civic offices, non air-conditioned	350	33
Civic offices, fully air-conditioned	410	38
Offices for letting, 5 to 10 storeys, air-conditioned	300	28
Offices for letting, high rise, air-conditioned	530	49
Offices for owner occupation high rise, air-conditioned	620	58
Prestige/headquarters office, 5 to 10 storeys, air-conditioned	550	51
Prestige/headquarters office, high rise, air-conditioned	805	75
Residential buildings		
Purpose designed single family housing 2 storey detached (single unit)	350	33
Social/economic apartment housing, low rise (no lifts)	400	37
Social/economic apartment housing, high rise (with lifts)	520	48

	Cost m^2 US$	Cost ft^2 US$
Private sector apartment building (standard specification)	575	53
Private sector apartment buildings (luxury)	705	65
Hotel, 5 star, city centre	1,500	139
Hotel, 3 star, city/provincial	900	84
Motel	750	70
Golf courses	370,000	per hole
Golf clubhouse	800	74
Health and education buildings		
General hospitals	750	70
Private hospitals	830	77
Health centres	650	60
Primary/Junior schools	420	39
Secondary/middle schools	525	49
University	530	49

Regional Variations

The approximate estimating costs are based on projects in Jakarta. Costs elsewhere can vary by up to plus or minus 20%.

Value Added Tax (VAT)

The standard rate of value added tax (VAT) is currently 10%, chargeable on general building work.

EXCHANGE RATES

The graph on the next page plots the movement of the Indonesian rupiah against the sterling, the euro, the US dollar and 100 Japanese yen since1998. The values used for the graph are quarterly and the method of calculating these is described and general guidance on the interpretation of the graph provided in section 2. The average exchange rate for the fourth quarter of 2008 was Rp17,214 to pound sterling, Rp14,446 to euro, Rp10,977 to US dollar and Rp11,456 to 100 Japanese yen.

THE INDONESIAN RUPIAH AGAINST STERLING, EURO, US DOLLAR AND 100 JAPANESE YEN

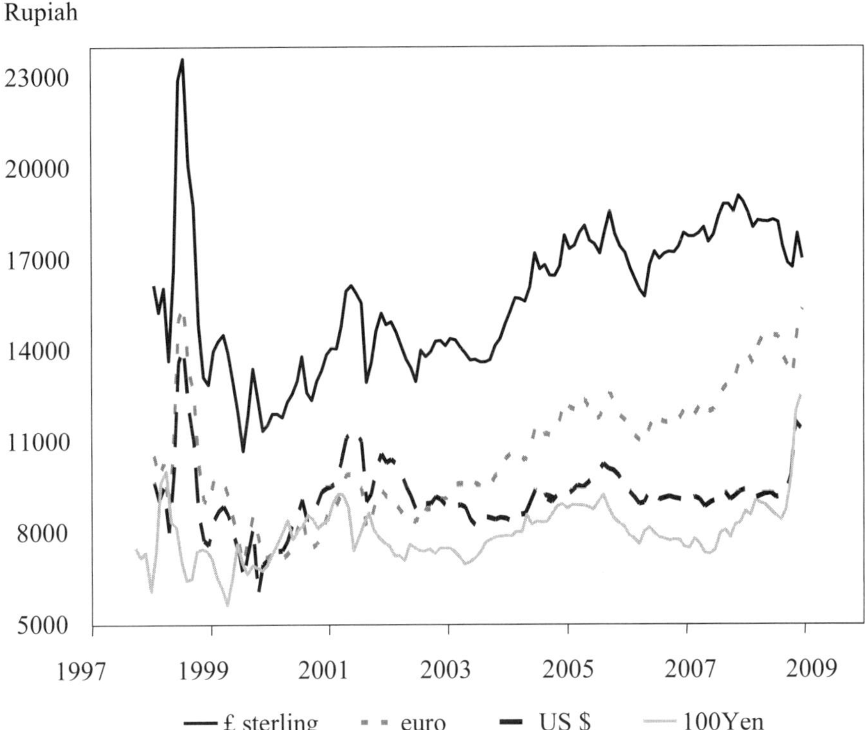

USEFUL ADDRESSES

Public Organizations

Agency of Assessment and Application of Technology
Badan Pengkajian dan Penerapan Teknologi (BPPT)
Jalan M.H. Thamrin 8
Jakarta 10340
Tel: (+62-21) 316-8200
Fax: (+62-21) 319-24319
E-mail: humas@bppt.go.id
Website: www.bppt.go.id

Indonesian Institute of Science
Lembaga Ilmu Pengetahuan Indonesia (LIPI)
Jalan Jendral Gatot Subroto 10
Jakarta 12710
Tel: (+62-21) 525-5711
Fax: (+62-21) 526-5457
Website: www.lipi.go.id

The Investment Coordinating Board
Badan Koordinasi Penanaman Modal (BKPM)
Jalan Jendral Gatot Subroto 44
Jakarta 12190
PO Box 3186
Tel: (+62-21) 525-2008 / 525-2649 / 525-4981
Fax: (+62-21) 525-4945
E-mail: sysadm@bkpm.go.id
Website: www.bkpm.go.id

Ministry of Culture and Tourism
Departemen Kebudayaan dan Pariwisata
Sapta Pesona Building
Jalan Medan Merdeka Barat 17
Jakarta 10110
Tel: (+62-21) 383-8167
Fax: (+62-21) 384-9715
Website: www.budpar.go.id

Ministry of Energy and Mineral Resources
Departemen Energi dan Sumber Daya Mineral
Jalan Medan Merdeka Selatan No 18
Jakarta 10110
Tel: (+62-21) 3804242 / 3813232 / 3446542
Fax: (+62-21) 3450846 / 3440649
E-mail: pusdatin@esdm.go.id
Website: www.esdm.go.id

Ministry of Industry
Departemen Perindustrian
Jalan Jendral Gatot Subroto Kav. 52-53
Jakarta 12950
Tel: (+62-21) 522-5194 / 527-1382 / 527-1387 / 527-1388
Fax: (+62-21) 526-1086
E-mail: karomas@depperin.go.id
Website: www.depperin.go.id

Ministry of Public Works
Departemen Pekerjaan Umum
Jalan Pattimura 20
Jakarta Selatan 12110
Tel: (+62-21) 739-5588
Fax: (+62-21) 722-0219
E-mail: pusdata.pu.go.id
Website: www.pu.go.id

The Ministry of Public Works is subdivided into:
Direktorat Binamarga
a directorate covering roads and bridges and construction work
Direktorat Cipta Karya
a directorate covering general building works
Direktorat Pengairan
a directorate covering hydrologic construction works
Direktorat Air Bersih
a directorate covering sanitation works

National Development Planning Agency
Badan Perencana Pembangunan Nasional (BAPPENAS)
Jalan Taman Suropati 2
Jakarta 10310
Tel: (+62-21) 390-5650
Fax: (+62-21) 314-5374
Website: www.bappenas.go.id

Regional Goverment Construction Ministries:
Kantor Wilayah Pekerjaan Umum
Public Works; Ministry Regional Offices under coordination of Central Government
Dinas Pekerjaan Umum
Public Works Provincial Offices under coordination of Regional Government

Science and Technology Research Centre
Pusat Penelitian Ilmu Pengetahuan dan Teknologi (PUSPITEK)
Building TMC-1, Puspitek Serpong, Tangerang
Banten 15314
Tel: (+62-21) 756-0001
Fax: (+62-21) 756-0071
Website: www.puspitek.net

State Ministry of Public Housing
Kementeri Negara Perumahan Rakyat
Jalan Raden Patah 1 No. 1, Level 2, Wing 4, Kebayoran Baru
Jakarta Selatan
Tel: (+62-21) 739-7727
Fax: (+62-21) 739-7777
Website: www.kemenpera.go.id

Statistics Indonesia of The Republic of Indonesia
Biro Pusat Statistik
Jalan Dr Sutomo 6-8
Jakarta 10710
Tel: (+62-21) 384-1195 / 384-12508 / 381-0291
Fax: (+62-21) 385-7046
E-mail: bpshq@bps.go.id
Website: www.bps.go.id

Trade And Professional Associations

Batam Industrial Estate Development Authority
Otorita Pengembangan Daerah Industri Pulau Batam
BIDA Building, Batam Centre
Batam 29400
Tel: (+62-778) 462-047 / 462-048
Fax: (+62-778) 462-456 / 462-240 / 462-492
Website: www.batam.go.id

DKI Jakarta City Development Coordinator
Dinas Penataan dan Pengawasan Bangunan DKI Jakarta
Jalan Taman Jatibaru No 1
Jakarta
Tel: (+62-21) 385-7093
Website: www.dppb.jakarta.go.id

Indonesian Architect Association
Ikatan Arsitek Indonesia (IAI)
Jakarta Design Centre, Level 7
Jalan Jendral Gatot Subroto
Jakarta 10260
Tel: (+62-21) 530-4715
Fax: (+62-21) 530-4722
E-mail: iai@iai.or.id
Website: www.iai.or.id

Indonesian Contractors' Association
Asosiasi Kontraktor Indonesia
Wijaya Graha Puri Blok D-1
Jl. Darmawangsa Raya No. 2
Jakarta 12160
Indonesia
Tel: (+62-21) 720 0794 / 727 90672
Fax: (+62-21) 720 6805
E-mail: akinet@cbn.net.id / akinet@aki.or.id
Website: www.aki.or.id

Indonesian Construction Expert Association
Himpunan Ahli Konstruksi Indonesia (HAKI)
Jalan Tebet Barat Dalam 10 No 5
Tebet
Jakarta 12810
Tel: (+62-21) 835-1186 / 829- 8518
Fax: (+62-21) 831-6451 / 835-1186
E-mail: haki@cbn.net.id
Website: www.haki-konstruksi.com

Indonesian Engineers Association
Persatuan Insinyur Indonesia
Jl. Halimun No. 39
Jakarta 12980
Tel: (+62-21) 835-2180 / 835-2181
Fax: (+62-21) 837-00663
E-mail: info@pii.or.id
Website: www.pii.or.id

Indonesian National Consultant Association
Ikatan Konsultan Nasional Indonesia (INKINDO)
Jalan Bendungan Hilir Raya No 29
Jakarta Pusat 10235
Tel: (+62-21) 573-8577
Fax: (+62-21) 573-3474
E-mail: inkindo@inkindo.org
Website: www.inkindo.org

National Chamber of Trade and Industry
Kamar Dagang dan Industri Nasional
Menara Kadin Indonesia Level 29
Jalan HR Rasuna Said X-5 Kav 2-3
Jakarta 12950
Indonesia
Tel: (+62-21) 527-4484
Fax: (+62-21) 527-4331 / 527-4332
E-mail: infokadin@kadin-indonesia.or.id
Website: www.kadin-indonesia.or.id

National Contractors' Association of Indonesia
Gabungan Pelaksana Konstruksi Nasional Indonesia (GAPENSI)
Graha GAPENSI
Jl. Raya Ragunan No. C/1
Jatipadang, Pasar Minggu
Jakarta 12540
Indonesia
Tel: (+62-21) 788-47247
Fax: (+62-21) 780-6119
E-mail: bppgapensi@link.net.id
Website: www.gapensi.or.id

Real Estate Indonesia
Persatuan Perusahaan Realestat Indonesia
Rukan Simprug Indah
Jl. Teuku Nyak Arief No.9B Kebayoran Lama
Jakarta Selatan 12210
Tel: (+62-21) 727-89105
Fax: (+62-21) 727-89155
Website: www.reindonesia.org

JAPAN

All data relate to 2008 unless otherwise indicated.

Population	
Population	127.7 mn
Urban population (2000)	64.6%
Population under 15 (2000 est.)	14.6%
Population 65 and over (2004 est.)	20.8%
Average annual growth rate (2005)	-0.06%
Geography	
Land area	377,835 km^2
Agricultural area	12.6%
Capital city	Tokyo
Population of capital city (2005)	12.87 mn
Economy	
Monetary unit	Yen (¥)
Exchange rate (average fourth quarter 2008) to:	
the pound sterling	¥ 150.00
the US dollar	¥ 96.00
the euro	¥ 126.00
Average annual inflation (2001 to 2008)	+0.1%
Inflation rate (2007)	1.4%
Gross Domestic Product (GDP)	¥ 557,219.4 bn
GDP per capita	¥ 3,096,093
Average annual real change in (GDP) (2007)	0.10%
Private consumption as a proportion of GDP	55%
Public consumption as a proportion of GDP	15%
Investment as a proportion of GDP	29.6%
Construction	
Gross value of construction output	¥ 49,174 bn

THE CONSTRUCTION INDUSTRY

Construction Output

The gross value of construction output in 2008 was ¥49,174 billion.

The table below shows the breakdown of work by type of construction for 2008.

VALUE OF WORK DONE BY TYPE OF CONSTRUCTION, 2008

Type of construction	*Public (¥ billion)*	*Private (¥ billion)*	*Total (¥ billion))*	*% of total*
Residential	476	16,892	17,368	35
Non-residential	1,777	8,930	10,707	22
Civil Engineering	14,670	6,429	21,099	43
Total	16,923	32,251	49,174	100

Source: *Ministry of Internal Affairs and Communications (Statistics Bureau, Director-General for Policy Planning & Statistical Research and Training Institute)*

Of the total value of construction completed in 2008, 66% was undertaken by the private sector. Residential building made up 52% of the total private construction.

Characteristics and Structure of the Industry

As of March 2007, there were about 484,649 construction contractors licensed either by the Ministry of Construction or by the Governors of Prefectures employing over 3.3 million people, a decline of 6.4% from the previous year. Of the total number, 99 are foreign firms licensed to engage in construction activities in Japan. There are 41 American constructions or engineering companies operating either independently or through partnerships or joint ventures with their local counterparts. Most of the contracting firms are small and medium size with more than 97% having capitalization of less than ¥50 million.

The Japanese economic recession and the contraction of the industry have resulted in an increase in the number of bankruptcies – some of which are public listed companies. In 2008, there were about 4,467 companies that became insolvent.

The slowdown in the domestic construction market has spurred Japanese construction firms to secure overseas work. Despite the global economic crisis, construction orders from overseas works have increased from ¥1,681 billion in 2007 to ¥2,042 billion in 2008, an increase of approximately 21%. However, such overseas works only account approximately 3.5% of the total amount of construction costs in Japan.

The table below shows the list of International Japanese contractors that are primarily involved in building, as ranked in the *Engineering News Record's 2008 Top 225 International Contractors.*

INTERNATIONAL JAPANESE CONTRACTORS PRIMARILY INVOLVED IN BUILDING, 2008

Major contractors	*ENR Rank 2008*
Obayashi Corporation	23
Kajima Corporation	24
Taisei Corporation	36
Takenaka Corporation	49
Shimizu Corporation	52
Penta-Ocean Construction Co. Ltd	92
Maeda Corporation	104
Hazama Corporation	111
Sumitomo Mitsui Construction Co. Ltd	115
Nishimatsu Construction Co. Ltd	138
Toda Corporation	182

Source: Engineering News Record 2008 Top 225 International Contractors

The *Construction Industry Law* requires contractors to obtain a licence to start a construction business. Nearly all site work in Japan is undertaken by specialty trade contractors who maintain a special relationship with a general contractor, known as a *ZENECON*. Under this relationship, the general contractor will endeavour to provide continuous employment for his subcontractors, in return for which each subcontractor will allow the general contractors to stipulate a contract price, and to monitor both his financial and project performance. The very large companies do not have a permanent workforce, but a family of subcontractors who are loosely connected to them.

One of the features of contracting organizations in Japan is that they undertake a considerable amount of research and development work. The range of research is very wide, from soil testing, environmental technology, information technology to air-supported domes. Earthquake engineering is important and the Japanese are generally regarded as world leaders in both research on the use of robots in construction and the development of intelligent buildings. Direct expenditure on research and development has been on the decline but amongst the top five construction firms, it continues to account for about 1% of total turnover.

For private sector projects, the contractor would normally checks all designs and products to be used in the project and reports to the client of any possible failures. The high level of responsibility placed on the contractor for the success of projects is one of the reasons why in-house research and development departments are needed.

The Ministry of Land and Infrastructure, Transport and Tourism (MLIT) oversees all aspects of construction. Research institutes and other organizations are under its control although each research institute has a large degree of autonomy.

Construction is also monitored by other ministries such as the Ministry of Agriculture.

There are three types of architectural engineer in Japan: first-class architectural engineer, second-class architectural engineer and wooden building architectural engineer. With effect from May 2009, qualification of Architectural Engineers specializes in structure and M&E will be required for designated buildings. First-class architectural engineer who have worked for over 5 years as structural or facility engineers must attend a set of lectures provided by the registration and lecture institutions listed by the MLIT before a licence can be issued. The other two categories are dealt with on a similar basis by prefecture governors. About 50% of first-class architects and most second-class architects work for contractors.

The table below shows the breakdown of licensed architects as at March 2008.

NUMBER OF LICENSED ARCHITECTS, MARCH 2008

Class	*No of architects*
First-class Architect	322,248
Second-class Architect	692,968
Wooden building Architect	4,562
Total	1,019,778

Source: Ministry of Land and Infrastructure, Transport and Tourism

Foreign architects seeking to register as first-class architects can obtain a licence (Apec Architect) without having to sit for the examination if they are recognized by the Minister of Construction as possessing equivalent qualifications to that of a first-class architect.

Clients and Finance

To encourage private sector involvement in public works projects, the government has adopted a scheme modelled along the lines of the UK's Private Finance Initiative (PFI). Prefecture and local governments are also exploring the use of such scheme for the construction of toll roads, government buildings and other infrastructure projects.

Selection of Design Consultants

For projects that are not on design-and-build procurement route, architects, engineers and cost consultants are usually appointed by the client either directly or after some form of competition. In some cases, other consultants are chosen by one of the main consultants. The most important basis for selection is track record with price as a secondary factor. Personal contacts and recommendations are sometimes relevant in the private sector but rarely in the public sector. In cases where the contractor is being appointed prior to other consultants, the client will be requested by the contractor to appoint an architect – often one of his selections. The architect would however, still be paid by the client.

Contractual Arrangements

In the public sector, construction companies of the appropriate category and experience are invited to bid. Central and local governments rank construction firms according to current and past track records, sales, financial status and technological capabilities when pre-qualifying those who will be on the tender list. The contract is then awarded to the lowest bidder. In the private sector, the client may appoint a specific contractor or invite selected contractors to bid – the latter being the more common system. Construction management system has recently generated much interest in the industry among the consultants and contractors. A recommendation to adopt construction management system was included in the paper (New action agenda concerning cost reduction members for public construction works) issued by the Committee on Administrative Reforms in 2000.

Generally, the Japanese contracting system is based on trust and mutual understanding. It is considered very important for both parties to maintain a good and long term relationship. Lawyers are rarely present during negotiations as it implies mistrust, and litigation is only used as the last resort. The contractor generally prepares the shop drawings, except for building services which are the responsibility of the specialist contractor.

The following are the two most commonly used standard contract forms:

- *Standard Form of Agreement and General Conditions of Government Contract for Works of Building and Civil Engineering*, prepared and recommended by the Construction Industry Council of Japan.
- *General Conditions of Construction Contract (GCCC)* approved jointly by the Architectural Institute of Japan, Architectural Association of Japan, Japan Institute of Architects, National General Contractors Association of Japan, Building Contractors Society, Japan Federation of Architects and Building Engineers Association and the Japan Federation of Architect Offices Association.

There is no bill of quantities for most cases but the contractor submits an itemized list of prices (including quantities) with his tender. Liquidated damages are payable if a project is delayed, and there is a defects liability period of two years for brick or concrete buildings and one year for timber structures which are extended to ten years and five years respectively if the defects have been wilfully caused or were due to negligence of the Contractor. The employer is given express rights to rectify the work and negotiations take place on dates and costs.

Liability and Insurance

Although the designer has primary responsibility for defects attributable to design, in some cases, the contractor corrects the defects in order to retain the confidence of the client. Some architects do not carry Professional Indemnity.

The Registration Organization for Warranties Houses, administered by the MLIT, provides a warranty scheme. This gives a ten year guarantee on the durability of structural components, including foundations, floors, walls and roof plus a five year warranty on the weather resistance of roof. In 2000, a ten year warranty for all areas to prevent water penetration was launched. The scheme is available to detached house builders using traditional Japanese timber construction techniques. Prefabricated house builders and condominium builders who compete with the single unit homebuilders, also provide a ten year protection on structural components.

CONSTRUCTION COST DATA

Cost of Labour

The figures that follow are typical of labour costs in the Tokyo area as at the fourth quarter of 2008. The wage rate is the basis of an employee's income.

	Wage rate (per day) ¥	*Number of hours worked per year*
Site operatives		
Mason/bricklayer	20,700	2,230
Carpenter	19,900	2,230
Plumber	18,000	2,230
Electrician	18,100	2,230
Structural steel erector	17,600	2,230
HVAC installer	18,100	2,230
Semi-skilled worker	14,000	2,230
Unskilled labourer	10,700	2,230
Equipment operator	17,400	2,230
Watchman/security	9,200	2,230
Site supervision		
General foreman	20,100	2,230
Trades foreman	21,000	2,230
Contractors' personnel		
Site manager	45,200	2,300
Resident engineer	34,900	2,300
Resident surveyor	34,900	2,300
Junior engineer	15,300	2,300
Junior surveyor	15,300	2,300

	Wage rate (per day) ¥	*Number of hours worked per year*
Planner	15,300	2,300
Consultants' personnel		
Senior architect	45,900	2,020
Senior engineer	45,900	2,020
Senior surveyor	31,100	2,020
Qualified architect	38,300	20,20
Qualified engineer	38,300	2,020
Qualified surveyor	24,700	2,020

Cost of Materials

The figures that follow are the costs of main construction materials, delivered to site in the Tokyo area, as incurred by contractors in the fourth quarter of 2008. These assume that the materials would be in quantities as required for a medium sized construction project and that the location of the works would be neither constrained nor remote.

	Unit	*Cost ¥*
Cement and aggregates		
Ordinary portland cement in 25kg bag	tonne	40,000
Coarse aggregates for concrete	m^3	3,450
Fine aggregates for concrete	m^3	4,050
Ready mixed concrete (210kg cement/cm^2)	m^3	11,750
Steel		
Mild steel reinforcement	tonne	109,000
High tensile steel reinforcement	tonne	105,000
Pre/Post compressing tendons	tonne	323,000
Structural steel sections	tonne	130,000
Bricks and blocks		
Common bricks (210 x 100 x 60mm)	1,000	195,000
Good quality facing bricks (210 x 100 x 60mm)	1,000	257,000
Hollow concrete blocks (190 x 190 x 390mm)	1,000	215,000
Solid concrete blocks (190 x 190 x 200mm)	each	NA
Precast concrete cladding units with exposed aggregate finish	m^2	10,300
Timber and insulation		
Softwood sections for carpentry	m^3	50,000
Softwood for joinery	m^3	80,000
Hardwood for joinery	m^3	185,000

	Unit	Cost ¥
Exterior quality plywood (12mm)	m²	840
Plywood for interior joinery (5mm)	m²	440
Softwood strip flooring (15mm)	m²	12,000
Chipboard sheet flooring (15mm)	m²	650
Softwood internal door complete with frames and ironmongery	each	81,900
Glass and ceramics		
Float glass (5 mm)	m²	1,330
Sealed double glazing units (FL3+A6+FL3) 12mm thick	m²	36,000
Good quality ceramic wall tiles	m²	3,650
Plaster and paint		
Plaster in 2 kg bags	tonne	45,500
Plasterboard (9mm thick)	m²	145
Emulsion paint in 5 litre tins	kg	310
Gloss oil paint in 5 litre tins	kg	400
Tiles and paviors		
Clay floor tiles (200 x 200mm)	m²	4,300
Vinyl floor tiles (2 x 300 x 300mm)	m²	700
Precast concrete paving slabs (300 x 300 x 60mm)	m²	4,440
Clay roof tiles	m²	3,100
Precast concrete roof tiles	m²	2,300
Drainage		
WC suite complete	each	44,000
Lavatory basin complete	each	26,600
100mm diameter clay drain pipes	m	2,230
150mm diameter stainless steel drain pipes	m	8,825

Unit Rates

The descriptions that follow are generally shortened versions of standard descriptions listed in full in section 4. Where an item has a two digit reference number (e.g. 05 or 33), this relates to the full description against that number in section 4. Where an item has an alphabetic suffix (e.g. 12A or 34B) this indicates that the standard description has been modified. Where a modification is major the complete modified description is included here and the standard description should be ignored; where a modification is minor (e.g. the insertion of a named hardwood) the shortened description has been modified here but, in general, the full description in section 4 prevails.

The unit rates below are main work items on a typical construction project in the Tokyo area in the fourth quarter of 2008. The rates include all necessary labour, materials and equipment. Allowances to cover preliminary and general items and contractors' overheads and profit have been added to the rates.

		Unit	*Rate ¥*
Excavation			
01	Mechanical excavation of foundation trenches	m^3	900
02	Hardcore filling making up levels	m^3	1,200
03	Earthwork support	m^2	37,400
Concrete work			
04	Plain in situ concrete in strip foundations in trenches	m^3	11,800
05	Reinforced in situ concrete in beds	m^3	12,100
06	Reinforced in situ concrete in walls	m^3	12,100
07	Reinforced in situ concrete in suspended floors or roof slabs	m^3	12,100
08	Reinforced in situ concrete in columns	m^3	12,500
09	Reinforced in situ concrete in isolated beams	m^3	12,500
10	Precast concrete slabs	m^2	8,400
Formwork			
11	Softwood formwork to concrete walls	m^2	3,950
12	Softwood formwork to concrete columns	m^2	3,950
13	Softwood or metal formwork to horizontal soffits of slabs	m^2	3,950
Reinforcement			
14	Reinforcement in concrete walls	tonne	140,000
15	Reinforcement in suspended concrete slabs	tonne	140,000
Steelwork			
17	Fabricate, supply and erect steel framed structure	tonne	174,000
18	Framed structural steelwork in universal joists sections	tonne	104,200
19	Structural steelwork lattice roof trusses	tonne	170,000
Brickwork and blockwork			
20	Precast lightweight aggregate hollow concrete block walls	m^2	7,300
21A	Solid (perforated) common bricks	m^2	N/A
Roofing			
24	Concrete interlocking roof tiles 430 x 380mm	m^2	6,600
25	Plain clay roof tiles 260 x 160mm	m^2	9,000
26	Fibre cement roof slates 600 x 300mm	m^2	2,250
27	Sawn softwood roof boarding	m^2	N/A

		Unit	*Rate ¥*
28	Particle board roof coverings	m^2	2,100
29	3 layers glass-fibre based bitumen felt roof covering	m^2	4,050
30	Bitumen based mastic asphalt roof covering	m^2	4,600
31	Glass-fibre mat roof insulation 160mm thick	m^2	4,200
33	Troughed galvanized steel roof cladding	m^2	4,080
Woodwork and metalwork			
34	Preservative treated sawn softwood 50 x 100mm	m	590
37	Two panel glazed door in hardwood size 850 x 2000mm	each	183,900
38	Solid core half hour fire resisting hardwood internal flush door, size 800 x 2000mm	each	79,000
39	Aluminium double glazed window, size 1200 x 1200mm	each	62,400
40	Aluminium double glazed door, size 850 x 2100mm	each	84,200
41	Hardwood skirtings	m	1,950
Plumbing			
42	UPVC half round eaves gutter	m	2,500
43	UPVC rainwater pipes	m	1,850
44	Light gauge copper cold water tubing	m	2,010
45	High pressure plastic pipes for cold water supply	m	1,230
46	Low pressure plastic pipes for cold water distribution	m	1,480
47	UPVC soil and vent pipes	m	N/A
48	White vitreous china WC suite	each	65,400
49	White vitreous china lavatory basin	each	58,400
51	Stainless steel single bowl sink and double drainer	each	74,800
Electrical work			
52	PVC insulated and copper sheathed cable	m	280
53	13 amp unswitched socket outlet	each	3,490
54	Flush mounted 20 amp, 1 way light switch	each	3,590
Finishings			
55A	2 coats gypsum based plaster on concrete walls 20mm thick	m^2	4,220
56	White glazed tiles on plaster walls	m^2	5,900
57	Red clay quarry tiles on concrete floors	m^2	14,270
58A	Cement and sand screed to concrete floors 30mm thick	m^2	1,950
59	Thermoplastic floor tiles on screed	m^2	2,350
60	Mineral fibre tiles on concealed suspension system	m^2	4,250

		Unit	*Rate ¥*
Glazing			
61	Glazing to wood	m²	5,200
Painting			
62	Emulsion on plaster walls	m²	910
63	Oil paint on timber	m²	1,230

Approximate Estimating

The building costs per unit area given below are averages incurred by building clients for typical buildings in the Tokyo area as at the fourth quarter of 2008. They are based upon the total floor area of all storeys, measured between external walls and without deduction for internal walls. Approximate estimating costs generally include mechanical and electrical installations but exclude furniture, loose or special equipment, and external works; they also exclude fees for professional services. The costs shown are for specifications and standards appropriate to Japan and this should be borne in mind when attempting comparisons with similarly described building types in other countries. A discussion of this issue is included in section 2. Comparative data for countries covered in this publication, including construction cost data, are presented in Part Three.

Approximate estimating costs must be treated with reserve; they cannot provide more than a rough guide to the probable cost of building.

	Cost m² ¥	*Cost ft² ¥*
Industrial buildings		
Factories for letting	150,000	14,000
Factories for owner occupation (light industrial use)	160,000	15,000
Factories for owner occupation (heavy industrial use)	170,000	16,000
Factory/office (high tech) for letting (shell and core only)	200,000	19,000
Factory/office (high tech) for letting (ground floor shell, first floor offices)	220,000	21,000
Factory/office (high tech) for owner occupation (controlled environment, fully finished)	250,000	23,000
High tech laboratory workshop centres (air-conditioned)	250,000	23,000
Warehouses, low bay (6 to 8m high) for letting	100,000	93,000
Warehouses, low bay for owner occupation	110,000	10,000
Warehouses, high bay for owner occupation	120,000	11,000
Administrative and commercial buildings		
Offices for letting, 5 to 10 storeys, non air-conditioned	200,000	19,000
Offices for letting, 5 to 10 storeys, air-conditioned	230,000	21,000

	Cost m² ¥	Cost ft² ¥
Offices for letting, high rise, air-conditioned	250,000	23,000
Offices for owner occupation high rise, air-conditioned	260,000	24,000
Prestige/headquarters office, 5 to 10 storeys, air-conditioned	300,000	28,000
Prestige/headquarters office, high rise, air-conditioned	350,000	33,000
Health and education buildings		
General hospitals (300 beds)	300,000	28,000
Private hospitals (100 beds)	280,000	26,000
Health centres	260,000	24,000
Primary/junior schools	230,000	21,000
Secondary/middle schools	220,000	20,000
University	280,000	26,000
Recreation and arts buildings		
Theatres (over 500 seats)	400,000	37,000
Sports halls including changing and social facilities	220,000	20,000
Swimming pools (international standard) (Olympic size)	each	98,000,000
Swimming pools (schools standard) including changing facilities	each	75,000,000
City centre/central libraries	350,000	33,000
Branch/local libraries	300,000	28,000
Residential buildings		
Private/mass market single family housing 2 storey detached/semi detached (multiple units)	190,000	18,000
Purpose designed single family housing 2 storey detached (single unit)	210,000	20,000
Social/economic apartment housing, high rise (with lifts)	220,000	20,000
Private sector apartment building (standard specification)	210,000	20,000
Private sector apartment buildings (luxury)	300,000	28,000
Student/nurses halls of residence	200,000	19,000
Homes for the elderly (shared accommodation)	200,000	19,000
Hotel, 5 star, city centre	400,000	37,000
Hotel, 3 star, city/provincial	300,000	28,000

Regional Variations

The approximate estimating costs are based on projects in Tokyo. These costs should be adjusted by the following factors to take account of regional variations:

Nagoya	:	-8%	Fukuoka	:	-9%
Osaka	:	-3%	Sapporo	:	-5%
Hiroshima	:	-7%			

EXCHANGE RATES AND INFLATION

The combined effect of exchange rates and inflation on prices within a country and price comparisons between countries is discussed in section 2.

Exchange Rates

The graph below plots the movement of the Japanese yen against the sterling, the euro and the US dollar since 1998. The values used for the graph are quarterly and the method of calculating these is described and general guidance on the interpretation of the graph provided in section 2. The average exchange rate in the fourth quarter of 2008 was ¥150 to pound sterling, ¥126 to euro and ¥96 to US dollar.

THE JAPANESE YEN AGAINST STERLING, EURO AND US DOLLAR

Price Inflation

The table below presents consumer price and building cost indices in Japan since 1980.

CONSUMER PRICE AND BUILDING COST INDICES

	Consumer price index		*Building cost index*	
Year	*average index*	*average change %*	*average index*	*average change %*
1980	100		100	
1981	105	5.0	103	3.0
1982	108	2.9	104	1.0
1983	110	1.9	103	-1.0
1984	112	1.8	103	0.0
1985	115	2.7	103	0.0
1986	115	0.0	102	-1.0
1987	115	0.0	104	2.0
1988	116	0.9	111	6.7
1989	119	2.6	118	6.3
1990	122	2.5	127	7.6
1991	126	3.3	135	6.3
1992	128	1.6	136	0.8
1993	130	1.3	132	-3.4
1994	131	0.7	126	-4.7
1995	130	-0.1	122	-3.1
1996	130	0.1	120	-1.5
1997	133	1.8	119	-0.6
1998	134	0.6	117	-2.0
1999	133	-0.3	116	-0.9
2000	132	-0.7	116	+0.1
2001	131	-0.7	114	-1.7
2002	130	-0.9	113	-1.0
2003	130	-0.3	113	+0.6
2004	130	-0.0	115	+1.1
2005	129	-0.3	116	+1.0
2006	129	-0.0	118	+2.1
2007	129	-0.0	120	+1.9
2008	129	-0.0	124	+3.0

USEFUL ADDRESSES

Public Organizations

The Architectural Institute of Japan
26-20, Shiba 5 Chome,Minato-ku
Tokyo 108-8414
Tel: (81) 3 3456 2051
Fax: (81) 3 3456 2058
E-mail:info@aij.or.jp

The Associated General Contractors of Japan, Inc
2-5-1 Hacchobori
Chuo-ku
Tokyo 104-0032
Tel: (81) 3 3551 9396
Fax: (81) 3 3555 3218
E-mail: kcho@zenken-net.or.jp
Website: www.zenken-net.or.jp

The Building Centre of Japan
6-1-8 Sotokanda Chiyoda-ku
Tokyo 101-8986
Tel: (81) 3 5816 7511
Fax: (81) 3 5816 7541
E-mail: soumu@bcj.or.jp
Website: www.bcj.or.jp

The Building Surveyors' Institute of Japan
Sunrise Mita building 7F
3-16-12 Shiba, Minato-ku
Tokyo 105-0014
Tel: (81) 3 3453 9591
Fax: (81) 3 3453 9597
E-mail: honbu@sekisan-kyokai.or.jp
Website: www.bsij.or.jp

The Japan Chamber of Commerce and Industry
3-2-2 Marunouchi
Chiyoda-ku
Tokyo 100-0005
Tel: (81) 3 3283 7824
Fax: (81) 3 3211 4859
E-mail: info@jcci.or.jp
Website: www.jcci.or.jp

Japan Civil Engineering Consultants Association
KY Sanbanchou building 8F
Sanbanchou
1-chome, Chiyoda-ku
Tokyo 102-0075
Tel: (81) 3 3229 7992
Fax: (81) 3 3239 1869
E-mail: info@jcca.or.jp
Website: www.jcca.or.jp

Japan Civil Engineering Contractors' Association, Inc
Tokyo Kensetsu Building
5-1-2-chome, Hacchobori
Chuo-ku
Tokyo 104-0032
Tel: (81) 3 3552 3201
Fax: (81) 3 3552 3206
E-mail: dokokyo@st.alpha-web.or.jp
Website: www.dokokyo.or.jp

Japan Federation of Construction Contractors, Inc
Tokyo Kensetsu Building 8F
2-5-1 Hacchobori
Chuo-ku
Tokyo 104-0032
Tel: (81) 3 3553 0701
Fax: (81) 3 3552 2360
E-mail: jfcc@mx1.alpha-web.ne.jp
Website: www.nikkenren.com

Japan Institute of Architects
JIA Kan
2-3-18 Jingumae
Shibuya-ku
Tokyo 150-0001
Tel: (81) 3 3408 7125
Fax: (81) 3 3408 7129
Website: www.jia.or.jp

Japan Structural Consultants Association
Hayashi Sanbancho building
Sanbancho 24
Chiyoda-ku
Tokyo 102-0075
Tel: (81) 3 3262 8498
Fax: (81) 3 3262 8486
E-mail: info@jsca.or.jp
Website: www.jsca.jp

Japanese Industrial Standards Committee
1-3-1 Kasumigaseki
Chiyoda-ku
Tokyo 100-890
Tel: (81) 3 3501 9471
E-mail: jisc@meti.go.jp
Website: www.jisc.go.jp

Management Research Society (Construction Industry)
11-8 Nihonbashi – Odenmachou 7F
Chuo-ku
Tokyo 103-0011
Tel: (81) 3 3663 2411
E-mail: info@kensetu-bukka.or.jp
Website: www.kensetu-bukka.or.jp

Ministry of Internal Affairs and Communication
Statistics Bureau
19-1 Wakamatsu-cho
Shinjuku-ku
Tokyo 162-8668
Tel: (81) 3 5273 2020
E-mail: webmaster@stat.go.jp
Website: www.stat.go.jp

Ministry of Land and Infrastructure, Transport and Tourism
2-1-3 Kasumigaseki
Chiyoda-ku
Tokyo 100-8944
Tel: (81) 3 5253 8111
Website: www.mlit.go.jp

MALAYSIA

All data relate to 2007 unless otherwise indicated.

Population	
Population	27.2 mn
Urban population (2005)	65%
Population under 15 (2007 est.)	32%
Population 65 and over (2007 est.)	5%
Average annual growth rate (2004 to 2007)	2.1%
Geography	
Land area	330,252 km^2
Agricultural area	18%
Capital city	Kuala Lumpur
Population of capital city (June 2006)	1.6 mn
Economy	
Monetary unit	Malaysian Ringgit (RM)
Exchange rate (average fourth quarter 2008) to:	
the pound sterling	RM 5.60
the US dollar	RM 3.56
the euro	RM 4.68
the yen x 100	RM 3.71
Average annual inflation (1998 to 2007)	2.4%
Inflation rate (2007 est.)	2.1%
Gross Domestic Product (GDP) at 2000 market prices (2006)	RM 474.4 bn
GDP per capita (2007 est.)	RM 22,951
Average annual real change in (GDP) (1998 to 2007)	4.4%
Private consumption as a proportion of GDP	51.1%
Public consumption as a proportion of GDP	12.9%
Investment as a proportion of GDP	23.4%
Construction	
Gross value of construction output	N/A
Net value of construction output	RM 7,447 mn
Net value of construction output as a proportion of GDP	1.6%

THE CONSTRUCTION INDUSTRY

Construction Output

The Malaysian construction industry has consistently contributed to the economy. Its growth rate fluctuates between extremities that vary from as high as 21.1% in 1995 to as low as -24% in 1998. In the whole year of 2007, the construction sector expanded 4.6% which is the highest growth since 1999. The construction sector strengthened further in the first quarter of 2008 by achieving a growth of 5.3% as compared to 4.1% in the first quarter of 2007. However, it eased to 3.9% in the second quarter 2008 and moderated further to 1.2% in the third quarter 2008. The moderation was influenced by the slower growth in Civil Engineering sub-sector and weaker activity in residential segments amidst higher prices of building materials.

Measures were taken by the Government to drive growth in the construction industry, which has been contracting over the last five years. Amongst them was the introduction of Private Financing Initiative (PFI) under the 9th Malaysia Plan as a new measure under the privatisation programme to increase opportunities for the private sector to participate in infrastructure and utilities development.

On 6 August 2008, the Government announced several provisions to allow variations in the contract value of government projects to assist contractors to cope with escalating prices of construction materials. Several measures to improve the delivery system for the construction sector, including expediting approvals, streamlining processes and introducing self-regulatory system were also taken.

The breakdown of work done by type of construction for 2004 (based on the latest official statistics available) is shown below.

VALUE OF NEW WORK DONE BY TYPE OF CONSTRUCTION, 1996-2004

		Government (RM/billion)			*Private (RM/billion)*		
Year	*Total*	*Residential*	*Non-Residential*	*Civil Engineering*	*Residential*	*Non-Residential*	*Civil Engineering*
1996	43.20	0.60	2.70	5.70	8.20	14.70	11.30
1998	39.50	0.50	3.20	4.50	8.20	13.10	10.00
2000	39.30	1.30	5.50	4.70	8.10	9.80	9.90
2002	40.80	1.80	6.30	5.80	8.80	8.30	9.80
2004	44.60	1.60	4.50	5.80	11.10	11.20	10.40

Source: Survey of Construction Industries 2005 – Department of Statistics, Malaysia

The residential sub-sector grew 3.5% during the first six months of 2008. However, due to escalating prices of construction material and global recession, new housing unit launched in the third quarter 2008 dropped by 46.1% from 9,210 units recorded in the second quarter 2008. Similarly, the sales performance declined by 6.6% from 19.7%. Developers were generally adopting a more cautious stance by holding back major projects launches and buyers adopting a wait-and-see attitude.

Office and commercial property developments in 2007 were fuelled by strong interest from funds. As at third quarter 2008, demand for space in the commercial buildings remained strong. High occupancy rate were seen in purpose-built office and shopping complex sub-sector. There are currently a total of 9,075,752m^2 existing retail space and 15,158,992m^2 of office space across the country. However, the market is likely to adjust when a combination of new supply of houses and slower growth in the service sector kicks in.

There are currently 154,659 rooms offered by 2,230 hotels across the country. As at third quarter 2008, higher tourist arrivals were recorded at 5.729 million people compared to 5.160 million in third quarter 2007. The average occupancy rate for three to five star hotels also recorded slight increase from 63.4% in second quarter 2008 to 66.5% in third quarter 2008. Following the success of Visit Malaysia Year 2007-2008, the Tourism Ministry of Malaysia has planned many new events in year 2009 in hope to have a steady growth in tourist arrivals and improved occupancies.

According to the labour force survey report, there was an increase of approximately 6.3% (58,600) workforce into the construction sector from 2nd half of 2007 to 1st half of 2008. The number of workforce in the construction sector as at 1st half of 2008 is recorded at 989,800 people. It is approximately 9.3% of the total workforce in Malaysia.

Characteristics and Structure of the Industry

The main regulatory agency for the construction industry is the Construction Industry Development Board Malaysia (CIDB). Since July 1995, all local and foreign contractors have to register with CIDB before undertaking any construction works in Malaysia.

Local construction companies are registered under seven grades (G1 to G7), based on three main criteria: tendering capacity, financial capacity and availability of human resources. As of June 2008, there are 63,465 registered companies classified as given in the table below.

NUMBER OF REGISTERED COMPANIES BY GRADE, 2008

Registration grade	*Limit of project size*	*Number of registered companies*
Grade 1	< RM100,000	34,359
Grade 2	< RM500,000	7,423
Grade 3	< RM1,000,000	10,700
Gradc 4	< RM3,000,000	2,361
Grade 5	< RM5,000,000	3,222
Grade 6	< RM10,000,000	1,119
Grade 7	No limit	4,281

Source: Construction Industry Development Board Malaysia

For government funded projects, local and foreign construction companies are required by the Ministry of Finance and Public Works Department (PWD) to register with *Pusat Khidmat Kontraktor (Contractor Service Centre or PKK),* which is under the Ministry of Entrepreneur and Cooperative Development.

The top ten major construction companies in terms of turnover are as follows.

MAJOR CONSTRUCTION COMPANIES IN MALAYSIA, 2007

Contractor	*Turnover (RM million)*
YTL Corporation Bhd	6,015.0
UEM World Bhd	6,979.0
Sunway Holding Bhd	1,897.0
UEM Builder Bhd	2,441.0
IJM Corporation Bhd	2,741.0
Ranhill Bhd	1,470.0
WCT Engineering Bhd	2,782.0
PECD Bhd	373.0
Gamuda Bhd	1,516.0
Muhibbah Engineering (M) Bhd	1,412.0

Source: Company Annual Reports

The professions are regulated by the respective professional bodies – *Pertubuhan Akitek Malaysia (PAM)*, Institution of Engineers Malaysia, Association of Consulting Engineers Malaysia, and the Institution of Surveyors Malaysia. Individual professional consultants have to register with their respective professional boards.

The table below shows the breakdown of professional consulting firms in 2005.

PROFESSIONAL CONSULTING FIRMS IN MALAYSIA, 2005

Location	*Architectural firms*	*Engineering firms*	*Surveying firms*
Peninsular Malaysia	964	967	665
Sabah	42	36	32
Sarawak	84	86	69
Total	1,090	1,089	766

Source: Yearbook of Statistics 200 – Department of Statistics, Malaysia

The title of Architect is protected by the *Architects Act 1967 (Revised 1972)* and restricted to an individual registered with the Board of Architects Malaysia.

Registration with the Board of Quantity Surveyors Malaysia is a condition precedent for a quantity surveyor to practise in Malaysia and use the designation.

The Institution of Surveyors Malaysia (ISM) is the official professional body of the surveying profession in Malaysia and comprises four sections: quantity surveying, property, consultancy and valuation surveying, land surveying and building surveying. As of 17 March 2008, ISM has a total of 4,387 members; 2,161 in the quantity surveying section, 916 in the property, consultancy and valuation surveying section, 1,015 in the land surveying and 295 in the building surveying section.

Clients and Finance

Investment in the construction industry is dominated by the private sector as a result of the privatisation programme promulgated by the government in 1983. Projects are procured through the sale and lease of assets, management contracts and build-operate-transfer (BOT) and its variants, build-operate (BO) and build-transfer (BT). The private sector will continue to play a key role in the country's economic growth. As at mid of 2007, there were RM20.3 billion new local jobs being identified, of which 37% were government and the balance were private sector/privatisation jobs.

Private funding is arranged through banks, trust funds and insurance companies. Increased foreign participation has seen some large multi-national companies providing a financing package for new projects.

Selection of Design Consultants

In the private sector, most consultants are selected and appointed by the developers based on track record and personal relationships besides cost consideration.

Project consultants for public sector projects are appointed through either open or selective bidding.

The push for privatisation has spurred design consultants to work with contractors in tendering for design and build projects and BOT projects.

Contractual Arrangements

Public projects are commonly procured through open tendering where advertisements are placed in the major newspapers. Selective tendering is only used for projects which satisfy certain criteria. Approval from the Ministry of Finance must be obtained for this method of tendering to be used, and also on the names of the shortlisted tenderers. Similarly, for direct negotiations, the procurement agency must seek the approval of the Ministry of Finance.

Only contractors registered with PKK and CIDB can tender for public projects exceeding RM50,000 in value. Projects funded by the World Bank and the Asian Development Bank are subject to the respective organization's own tendering procedures.

In the private sector, tenders are invited through open and selective tendering, and by direct negotiations. For major projects, it is common to have a pre-qualification exercise to shortlist tenderers.

The Jabatan Kerja Raya (JKR) or Public Works Department and the CIDB have their own standard form of contract. In the private sector, the *PAM (Pertubuhan Akitek Malaysia)* Agreement and Conditions of Building Contract is widely adopted. The standard form was extensively revised and launched in April 2007 to replace the earlier *PAM* Forms. There are three versions: without quantities, with quantities and nominated subcontract forms.

Development Control and Standards

The law on building control consists of the various *Planning Acts, Uniform Building By-Laws and Street, Drainage and Building Act 1974.* Planning permission is a pre-requisite to any application for local authority approval for projects involving development or change of use.

The *Kuala Lumpur Building By-Law 1985* requires a Qualified Person (QP) to be appointed for both design and the supervision of the execution of a development. A QP is defined as 'any architect, registered building draughtsman or engineer'. Occupation of a completed building requires a Temporary Certificate of Fitness (TCOF) to be obtained. A Certificate of Fitness (COF) is granted when all building works are completed and reports and certificates submitted to the authorities. As of April 2007, the COF is being replaced by the Certificate of Completion and Compliance (CCC). The CCC is issued by the project's Principal Submitting Person (PSP) who is either a Professional Architect, Professional Engineer or a Registered Building Draughtsman allowed by the Architects Act to issue a CCC for buildings not exceeding two storey and an area less than 300 square meters.

Malaysian standards (MS) are developed and promulgated by SIRIM Berhad (formerly known as Standards and Industrial Research Institute of Malaysia before its corporatisation in September 1996). There are currently over 5,008 Malaysian standards.

CONSTRUCTION COST DATA

Cost of Labour

The figures on the next page are typical of labour costs in Kuala Lumpur as at the fourth quarter of 2008. Cost of labour indicates the cost to a contractor of employing that employee.

Labour rate (per day = 8 hr)

	Wage rate (per day) RM	*Number of hours worked per year*
Site operatives		
Mason/bricklayer	90	2,304
Carpenter	90	2,304
Plumber	120	2,304
Electrician	120	2,304
Structural steel erector	100	2,304
HVAC installer	100	2,304
Semi-skilled worker	60	2,304
Unskilled labourer	50	2,304
Equipment operator	100	2,304
Watchman/security	70	2,304
	(per month)	
Site supervision		
General foreman	4,500	2,304
Trades foreman	5,000	2,304
Clerk of works	3,500	2,304
Contractors' personnel		
Site manager	8,000	2,304
Resident engineer	7,000	2,304
Resident surveyor	7,000	2,304
Junior engineer	4,000	2,304
Junior surveyor	4,000	2,304
Consultants' personnel		
Senior architect	10,000	2,304
Senior engineer	9,000	2,304
Senior surveyor	7,000	2,304
Qualified architect	6,500	2,304
Qualified engineer	6,000	2,304
Qualified surveyor	5,500	2,304

Cost of Materials

The figures on the next page are the costs of main construction materials, delivered to site in the capital area, as incurred by contractors in the fourth quarter of 2008. These assume that the materials would be in quantities as required for a medium sized construction project and that the location of the works would be neither constrained nor remote.

	Unit	*Cost RM*
Cement and aggregate		
Ordinary portland cement in 50kg bags	tonne	290.00
Coarse aggregates for concrete in 20mm granite	m^3	40.00
Ready mixed concrete (mix 1:2:4)	m^3	200.00
Ready mixed concrete (mix 1:1:2)	m^3	215.00
Steel		
Mild steel reinforcement 10mm-40mm diameter	tonne	2,200.00
High tensile steel reinforcement 10mm-40mm diameter	tonne	2,100.00
Structural steel sections	tonne	4,300.00
Bricks and blocks		
Common bricks (215 x 102 x 65mm)	pc	0.30
Good quality facing bricks (210 x 100 x 70mm)	pc	0.80
Hollow concrete blocks (190 x 390 x 190mm)	pc	2.30
Solid concrete blocks integrated (190 x 390 x 190mm)	pc	2.90
Timber and insulation		
Exterior quality plywood	m^2	25.00
Softwood strip flooring	m^2	250.00
Softwood internal door complete with frames and ironmongery	each	550.00
Glass and ceramics		
Float glass (5 mm)	m^2	40.00
Good quality ceramic wall tiles (150 x 150mm)	m^2	30.00
Plaster and paint		
Plasterboard (15mm thick)	m^2	18.00
Emulsion paint in 5 litre tins	litre	17.50
Gloss oil paint in 5 litre tins	litre	20.00
Tiles and paviors		
Clay floor tiles (200 x 200 x 13mm)	m^2	60.00
Vinyl floor tiles (300 x 300 x 2mm)	m^2	32.00
Precast concrete paving slabs (300 x 300 x 60mm)	m^2	65.00
Clay roof tiles	pc	4.20
Precast concrete roof tiles	pc	1.85
Drainage		
100mm diameter clay drain pipes	m	20.00
150mm diameter UPVC drain pipes	m	23.00

Unit Rates

The descriptions that follow are generally shortened versions of standard descriptions listed in full in section 4. Where an item has a two digit reference number (e.g. 05 or 33), this relates to the full description against that number in section 4. Where an item has an alphabetic suffix (e.g. 12A or 34B) this indicates that the standard description has been modified. Where a modification is major the complete modified description is included here and the standard description should be ignored; where a modification is minor (e.g. the insertion of a named hardwood) the shortened description has been modified here but, in general, the full description in section 4 prevails.

The unit rates below are for main work items in a typical construction project in the Kuala Lumpur area in the fourth quarter of 2008. The rates include all necessary labour, materials and equipment. Allowances of 6% - 8% to cover preliminary and general items and 15% to cover contractors' overheads and profit have been included in the rates.

		Unit	*Rate RM*
Excavation			
01	Mechanical excavation of foundation trenches	m^3	28.00
02	Hardcore filling making up levels	m^3	50.00
Concrete work			
04A	Plain in situ concrete in strip foundations in trenches (C25)	m^3	260.00
05A	Reinforced in situ concrete in beds (C35)	m^3	280.00
06A	Reinforced in situ concrete in walls (C35)	m^3	280.00
07A	Reinforced in situ concrete in suspended floors or roof slabs (C35)	m^3	280.00
08A	Reinforced in situ concrete in columns (C35)	m^3	280.00
09A	Reinforced in situ concrete in isolated beams (C35)	m^3	280.00
Formwork			
11	Softwood formwork to concrete walls	m^2	35.00
12	Softwood formwork to concrete columns	m^2	35.00
13	Softwood to horizontal soffits of slabs	m^2	35.00
Reinforcement			
14	Reinforcement in concrete walls	tonne	3,800.00
15	Reinforcement in suspended concrete slabs	tonne	3,800.00
16	Fabric reinforcement in concrete beds ($3kg/m^2$)	m^2	13.00
Steelwork			
17	Fabricate, supply and erect steel framed structure	tonne	6,000.00
18	Framed structural steelwork in universal joist sections	tonne	6,000.00

		Unit	*Rate RM*
19	Structural steelwork lattice roof trusses	tonne	6,000.00
Brickwork and blockwork			
20	Precast lightweight aggregate hollow concrete block walls	m^2	43.00
21	Solid (perforated) common bricks	m^2	35.00
22	Sand lime bricks	m^2	30.00
23	Facing bricks	m^2	55.00
Roofing			
24	Concrete interlocking roof tiles 430 x 380mm	m^2	43.00
25	Plain clay roof tiles 260 x 160mm	m^2	80.00
26	Fibre cement roof slates 600 x 300mm	m^2	30.00
33	Troughed galvanized steel roof cladding	m^2	N/A
Woodwork and metalwork			
34	Preservative treated sawn softwood 50 x 100mm	m^3	N/A
35	Preservative treated sawn softwood 50 x 150mm	m^3	N/A
36	Single glazed casement window in hardwood, size 650 x 900mm	each	55.00
38	Solid core half hour fire resisting hardwood internal flush doors, size 800 x 2000mm	each	950.00
39	Aluminium double glazed window, size 1200 x 1200mm	each	1,120.00
40	Aluminium double glazed door, size 850 x 2100mm	each	1,300.00
41	Hardwood skirtings	m	15.00
Plumbing			
42	UPVC half round eaves gutter	m	25.00
43	UPVC rainwater pipes 100mm	m	40.00
47	UPVC soil and vent pipes 100mm	m	50.00
48	White vitreous china WC suite	each	650.00
49	White vitreous china lavatory basin	each	400.00
51	Stainless steel single bowl sink and double drainer	each	350.00
Electrical Work			
52A-54A	PVC insulated and copper sheathed cable with 13 amp switched socket outlet, flush mounted 20 amp, 1 way light switch (within 10m range)		
	Per power point	each	100 - 120
	Per light point	each	80 - 100

		Unit	*Rate RM*
Finishings			
55A	2 coats gypsum based plaster on concrete walls 20mm thick	m²	18.00
56	White glazed tiles on plaster walls	m²	55.00
58A	Cement and sand screed to concrete floors 30mm thick	m²	15.00
60	Mineral fibre tiles on concealed suspension system	m²	60.00
Glazing			
61A	6mm clear float glass; glazing to wood	m²	60.00
Painting			
62	Emulsion on plaster walls	m²	4.50
63	Oil paint on timber	m²	6.50

Approximate Estimating

The building costs per unit area that follow are averages incurred by building clients for typical buildings in the capital area as at the fourth quarter of 2008. They are based upon the total floor area of all storeys, measured between external walls and without deduction for internal walls.

Approximate estimating costs generally include mechanical and electrical installations but exclude furniture, loose or special equipment, and external works; they also exclude fees for professional services. The costs shown are for specifications and standards appropriate to Malaysia and this should be borne in mind when attempting comparisons with similarly described building types in other countries. A discussion of this issue is included in section 2. Comparative data for countries covered in this publication, including construction cost data, are presented in Part Three.

Approximate estimating costs must be treated with reserve; they cannot provide more than a rough guide to the probable cost of building.

	Cost m² RM	*Cost ft² RM*
Industrial buildings		
Factories for letting	1,410	131
Factories for owner occupation (light industrial use)	1,500	139
Factories for owner occupation (heavy industrial use)	1,870	174
Factory/office (high-tech) for letting (shell and core only)	1,650	153
Factory/office (high-tech) for letting (ground floor shell, first floor offices)	2,020	188
Factory/office (high tech) for owner occupation (controlled environment, fully finished)	3,330	309

	Cost *m^2 RM*	*Cost* *ft^2 RM*
High tech laboratory workshop centres (air-conditioned)	1,670	155
Administrative and commercial buildings		
Civic offices, non air-conditioned	1,290	120
Civic offices, fully air-conditioned	1,650	153
Offices for letting, 5 to 10 storeys, non air-conditioned	1,550	144
Offices for letting, 5 to 10 storeys, air-conditioned	1,780	165
Offices for letting, high rise, air-conditioned	2,630	244
Offices for owner occupation, 5 to 10 storeys, non air-conditioned	1,940	180
Offices for owner occupation, 5 to 10 storeys, air-conditioned	2,160	201
Offices for owner occupation, high rise, air-conditioned	3,160	294
Prestige/headquarters office, 5 to 10 storeys, air-conditioned	2,870	267
Prestige/headquarters office, high rise, air-conditioned	3,760	349
Health and education buildings		
General hospitals (excluding specialist equipment and installation) (main hospital)	bed	300,000
Private hospitals (excluding specialist equipment and installation)	bed	190,000
Primary/junior schools	1,010	94
Secondary/middle schools	1,160	108
Recreation and arts buildings		
Theatres (over 500 seats) including seating and stage equipment	seat	12,100
Theatres (less than 500 seats) including seating and stage equipment	seat	15,000
Concert halls including seating	4,450	413
Sports hall including changing and social facilities	1,880	175
Swimming pools (international standard) including changing and social facilities	each	1,350,000
Swimming pools (schools standard) including changing facilities	each	500,000
National museums including full air-conditioning and standby generator	3,800	353
Local museums including air-conditioning	3,220	299

	Cost m^2 RM	*Cost ft^2 RM*
Residential buildings		
Social/economic single family housing (multiple units)	940	87
Private/mass market single family housing 2 storey detached/semi detached (multiple units)	2,430	226
Purpose designed single family housing 2 storey detached (single unit)	2,990	278
Social/economic apartment housing, low rise (no lifts)	740	69
Social/economic apartment housing, high rise (with lifts)	1,000	93
Private sector apartment building (standard specification)	1,670	155
Private sector apartment buildings (luxury)	3,720	346
Student/nurses halls of residence	1,110	103
Homes for the elderly (shared accommodation)	1,020	95
Homes for the elderly (self contained with shared communal facilities)	1,080	100
Hotel, 5 star, city centre (inclusive of FF & E)	7,700	715
Hotel, 3 star, city/provincial (ditto)	5,520	513
Motel (ditto)	2,980	277

Regional Variations

The approximate estimating costs are based on projects in the capital. Adjust these costs by the following factors to take account of regional variations:

Selangor	:	0%
Penang	:	+5%
Johore	:	+15%
Kota Kinabalu	:	+15%
Kuching	:	+15%

EXCHANGE RATES AND INFLATION

The combined effect of exchange rates and inflation on prices within a country and price comparisons between countries is discussed in section 2.

Exchange Rates

The graph below plots the movement of the Malaysian ringgit against the sterling, the euro, the US dollar and 100 Japanese yen since 1998. The values used for the graph are quarterly and the method of calculating these is described and general guidance on the interpretation of the graph provided in section 2. The average exchange rate in the fourth quarter of 2008 was RM5.60 to pound sterling, RM4.68 to euro, RM3.56 to US dollar and RM3.71 to 100 Japanese yen.

THE MALAYSIAN RINGGIT AGAINST STERLING, EURO, US DOLLAR AND 100 JAPANESE YEN

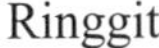

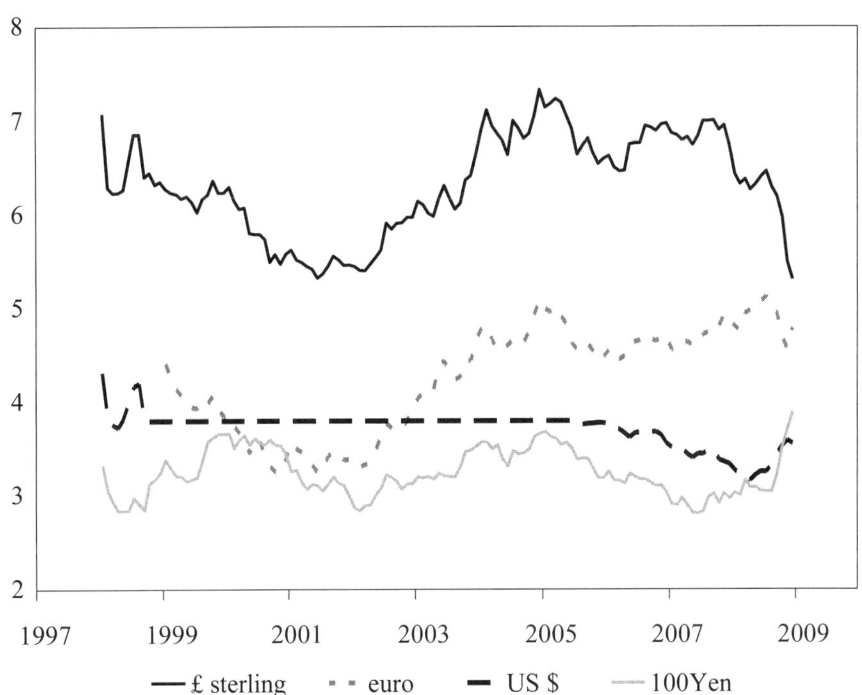

Price Inflation

The table below shows consumer price and house price inflation in Malaysia since 1998.

CONSUMER PRICE AND HOUSE PRICE INDEX

Year	*Consumer index*	*Change from previous year*	*House price index*	*Change from previous year*
1998	95.8	-	96.6	-
1999	98.5	2.8	94.3	-2.3
2000	100.0	1.5	100.0	5.7
2001	101.5	1.4	101.1	1.1
2002	103.5	1.8	103.6	2.5
2003	104.7	1.2	107.9	4.3
2004	106.2	1.5	113.3	5.4
2005	109.4	3.0	115.9	2.6
2006	113.3	3.8	118.3	2.4
2007	113.3	3.8	123.8	5.5
2008	111.4	5.5	128.8*	5.0

*Note : * 2nd Quarter, Preliminary*
Source: Department of Statistics, Malaysia &
The Malaysia House Price Index, Valuation & Property Service Department

USEFUL ADDRESSES

Public Organizations

Construction Industry Development Board Malaysia
Ibu Pejabat CIDB
Tingkat 7, Grand Seasons Avenue
No 72, Jalan Pahang
53000 Kuala Lumpur
Tel: (60) 3 2617 0200
Fax: (60) 3 2617 0220
Website: www.cidb.gov.my

Department of National Housing
Jabatan Perumahan Negara
Kementerian Perumahan dan Kerajaan Tempatan
Paras 6-7, Block K
Pusat Bandar Damansara
Peti Surat 12579
50782 Kuala Lumpur
Tel: (60) 3 2094 7033
Fax: (60) 3 2093 0709
Website: www.kpkt.gov.my/jpn

The Local Authority
Pihak Berkuasa Tempatan
(in every state there is a Local Authority)

The Malaysian Highway Authority
Lembaga Lebuhraya Malaysia
KM 6 Jalan Kajang Serdang
Peti Surat 22, 43000 Kajang, Selangor Darul Ehsan
Tel: (60) 3 8737 3000 / 8738 3000
Fax: (60) 3 8737 3555
Website: www.llmnet.gov.my

The Ministry of Housing and Local Government
Kementerian Perumahan dan Kerajaan Tempatan
Level 3-7, Block K, P.O.Box 12579
Damansara Town Centre, 50782 Kuala Lumpur
Tel: (60) 3 2094 7033
Fax: (60) 3 2094 9720
Website: www.kpkt.gov.my

The Ministry of Science, Technology and Innovation
Kementerian Sains, Teknologi dan Inovasi
Aras 1-7, Block C4 & C5, Kompleks C
Pusat Pentadbiran Kerajaan Persekutuan
Administrative Centre
62662 Putrajaya, Wilayah Persekutuan
Tel: (60) 3 8885 8000
Fax: (60) 3 8888 9070
E-mail: pro@mosti.gov.my
Website: www.mosti.gov.my

The Ministry of Works Malaysia
Kementerian Kerjaraya Malaysia
Blok B, Tingkat 5, Kompleks Kerja Raya
Jalan Sultan Salahuddin
50580 Kuala Lumpur
Tel: (60) 3 2711 1100
Fax: (60) 3 2711 1590 / 1592
Website: www.kkr.gov.my

Pusat Khidmat Kontraktor
Kementerian Kerja Raya Malaysia
Aras 5, Block Menara
No. 18 Persiaran Perdana, Presint 2
62652 Putrajaya
Tel: (60) 3 8880 5000
Fax: (60) 3 8880 5204 / 5300
E-mail: pkk@kkr.gov.my
Website: pkk.kkr.gov.my

SIRIM Berhad
No.1 Persiaran Dato' Menteri
Seksyen 2, Peti Surat 7035
40911 Shah Alam
Tel: (60) 3 5544 6000
Fax: (60) 3 5510 8095
E-mail: web@sirim.my
Website: www.sirim.my

Town and Country Planning Department
Jabatan Perancang Bandar dan Desa
Jabatan Perancang Department
Jalan Cenderasari
50464 Kuala Lumpur
Tel: (60) 3 2698 9211
Fax: (60) 3 2698 9994
Website: www.townplan.gov.my

Urban Development Authority (UDA)
UDA Holdings Berhad
Menara Bukit Bintang
Lot 111 Jalan Bukit Bintang
Peti Surat 10080
50704 Kuala Lumpur
Tel: (60) 3 2730 8500
Fax: (60) 3 2713 8500 / 8555
E-mail: uhb@udanet.com
Website: www.udaholdings.com.my

Trade And Professional Organizations

Association of Consulting Engineers Malaysia
63-2 & 65-2 Medan Setia 1
Damansara Heights
50490 Kuala Lumpur
Tel: (60) 3 2095 0031 / 0079 / 0158
Fax: (60) 3 2095 3499
Website: www.acem.com.my

Board of Architects Malaysia
Lembaga Akitek Malaysia
Tingkat 17, Block F, Ibu Pejabat JKR
Jalan Sultan Salahuddin
Peti Surat 12695
50786 Kuala Lumpur
Tel: (60) 3 2698 2878 / 2696 7087
Fax: (60) 3 2693 6881
Website: www.lam.gov.my

Institute of Architects Malaysia
4 & 6 Jalan Tangsi
50480 Kuala Lumpur
Tel: (60) 3 2693 4182
Fax: (60) 3 2692 8782
E-mail: info@pam.com.my
Website: www.pam.org.my

Institution of Engineers Malaysia
Bangunan Ingenieur
Lots 60 / 62 Jalan 52/4
Peti Surat 223 (Jalan Sultan)
46720 Petaling Jaya, Selangor Darul Ehsan
Tel: (60) 3 7968 4001
Fax: (60) 3 7957 7678
E-mail: sec@iem.org.my
Website: www.iem.org.my

Institution of Surveyors Malaysia
3rd Floor Bangunan Juruukur
64-66 Jalan 52/4
46200 Petaling Jaya
Selangor
Tel: (60) 3 7955 1773
Fax: (60) 3 7955 0253
E-mail: secretariat@ism.org.my
Website: www.ism.org.my

Master Builders Association Malaysia
2-1 First Floor Jalan 2/109E
Desa Business Park
Taman Desa
Off Jalan Klang Lama
58100 Kuala Lumpur
Tel: (60) 3 7984 8636
Fax: (60) 3 7982 6811
E-mail: mbam01@mbam.org.my
Website: www.mbam.org.my

NEW ZEALAND

All data relate to 2008 unless otherwise indicated.

Population	
Population	4.29 mn
Urban population	72%
Population under 15	22%
Population 65 and over	12.5%
Average annual growth rate (2004 to 2008)	1.4%
Geography	
Land area	268,670 km^2
Agricultural area	60%
Capital city	Wellington
Population of capital city	379,100
Economy	
Monetary unit	New Zealand Dollar (NZ$)
Exchange rate (average fourth quarter 2008) to:	
the pound sterling	NZ$ 2.72
the US dollar	NZ$ 1.73
the euro	NZ$ 2.28
the yen x 100	NZ$ 1.79
Average annual inflation (1998 to 2008)	2.3%
Inflation rate	3.7%
Gross Domestic Product (GDP)	NZ$ 135.8 bn
GDP per capita	NZ$ 31,626
Average annual real change in (GDP) (1996 to 2008)	3.37%
Private consumption as a proportion of GDP	58.7%
Public consumption as a proportion of GDP	19.3%
Investment as a proportion of GDP	22%
Construction	
Gross value of construction output	NZ$ 13.2 bn
Net value of construction output	NZ$ 6.29 bn
Net value of construction output as a proportion of GDP	4.6%

THE CONSTRUCTION INDUSTRY

Construction Output

The gross output of construction in year ending September 2008 was NZ$13.2 billion or 9.7% of the GDP, equivalent of US$7.63 billion.

The net value of construction output in year ending September 2008 was NZ$6.29 billion equivalent to 4.66% of GDP.

The breakdown by type of work is shown below:

VALUE OF BUILDING WORK PUT IN PLACE, 2008

Type of building	*Value NZ$ million*	*Percentage of total building work (%)*
Residential building		
new dwellings	6,559	49.7
alterations additions and outbuildings	1,437	10.9
Total residential building	7,996	60.5
Non-residential building		
Accommodation buildings	417	3.2
Hospitals and nursing homes	455	3.4
Factories and industrial buildings	472	3.6
Commercial buildings	1,732	13.1
Education buildings	602	4.6
Miscellaneous* and multi purpose	1,531	11.6
Total non-residential building	5,210	39.5
Total	13,206	100.0

Source: Statistics New Zealand

**Social, cultural, religious, recreational and farm buildings*

Characteristics and Structure of the Industry

The Resource Management Act 1991 (RMA) is a framework for governing the planning and development of New Zealand.

The RMA sets out who has what responsibilities in local and central government, and the rules for carrying out the planning process. It applies to the construction of all infrastructures, including transport infrastructure such as roads, railway lines, ports and ferry / terminals.

The RMA is still identified by developers as a major area of uncertainty in early stage planning of projects. The principles of the Act are generally well regarded, however uncertain and contradictory interpretation of the legislation by the Territorial Authorities, create problems for projects which require notifiable resource consents.

The industry continues to operate in an environment of low profit margins and uncertain and cyclical workload with many contractors maintaining minimum workforces and using contract labour to reduce their risk exposure to a sudden drop in forward workload. On the other hand, this has led to experimentation with new construction techniques and methods of on-site operation which, when successful, will result in a more efficient industry.

Health and safety within the workplace has been a measure of success for many of the top construction firms in New Zealand. The development of the Site Safe Passport ensures that all workers are up-to-date with current standards and skills and is now compulsory for workers to have one to enter a site.

The top 10 contractors in New Zealand are shown in the table below:

TOP 10 BUILDERS, 2007-2008

Contractor	*Turnover (NZ$ million)*
Fletcher Construction	369.43
Hawkins	326.28
Mainzeal/Mainworks	204.12
Naylor Love	101.36
Arrow International	93.41
Calder Stewart	79.42
Ormat Systems	69.08
Aspec Construction	62.56
Watts & Hughes/Cobalt	59.61
Hayden & Rollet Ltd	58.97

Source: Whats On Report, Auckland (www.whatson.co.nz) – +0064 0800 WHATSON

Fletcher Construction had a total revenue of nearly US$2 billion in Year 2000 and will have over 100 years experience by Year 2009. The company employs over 2,660 staff and is the major driver in setting up the Site Safe industry health and safety organisation in 1999.

Development Control and Standards

The introduction of the Construction Contracts Act in 2002 has made a remarkable change in the construction industry's payment structure by providing processes for progress payments on construction contracts, for the enforcement of overdue payments and for the adjudication of disputes.

NZS 3604 (Code of practice for light timber frame buildings not requiring specific design) has been continually updated and continues to be the standard for timber framed building not exceeding two storeys above ground level.

CONSTRUCTION COST DATA

Cost of Labour

The figures below are typical of labour costs in Auckland as at the fourth quarter of 2008. The wage rate is the basis of an employee's income.

	Avg Wage rate (per hour) NZ$	*Number of hours worked per year*
Site operatives		
Mason/bricklayer	39	2,068
Carpenter	43	2,068
Plumber	48	2,068
Electrician	48	2,068
Structural steel erector	43	2,068
HVAC installer	48	2,068
Semi-skilled worker	38	2,068
Unskilled labourer	24	2,068
Equipment operator	29	2,068
Watchman/security	24	2,068
Site supervision		
General foreman	46	2,068
Trades foreman	35	2,068
Project Manager	57	2,068
Contractors' personnel		
Site manager	46	2,068
Resident engineer	51	1,762
Resident surveyor	51	1,762
Junior engineer	29	1,762
Junior surveyor	29	1,762
Planner	51	1,762
Consultants' personnel		
Senior architect	48	1,762
Senior engineer	51	1,762
Senior surveyor	55	1,762
Qualified architect	29	1,762
Qualified engineer	29	1,762
Qualified surveyor	48	1,762

Cost of Materials

The figures that follow are the costs of main construction materials, delivered to site in the Auckland area, as incurred by contractors in the fourth quarter of 2008. These assume that the materials would be in quantities as required for a medium sized construction project and that the location of the works would be neither constrained nor remote.

All the costs in this section exclude Goods and Services tax (GST).

	Unit	*Cost NZ$*
Cement and aggregate		
Ordinary portland cement in 40kg bags	tonne	355.00
Coarse aggregates for concrete	m^3	55.00
Fine aggregates for concrete	m^3	65.00
Ready mixed concrete (17.5 MPa)	m^3	165.00
Ready mixed concrete (30.0 MPa)	m^3	184.00
Steel		
Mild steel reinforcement (16mm)	tonne	1,900.00
High tensile steel reinforcement (16mm)	tonne	1,900.00
Structural steel sections	tonne	2,800.00
Bricks and blocks		
Common bricks (190 x 90 x 90mm)	1,000	1,498.00
Good quality facing bricks (230 x 119 x 70mm)	1,000	1,509.00
Hollow concrete blocks	1,000	345.00
Solid concrete blocks	m^2	431.00
Precast concrete cladding units with exposed aggregate finish (200mm thick)	m^2	350.00
Timber and insulation		
Softwood sections for carpentry	m^3	1,800.00
Softwood for joinery	m^3	2,300.00
Hardwood for joinery	m^3	2,700.00
Exterior quality plywood (9mm)	m^2	65.00
Plywood for interior joinery (9mm)	m^2	45.00
Softwood strip flooring (50mm)	m	12.00
Chipboard sheet flooring (20mm)	m^2	40.00
100mm thick quilt insulation	m^2	13.00
60mm thick rigid slab insulation	m^2	40.00
Softwood internal door complete with frames and ironmongery	each	500.00
Glass and ceramics		
Float glass (6mm)	m^2	200.00
Sealed double glazing units (two layers laminated glass)	m^2	700.00

	Unit	*Cost NZ$*
Plaster and paint		
Good quality ceramic wall tiles (150 x 150mm)	m²	40.00
Plaster in 50kg bags	tonne	598.00
Plasterboard (13mm thick)	m²	15.00
Emulsion paint in 5 litre tins	m²	7.50
Gloss oil paint in 5 litre tins	m²	10.50
Tiles and pavers		
Clay floor tiles (200 x 200mm)	m²	98.00
Vinyl floor tiles (2mm thick)	m²	38.00
Precast concrete paving slabs (100mm)	m²	85.00
Precast concrete roof tiles – single storey	m²	37.50
Drainage		
WC suite complete	each	1,000.00
Lavatory basin complete	each	450.00
100mm diameter clay drain pipes	m	77.00

Unit Rates

The descriptions below are generally shortened versions of standard descriptions listed in full in section 4. Where an item has a two digit reference number (e.g. 05 or 33), this relates to the full description against that number in section 4. Where an item has an alphabetic suffix (e.g. 12A or 34B) this indicates that the standard description has been modified. Where a modification is major the complete modified description is included here and the standard description should be ignored; where a modification is minor (e.g. the insertion of a named hardwood) the shortened description has been modified here but, in general, the full description in section 4 prevails.

The unit rates below are for main work items on a typical construction project in the Auckland area in the fourth quarter of 2008. The rates include all necessary labour, materials and equipment. Allowances to cover preliminary and general items and contractors' overheads and profit should be added to these rates. All the rates in this section exclude Goods and Services tax (GST).

		Unit	*Rate NZ$*
Excavation			
01	Mechanical excavation of foundation trenches	m³	45.00
02	Hardcore filling making up levels	m³	85.00
03	Earthwork support	m²	90.00
Concrete work			
04	Plain in situ concrete in strip foundations in trenches	m³	280.00
05	Reinforced in situ concrete in beds	m³	285.00
06	Reinforced in situ concrete in walls	m³	300.00

		Unit	Rate NZ$
07	Reinforced in situ concrete in suspended floors or roof slabs	m^3	285.00
08	Reinforced in situ concrete in columns	m^3	325.00
09	Reinforced in situ concrete in isolated beams	m^3	325.00
10	Precast concrete slab	m^2	180.00
Formwork			
11	Softwood formwork to concrete walls	m^2	140.00
12	Softwood formwork to concrete columns	m^2	145.00
13	Softwood formwork to horizontal soffits of slabs	m^2	175.00
Reinforcement			
14	Reinforcement in concrete walls	tonne	2,750.00
15	Reinforcement in suspended concrete slabs	tonne	2,500.00
16	Fabric reinforcement in concrete beds	m^2	12.00
Steelwork			
17	Fabricate, supply and erect steel framed structure	tonne	4,800.00
18	Framed structural steelwork in universal joist sections	tonne	6,000.00
19	Structural steelwork lattice roof trusses	tonne	6,500.00
Brickwork and blockwork			
20	Precast lightweight aggregate hollow concrete block walls	m^2	95.00
21A	Solid (perforated) concrete blocks	m^2	127.00
23	Facing bricks	m^2	135.00
Roofing			
24	Concrete interlocking roof tiles 430 x 380mm	m^2	55.00
26	Fibre cement roof slates 600 x 300mm	m^2	140.00
27	Sawn softwood roof boarding	m^2	180.00
29	3 layers glass-fibre based bitumen felt roof covering	m^2	130.00
30	Bitumen based mastic asphalt roof covering	m^2	85.00
31A	Glass-fibre mat roof insulation 100mm thick	m^2	16.00
33	Troughed galvanized steel roof cladding	m^2	55.00
Woodwork and metalwork			
34	Preservative treated sawn softwood 50 x 100mm	m	12
35	Preservative treated sawn softwood 50 x 150mm	m	15
36	Single glazed casement window in hardwood, 650 x 900mm	m^2	800.00
37	Two panel glazed door in hardwood, 850 x 2000mm including ironmongery	each	1,500.00

		Unit	Rate NZ$
38A	Solid core half hour fire resisting hardwood internal flush doors, size 800 x 2000mm including ironmongery	each	1,775.00
39	Aluminium double glazed window, size 1200 x 1200mm	m^2	650.00
40	Aluminium double glazed door, size 850 x 2100mm	each	3,500.00
41	Hardwood skirtings	m	20.00
Plumbing			
42	UPVC half round eaves gutter	m	30.00
43	UPVC rainwater pipes	m	30.00
44	Light gauge copper cold water tubing	m	35.00
45	High pressure plastic pipes for cold water supply	m	30.00
46	Low pressure plastic pipes for cold water distribution	m	25.00
47	UPVC soil and vent pipes	m	40.00
48	White vitreous china WC suite	each	1,000.00
49	White vitreous china lavatory basin	each	550.00
51	Stainless steel single bowl sink and double drainer	each	750.00
Electrical work			
52	PVC insulated and copper sheathed cable	m	6.00
53A	15 amp switched socket outlet	each	50.00
54	Flush mounted 20 amp, 1 way light switch	each	40.00
Finishings			
55	2 coats gypsum based plaster on brick walls	m^2	75.00
56	White glazed tiles on plaster walls	m^2	140.00
57	Red clay quarry tiles on concrete floors	m^2	125.00
58	Cement and sand screed to concrete floors	m^2	45.00
60	Mineral fibre tiles on exposed two way suspension system	m^2	45.00
Glazing			
61A	6mm clear float glass; glazing to wood	m^2	160.00
Painting			
62	Emulsion on plaster walls	m^2	12.00
63	Oil paint on timber	m^2	15.00

Approximate Estimating

The building costs per unit area given below are averages incurred by building clients for typical buildings in the Auckland area as at the fourth quarter of 2008. They are based upon the total floor area of all storeys, measured over external walls and without deduction for internal walls.

Approximate estimating costs generally include mechanical and electrical installations but exclude furniture, loose or special equipment, and external works; they also exclude fees for professional services. The costs shown are for specifications and standards appropriate to New Zealand and this should be borne in mind when attempting comparisons with similarly described building types in other countries. A discussion of this issue is included in section 2. Comparative data for countries covered in this publication, including construction cost data, are presented in Part Three.

Approximate estimating costs must be treated with caution; they cannot provide more than a rough guide to the probable cost of building. All the rates in this section exclude Goods and Services Tax (GST).

	Cost m^2 NZ\$	*Cost ft^2 NZ\$*
Industrial buildings		
Factories for letting	750	70
Factories for owner occupation (light industrial use)	750	70
Factories for owner occupation (heavy industrial use)	850	79
Factory/office (high-tech) for letting (shell and core only)	1,200	111
Factory/office (high-tech) for letting (ground floor shell, first floor offices)	1,400	130
Factory/office (high tech) for owner occupation (controlled environment fully finished)	1,600	149
High tech laboratory workshop centres (air-conditioned)	3,000	279
Warehouses, low bay (6 to 8m high) for letting (no heating)	600	56
Warehouses, low bay for owner occupation (including heating)	700	65
Warehouses, high bay for owner occupation (including heating)	750	70
Administrative and commercial buildings		
Civic offices, non air-conditioned	1,500	139
Civic offices, fully air-conditioned	1,700	158
Offices for letting, 5 to 10 storeys, non air-conditioned	1,600	149
Offices for letting, 5 to 10 storeys, air-conditioned	1,900	177
Offices for letting, high rise, air-conditioned	2,500	232
Offices for owner occupation 5 to 10 storeys, non air-conditioned	1,600	149
Offices for owner occupation high rise, air-conditioned	1,900	177
Prestige/headquarters office, 5 to 10 storeys, air-conditioned	2,750	255

	Cost m² NZ$	Cost ft² NZ$
Prestige/headquarters office, high rise, air-conditioned	3,000	279
Health and education buildings		
General hospitals (100 beds)	4,000	372
Teaching hospitals (100 beds)	4,000	372
Private hospitals (100 beds) aged persons	2,750	255
Health centres	2,250	209
Nursery schools	1,700	158
Primary/junior schools	1,900	177
Secondary/middle schools	2,200	204
University (arts) buildings	2,400	223
University (science) buildings	2,750	255
Management training centres	2,000	186
Recreation and arts buildings		
Theatres (over 500 seats) including seating and stage equipment	3,000	279
Theatres (less than 500 seats) including seating and stage equipment	3,250	302
Concert halls including seating and stage equipment	3,000	279
Sports halls including changing and social facilities	2,500	232
Swimming pools (international standard) including changing and social facilities	3,400	316
Swimming pools (schools standard) including changing facilities	3,200	297
National museums including full air-conditioning and standby generator	3,500	325
Local museums including air-conditioning	3,000	279
Residential buildings		
Social/economic single family housing (multiple units)	1,100	102
Private/mass market single family housing 2 storey detached / semi detached (multiple units)	1,400	130
Purpose designed single family housing 2 storey detached (single unit)	1,700	158
Social/economic apartment housing, low rise (no lifts)	2,000	186
Private sector apartment building (standard specification)	2,400	223
Private sector apartment buildings (luxury)	3,000	279
Student/nurses halls of residence	2,000	186
Homes for the elderly (shared accommodation)	2,200	204
Homes for the elderly (self contained with shared communal facilities)	2,400	223
Hotel, 5 star, city centre	3,900	362
Hotel, 3 star, city/provincial	3,020	281
Motel	2,500	232

Goods And Services Tax (GST)

The standard rate of Goods and Services tax (GST) is currently 12.5%, chargeable on all building work.

EXCHANGE RATES AND INFLATION

The combined effect of exchange rates and inflation on prices within a country and price comparisons between countries is discussed in section 2.

Exchange Rates

The graph below plots the movement of the New Zealand dollar against the sterling, the euro, the US dollar and 100 Japanese yen since 1998. The values used for the graph are quarterly and the method of calculating these is described and general guidance on the interpretation of the graph provided in section 2. The average exchange rate in the fourth quarter of 2008 was NZ$2.72 to pound sterling, NZ$2.28 to euro, NZ$1.73 to US dollar and NZ$1.79 to 100 Japanese yen.

THE NEW ZEALAND DOLLAR AGAINST STERLING, EURO, US DOLLAR AND 100 JAPANESE YEN

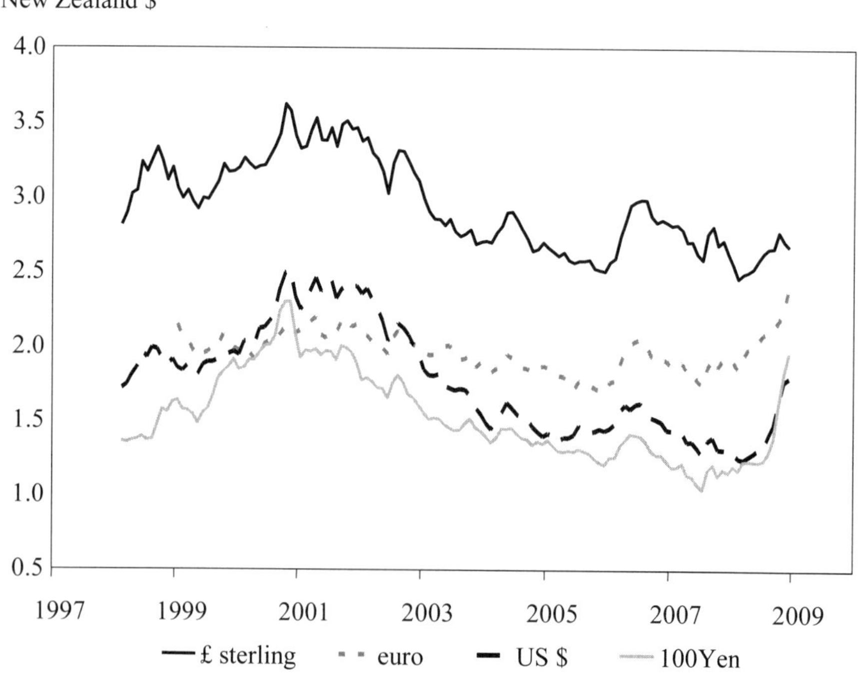

Price Inflation

The table below presents consumer price index and construction price inflation in New Zealand since 2001.

CONSUMER PRICE AND CONSTRUCTION COST INFLATION

	Consumer price index		*Construction cost index*	
Year	*Jun Qtr*	*change %*	*Dec Qtr*	*change %*
2001	870		4320	
2002	894	2.8	4450	3.0
2003	907	1.5	4500	1.1
2004	927	2.2	4710	4.7
2005	962	3.8	4960	5.3
2006	1000	4.0	5350	7.9
2007	1020	2.0	5650	5.6
2008	1061	4.0	5930	5.0

Changes in the calculation of the CPI figures have been recommended by the 1997 CPI Revision Advisory Committee which has recommended that the CPI be calculated but confirmed the essential soundness of the current CPI and the way it is calculated.

USEFUL ADDRESSES

Public Organizations

Building Industry Authority
P.O. Box 11-846
Wellington
Tel: (64) 4 471 0794
Fax: (64) 4 471 0798
E-mail: bia@bia.co.nz
Website: www.bia.govt.nz

Department of Building and Housing
P.O. Box 173
Wellington
Tel: (64) 4 474 2921
Fax: (64) 4 499 1791
Website: www.dbh.govt.nz

Ministry of Transport
P.O. Box 3175
Wellington
Tel: (64) 4 439 9000
Fax: (64) 4 439 9001
E-mail: info@transport.govt.nz
Website: www.transport.govt.nz

Standards New Zealand (SNZ)
Private Bag 2439
Wellington
Tel: (64) 4 498 5990
Fax: (64) 4 498 5994
E-mail: enquiries@standards.co.nz
Website: www.standards.co.nz

Statistics New Zealand
Statistics House
The Boulevard
Harbour Quays
P.O. Box 2922
Wellington 6140
Tel: (64) 4 931 4600
Fax: (64) 4 931 4610
E-mail: educationservices@stats.govt.nz
Website: www.stats.govt.nz

Trade And Professional Associations

Building Research Association of New Zealand
Private Bag 50908
Porirua 5240
Tel: (64) 4 237 1170
Fax: (64) 4 237 1171
E-mail: branz@branz.co.nz
Website: www.branz.co.nz

Cement and Concrete Association of New Zealand
P.O. Box 448
Wellington
Tel: (64) 4 499 8820
Fax: (64) 4 499 7760
E-mail: admin@cca.org.nz
Website: www.cca.org.nz

Designers Institute of New Zealand
P.O. Box 109423
Newmarket
Auckland
Tel: (64) 9 529 1713
Fax: (64) 9 529 1714
E-mail: info@dinz.org.nz
Website: www.dinz.org.nz

Electrical Contractors Association of New Zealand Inc
P.O. Box 12434
Wellington 6144
Tel: (64) 4 494 1540
Fax: (64) 4 494 1549
E-mail: carl@ecanz.org.nz
Website: www.ecanz.org.nz

Institute of Professional Engineers
P.O. Box 12-241
Wellington
Tel: (64) 4 473 9444
Fax: (64) 4 474 8933
E-mail: ipenz@ipenz.org.nz
Website: www.ipenz.org.nz

National Contractors' Federation
P.O. Box 12-013
Wellington
Tel: (64) 4 496 3270
Fax: (64) 4 496 3272
E-mail: nadia@nzcontractors.co.nz
Website: www.nzcontractor.co.nz

New Zealand Heavy Engineering Research Association
P.O. Box 76-134
Manukau City
Auckland
Tel: (64) 9 262 2885
Fax: (64) 9 262 2856
E-mail: marketing@hera.org.nz
Website: www.hera.org.nz

New Zealand Institute of Architects Inc
P O Box 2516
Shortland Street
Auckland 1140
Tel: (64) 9 623 6080
Fax: (64) 9 623 6081
Website: www.nzia.co.nz

New Zealand Institute of Building Inc
P.O. Box 303-159
North Harbour
Auckland
Tel: (64) 9 4481 911
Fax: (64) 9 4482 022
E-mail: adminnziob@nznet.gen.nz
Website: www.nziob.org.nz

New Zealand Institute of Quantity Surveyors
P.O. Box 10-469
The Terrace
Wellington
Tel: (64) 4 473 5521
Fax: (64) 4 473 2918
E-mail: office@nziqs.co.nz
Website: www.nziqs.co.nz

New Zealand Institute of Surveyors
5th Floor
St John House
114 The Terrace
Wellington
Tel: (64) 4 471 1774
Fax: (64) 4 471 1907
E-mail: nzis@surveyors.org.nz
Website: www.surveyors.org.nz

PAKISTAN

All data relate to 2007 unless otherwise indicated.

Population	
Population	164.741 mn
Urban population (2005)	35%
Population under 15	37%
Population 65 and over	4%
Average annual growth rate	1.98%
Geography	
Land area	803,940 km^2
Agricultural area (2005 to 2006 Provisional)	34%
Capital city	Islamabad
Population of capital city (2005 est.)	800,000
Economy	
Monetary unit	Pakistan Rupee (Rs)
Exchange rate (average fourth quarter 2008) to:	
the pound sterling	Rs 119.06
the US dollar	Rs 78.90
the euro	Rs 109.17
the yen x 100	Rs 88.00
Average annual inflation (1998 to 2007)	5.9%
Inflation rate	7.8%
Gross Domestic Product (GDP) at market price	US$ 143.65 bn
GDP per capita (PPP)	US$ 2,600
Average annual real change in (GDP) (1998 to 2007)	5.6%
Private consumption as a proportion of GDP	47.0%
Gross domestic Investment as a proportion of GDP	23.0%
Construction	
Gross value of construction output (2006)	Rs 143,916 mn

THE CONSTRUCTION INDUSTRY

Construction Output

Benefiting from both public and private investments, the construction industry of Pakistan is a prime source of employment generation on offering job opportunities to millions of unskilled, semi-skilled and skilled work force.

The construction sector has been one of the star performers of the fiscal year 2004-2005, as against a sharp downturn of 10.7% in 2003-2004. The construction sector recorded an equally sharp upturn of 17.9% in 2006-2007. In year 2007-2008, the trend of slow growth was recorded in construction industry as the growth rate curtailed from 17.9% (recorded in 2006-2007) to 15.2% (recorded in 2007-2008). The industry saw a decline of 2.7% in year 2007-2008.

The decline in 2007-2008 was mainly caused by devaluation of Pakistan Rupees and high inflation, which was observed globally.

In the last two years, the government has taken various budgetary and non-budgetary measures which yielded positive results and thus construction activities in Pakistan gathered momentum. The budget allocated for construction and transportation increased from Rs 6.2 billion (US$ 78.7 million) in fiscal year 2007-2008 to Rs 6.5 billion (US$ 82.5 million) in fiscal year 2008-2009. More funds have been allocated to the construction industry for its betterment and growth.

The sectoral share in GDP from the construction industry increased from 2.5% (in 2006-2007) to 2.7% (in 2007-2008). Similarly the demand for construction related materials has surged.

Many national and international real estate developers launched different construction projects in Pakistan in the year 2007-2008. But due to the global economic crisis, Pakistan also suffers loss in all sectors including the construction industry. Progress of many of the mega projects are either being slowed down or halted. The Government as well as private sector have been taking all the necessary steps to regain the boost in the industry.

Year	*Real GDP Growth Rate (%)*	*Construction Sectoral Share in GDP (%)*	*GDP at Constant Factor Cost (Rs Million)*
2000-2001	0.5	2.4	87,846
2001-2002	1.6	2.4	89,241
2002-2003	4.0	2.4	92,789
2003-2004	-10.7	2.0	82,818
2004-2005	18.6	2.1	98,190
2005-2006	5.7	2.1	103,750
2006-2007	17.9	2.5	127,616
2007-2008	15.2	2.7	146,962

Source: Tables 5, 12 & 13 of the National Accounts by Federal Bureau of Statistics, Government of Pakistan

Characteristics and Structure of the Industry

As of 2008, there are approximately 30,500 registered constructors and 150 operators in Pakistan. Even though the number of labour force in the construction industry has increased, it is still merely 6.56% of the overall labour force.

In Pakistan, engineering works can only be constructed/operated by a constructor/operator licensed by the Pakistan Engineering Council (PEC). There are six categories of registration which differ mainly by the limit of construction cost of project to be constructed or capital cost of project to be operated.

	Limit of construction cost of project (million rupees)	*Average annual value of work for last 3 years (million rupees)*	*Largest project value during last 3 years (million rupees)*	*Paid up capital or net/capital worth (million rupees)*	*Minimum requirement of professional credit points (pcp-credits)*
Constructor's Categories					
C-1	No limit	20	15	20	100
C-2	Up to 100	15	10	10	70
C-3	Up to 50	5	3.75	2.5	40
C-4	Up to 20	2	1.5	1	20
C-5	Up to 10	1.4	0.75	0.5	10
C-6	Up to 5	0.5	0.38	0.25	5

Note: Construction cost of a project shall exclude cost of land; plant and machinery permanently installed in the works but shall include cost of erection, installation, testing and commissioning.

Operator's Categories					
O-1	No limit	4	2	4	100
O-2	Up to 50	3	1.6	3	70
O-3	Up to 20	1	0.8	2	40
O-4	Up to 8	0.5	0.5	2	20
O-5	Up to 4	0.3	0.3	1	10
O-6	Up to 2	0.1	0.2	0.5	5

Note: Capital costs of projects and other values in the above table are based on the value of the operator's fees.

Source: Pakistan Engineering Council

Clients and Finance

A quarter of the total construction investment is made up of the public sector and the balance from the private sector.

Most of the building projects are privately funded where the financing is generally arranged through banks. Private funding is becoming more common due to increase in the privatisation of the public organizations. However, the majority of the civil engineering and infrastructure projects are still financed by the public funding.

Selection of Design Consultants

The professions are regulated by the appropriate professional bodies – Institute of Architects Pakistan (IAP), Institute of Engineers Pakistan (IEP) and Association of Consulting Engineers Pakistan (ACEP), Individual professional consultant has to register with his respective professional board.

PEC is the regulatory body for engineering works. It is compulsory for all practising engineers and engineering firms to be registered with PEC.

Architects and town planners shall register individually (not as a firm) with Pakistan Council of Architect and Town Planners (PCATP) which is the regulating body for architects and town planners.

There is no independent body for quantity surveyors. Quantity surveying works are normally done in-house as an integral part of the engineering and/or architectural practices.

In the private sector, most consultants are selected and appointed by the developers based on track record and personal relationships besides cost consideration. In the public sector, the selection criteria of project consultants are based on experience, quota, contracts and cost. The consultants must be registered with the various development authorities such as Capital Development Authority Islamabad, Rawalpindi Development Authority, Rawalpindi District Council, Karachi Building Control Authority (KBCA), Pakistan Housing Authority (PHA), Public Works Department (PWD), Defence Housing Authorities, etc.

Contractual Arrangements

In the public sector, there are various forms of contract being used for large tender bids which are essentially modified FIDIC Forms of Contracts. For smaller projects, simple forms of contract are normally used. The PWD usually invites contracting companies to tender through open advertisements in the major newspapers. Sometimes, tenderers are selected through a pre-qualification exercise for larger jobs. For private sector projects, the most common procurement method is by selective tendering.

In the private sector, the PEC Engineering Forms of contract are widely adopted. Bill of Quantities, drawings and specification are commonly used as the basis for tender.

PEC produced the following standard bidding/contract documents with modifications based on FIDIC and World Bank formats and specifically tailored to be in line with relevant PEC construction and consultancy by-laws and Government of Pakistan requirements:

a) Standard form of bidding documents (civil works) – to be used for construction contract over Rs 50 million
b) Standard form of tender documents for procurement of works (electrical and mechanical) – to be used for E&M procurement contract over Rs 50 million
c) Standard form of tender documents for procurement of works (for smaller contracts) – to be used for all type of procurement contract below Rs 50 million
d) Standard form of contracts for engineering consultancy for large projects (Time based/Lump sum assignments) and smaller projects

These documents are applicable to all projects to be executed in Pakistan.

Development Control

Presently the functions of development control are administered by several institutions. For instance, Karachi Building Control Authority (KBCA) which was created under the provision of Sindh Building Control Ordinance 1979, is one of the legal valid bodies undertaking this task.

KBCA is a regulatory and supervisory body whose prime function is to grant approval of building plans and "No Objection Certificates (NOC)", etc. and the conformation with the existing Building & Town planning regulations. However, quality, soundness and implementation of appropriate design/specifications are the responsibility of the concerned professionals licensed by KBCA under Karachi Building Control Licensing Regulations 1982.

City District Government Karachi (CDGK) also has its own claim under the Sindh Local government Ordinance.

Federally controlled and constituted bodies such as Cantonment Board of the Ministry of Defence, Karachi Port Trust, Pakistan Railways, Ministry of Works Pakistan, Board of Revenue, Sindh Katchi Abadies Authorities and Sindh Industrial Trading Estates Karachi have their own jurisdiction which stands untainted from the other local authorities. Similarly, other autonomous land owning agencies such as public universities are not controlled by the conventional building control practices.

In general all developments must adhere and conform to the Town Planning and Building Control in accordance with Master Plan and Environmental Control (Building and Town Planning Regulations 2002, given cover under Sindh Building Control Ordinance (SBCO) 1979) and the developers/builders must obtain the followings:

a) Building Plan approvals and NOCs from the various utilities authorities
b) Approval of Structural Designs of Buildings

c) Obtaining of NOCs for sale and advertisement for public sale projects i.e. fixation/approval of unit price, time period and specifications of construction and development
d) Submission of as-built plan upon completion
e) Obtaining occupation/completion certificate

Standards

Generally the specifications for construction works are based on the latest edition of Pakistan Standards, British Standards (BS), American Concrete Institute Standards (ACI) and American Society for Testing and Materials Standards (ASTM).

CONSTRUCTION COST DATA

Cost of Labour

The figures below are typical of labour costs in Pakistan as at the fourth quarter of 2008. Cost of labour indicates the cost to a contractor of employing that employee.

Labour rate (per day = 8 hr)

	Cost of labour (per day) Rs	*Number of hours worked per year*
Mason/bricklayer	600	2,139
Carpenter	575	2,139
Plumber	550	2,139
Electrician	500	2,139
Aluminium technician	500	2,139
Welder	550	2,139
Painter	500	2,139
Crane operator	550	2,139
Pipe fitter	500	2,139
Skilled workers	600	2,139
Semi-skilled workers	450	2,139
Unskilled workers	350	2,139
	(per hour)	
Foreman asphalt	75	2,139
Foreman concrete	75	2,139
Foreman earthwork	70	2,139
Supervisor	63	2,139
Site engineer	125	2,139

	Cost of labour (per hour) Rs	*Number of hours worked per year*
Asphalt plant engineer	125	2,139
Concrete plant engineer	125	2,139
Surveyor	105	2,139
Assistant surveyor	85	2,139
Steel binder/cutter	75	2,139

Cost of Materials

The figures that follow are the costs of main construction materials, delivered to site in the urban area, as incurred by contractors in the fourth quarter of 2008. These assume that the materials would be in quantities as required for a medium sized construction project and that the location of the works would be neither constrained nor remote.

All estimated amounts are rounded to nearest Rs 50 - 100.

	Unit	*Cost Rs*
Cement and aggregate		
Ordinary portland cement in 50kg bags	tonne	6,740
Coarse aggregates for concrete (3/4" down)	m^3	825
Fine aggregates for concrete (Local Sand)	m^3	670
Ready mixed concrete (in OP Cement)		
– 1000 psi	m^3	4,400
– 3000 psi	m^3	5,400
– 3750 psi	m^3	5,800
– 4500 psi	m^3	6,360
– 6000 psi	m^3	7,055
– 7000 psi	m^3	8,200
– 9000 psi	m^3	11,000
Ready mixed concrete (in SR Cement is on average Rs 125/m^3 higher)		
Steel		
Plain steel reinforcement 10 - 40mm diameter	tonne	70,000 - 71,000
Cold-worked deformed steel bars 10 - 40mm diameter	tonne	80,000 - 81,000
Structural steel sections	tonne	85,000 - 95,000
Bricks and blocks		
Hollow concrete blocks		
– 1200 psi (150 x 200 x 305mm)	pc	19
– 1200 psi (100 x 200 x 305mm)	pc	17
– 2000 psi (150 x 200 x 305mm)	pc	58

	Unit	*Cost Rs*
Solid concrete blocks		
– 1200 psi (150 x 200 x 305mm)	pc	25
– 1200 psi (100 x 200 x 305mm)	pc	23
– 3000 psi (150 x 200 x 305mm)	pc	80
Hollow concrete fairface blocks		
– 1050 psi (190 x 190 x 390mm)	pc	38
– 1050 psi (140 x 190 x 390mm)	pc	29
– 1050 psi (90 x 190 x 390mm)	pc	23
Solid concrete fairface blocks		
– 1050 psi (190 x 190 x 390mm)	pc	50
– 1050 psi (140 x 190 x 390mm)	pc	40
– 1050 psi (90 x 190 x 390mm)	pc	30
Timber and insulation		
Softwood sections for formwork	m^3	26,500
Hardwood for joinery (Deodar Wood)	m^3	113,000
Teak plywood 4mm (Prime) (4'x8')	m^2	520
Commercial plywood 4mm for joinery (4'8')	m^2	500
Formica ply sheets 12mm thick (Local)	m^2	390
100mm thick foam insulation	m^2	1,300
25mm thick foam insulation	m^2	375 - 430
Plaster and paint		
Plastic emulsion paint	litre	255
Distemper paint	litre	150
Matt enamel paint	litre	300
Paint for external surface	litre	300
Colour crete plaster for external surface	m^2	645
Cement tiles and pavers		
Clay floor tiles (200 x 200mm)	m^2	410
Kerb stone (150 x 300 x 450mm) (Local)	pc	220
Precast concrete paving slabs boston (300 x 300 x 60mm)	m^2	1,000
Clay roof tiles (400 x 225mm)	pc	110
Precast concrete roof tiles (203 x 406mm)	m^2	550
Drainage		
WC suite complete	each	10,000 - 14,000
Lavatory basin complete	each	7,000 - 80,000
100mm diameter UPVC drain pipes	m	755
150mm diameter UPVC drain pipes	m	1,635

Unit Rates

The descriptions that follow are generally shortened versions of standard descriptions listed in full in section 4. Where an item has a two digit reference number (e.g. 05 or 33), this relates to the full description against that number in section 4. Where an item has an alphabetic suffix (e.g.12A or 34B) this indicates that the standard description has been modified. Where a modification is major the complete modified description is included here and the standard description should be ignored; where a modification is minor (e.g. the insertion of a named hardwood) the shortened description has been modified here but, in general, the full description in section 4 prevails.

The unit rates below are for main work items on a typical construction project in the Karachi area in the fourth quarter of 2008. The rates include all necessary labour, materials and equipment. Allowances of 20% to cover preliminary and general items and 20% to cover contractors' overheads and profit have been included in the rates which are the normal industry allowances in Karachi. It is customary to induce preliminaries and overheads in the unit rates.

		Unit	*Rate Rs*
Excavation			
01	Mechanical excavation of foundation trenches	m^3	425
02	Hardcore filling making up levels	m^3	885
Concrete work			
04A	Plain in situ concrete in strip foundations in trenches (G25)	m^3	6,400 - 7,745
05A	Reinforced in situ concrete in beds, walls, suspended floors/roof slabs, columns, isolated beams (G35)	m^3	7,200 - 8,715
Formwork			
11	Softwood formwork to concrete walls, concrete columns, horizontal soffits of slabs	m^2	270 - 325
Reinforcement			
14	Reinforcement in concrete walls, suspended concrete	tonne	80,000
15	Reinforcement in slabs	tonne	80,000
16	Fabric reinforcement in concrete beds	m^2	240 - 405
Steelwork			
17	Fabricate, supply and erect framed structure	tonne	165,000
18	Framed structural steelwork in universal joist sections	tonne	165,000
19	Structural steelwork lattice roof trusses	tonne	165,000

		Unit	*Rate Rs*
Brickwork and blockwork			
20	Precast hollow concrete block walls	m^2	1,130
21	Solid (perforated) common bricks	m^2	810
23	Facing bricks	m^2	1,350
Roofing			
25	Plain clay roof tiles 260 x 160mm	m^2	740
27A	Sawn softwood roof boarding (12mm)	m^2	1,410
30	Bitumen based mastic asphalt roof covering	m^2	450
33	Troughed galvanized steel roof cladding	m^2	650 - 800
Woodwork and metalwork			
34	Preservative treated sawn softwood 50 x 150mm	m^3	106,000
36	Single glazed casement window in hardwood, size 650 x 900mm	each	74,300
37	Two panel glazed door in hardwood size 850 x 2000mm	each	14,000
38	Solid core half hour fire resisting hardwood internal flush doors, size 800 x 2000mm	each	15,000
39	Aluminium double glazed window, size 1200 x 1200mm	each	14,400 - 15,200
40	Aluminium double glazed door, size 850 x 2100mm	each	15,500 - 17,500
41	Hardwood skirtings	m	500 - 535
Plumbing			
42A	UPVC half round eaves gutter (6” diameter)	m	2,400
43	UPVC rainwater pipes (4” diameter)	m	2,300
44	Light gauge copper cold water tubing (1/2” diameter)	m	595
45A	High pressure plastic pipes for cold water supply (2” diameter)	m	1,235
46	Low pressure plastic pipes for cold water distribution (3/4” diameter)	m	315
47	UPVC soil and vent pipes (3” or 4” diameter)	m	1,150
48	White vitreous china WC suite	each	14,500
49	White vitreous china lavatory basin	each	8,500
51	Stainless steel single bowl sink and double drainer	each	11,700
Electrical Work			
52	PVC insulated and copper sheathed cable	m	1,900
53	13 amp unswitched socket outlet	each	490
54	Flush mounted 20 amp, 1 way light switch	each	350

		Unit	Rate Rs
Finishings			
55A	2 coats gypsum based plaster on concrete walls 20mm thick	m^2	500
56	White glazed tiles on plaster walls	m^2	1,290 - 1,395
57	Red clay quarry tiles on concrete floors	m^2	900
58A	Cement and sand screed to concrete floors 30mm thick	m^2	345 - 376
60	Mineral fibre tiles on concealed suspension system	m^2	950
Glazing			
61A	6mm clear float glass; glazing to wood	m^2	600 - 700
Painting			
62	Emulsion on plaster walls	m^2	215 - 235
63	Oil paint on timber	m^2	215 - 235

Approximate Estimating

The building costs per unit area that follow are averages incurred by building clients for typical buildings in the urban area as at the fourth quarter of 2008. They are based upon the total floor area of all storeys, measured between external walls and without deduction for internal walls.

Approximate estimating costs generally include mechanical and electrical installations but exclude furniture, loose or special equipment, and external works; they also exclude fees for professional services. The costs shown are for specifications and standards appropriate to Karachi and this should be borne in mind when attempting comparisons with similarly described building types in other countries. A discussion of this issue is included in section 2. Comparative data for countries covered in this publication, including construction cost data, are presented in Part Three.

Approximate estimating costs must be treated with caution; they cannot provide more than a rough guide to the probable cost of building.

	Cost m^2 Rs	Cost ft^2 Rs
Industrial buildings		
Factories for letting	12,912 - 13,988	1,200 - 1,300
Factories for owner occupation (light industrial use)	16,140 - 17,216	1,500 - 1,600
Factories for owner occupation (heavy industrial use)	20,444 - 23,672	1,900 - 2,200
High tech laboratory workshop centres (air-conditioned)	30,128 - 32,280	2,800 - 3,000

	Cost *m^2 Rs*	*Cost* *ft^2 Rs*
Administrative and commercial buildings (* air-conditioned)		
Civic offices*	32,280	3,000
Offices for letting, 5 to 10 storeys*	34,432 - 37,660	3,200 - 3,500
Offices for letting, high rise*	37,660 - 39,812	3,500 - 3,700
Offices for owner occupation, 5 to 10 storeys*	40,888 - 43,040	3,800 - 4,000
Offices for owner occupation, high rise*	43,040 - 47,344	4,000 - 4,400
Prestige/headquarters office, 5 to 10 storeys*	63,484 - 65,636	5,900 - 6,100
Prestige/headquarters office, high rise*	67,788 - 71,016	6,300 - 6,600
Health and education buildings		
General hospitals (excluding specialist equipment and installation)	69,940 - 81,776	6,500 - 7,600
Private hospitals (excluding specialist equipment and installation)	80,700 - 93,612	7,500 - 8,700
Primary/junior schools	35,508 - 43,040	3,300 - 4,000
Secondary/middle schools	35,508 - 43,040	3,300 - 4,000
University	52,724	4,900
Recreation and arts buildings		
Theatres (over 500 seats) including seating and stage equipment	seat	70,000 - 75,000
Concert halls including seating	27,976 - 32,280	2,600 - 3,000
Sports hall including changing and social facilities	38,736	3,600
Swimming pools (schools standard) including changing facilities	each	3,000,000
National museums including full air-conditioning and standby generator	32,280 - 43,040	3,000 - 4,000
Residential buildings		
Social/economic single family housing (B type)	10,760 - 12,912	1,000 - 1,200
Private/mass market single family housing 2 storey detached/semi detached	15,064	1,400
Terrace houses	24,748 - 32,280	2,300 - 3,000
Social/economic apartment housing, low rise (no lifts)	13,988	1,300
Social/economic apartment housing, high rise (with lifts)	16,140	1,500

	Cost m^2 Rs	Cost ft^2 Rs
Private sector apartment building (standard specification)	35,508 - 39,812	3,300 - 3,700
Private sector apartment buildings (luxury)	59,180 - 67,788	5,500 - 6,300
Student/nurses halls of residence	19,368 - 21,520	1,800 - 2,000
Homes for the elderly (shared accommodation)	13,988 - 16,140	1,300 - 1,500
Homes for the elderly (self contained with shared communal facilities)	16,140 - 18,292	1,500 - 1,700
Hotel, 5 star, city centre (inclusive of F.F. & E.)	118,360 - 139,880	11,000 - 13,000
Hotel, 3 star, city/provincial (inclusive of F.F. & E.)	75,320 - 94,688	7,000 - 8,800
Motel (inclusive of F.F. & E.)	33,356 - 39,812	3,100 - 3,700

Regional Variations

The approximate estimating costs are based on projects in Karachi. For other parts of Pakistan, adjust these costs by the following factors to take account of regional variations:

Karachi	:	0%
Lahore	:	+ 0 to 5%
Islamabad	:	+ 5% to 10%

EXCHANGE RATES AND INFLATION

The combined effect of exchange rates and inflation on prices within a country and price comparisons between countries is discussed in section 2.

Exchange Rates

The graph on the next page plots the movement of the Pakistan Rupee against the sterling, the euro, the US dollar and 100 Japanese yen since 1999. The values used for the graph are quarterly and the method of calculating these is described and general guidance on the interpretation of the graph provided in section 2. The average exchange rate in the fourth quarter of 2008 was Rs 119.06 to pound sterling, Rs 109.17 to euro, Rs 78.90 to US dollar and Rs 88.00 to 100 Japanese yen.

THE PAKISTAN RUPEE AGAINST STERLING, EURO, US DOLLAR AND 100 JAPANESE YEN

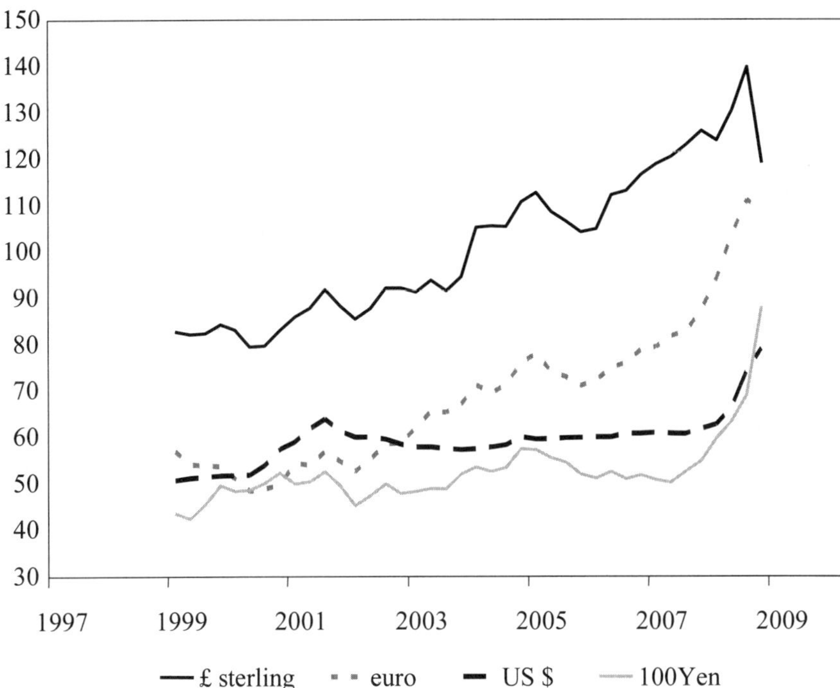

Price Inflation

The table on the next page represents inflation rates based on Sensitive Price Indicator (SPI), Consumer Price Index (CPI) and Wholesale Price Index (WPI) in Pakistan since 1991.

SENSITIVE PRICE INDICATOR, CONSUMER PRICE INDEX AND WHOLESALE PRICE INDEX

Year	*SPI*	*CPI*	*WPI*
1991-1992	10.50	10.60	9.80
1992-1993	10.70	9.80	7.40
1993-1994	11.10	11.30	16.40
1994-1995	15.00	13.00	16.00
1995-1996	10.70	10.80	11.10
1996-1997	12.50	11.80	13.00
1997-1998	7.40	7.80	6.60
1998-1999	6.40	5.70	6.40
1999-2000	1.80	3.60	1.80
2000-2001	4.80	4.40	6.20
2001-2002	3.40	3.50	2.10
2002-2003	3.60	3.10	5.90
2003-2004	6.80	4.60	7.90
2004-2005	11.60	9.30	6.80
2005-2006	7.00	7.90	10.10
2006-2007	10.80	7.80	6.90
2007-2008	16.80	12.00	16.40

Source: Statistic Division, Government of Pakistan
Note: Yearly Inflation rate of Pakistan from the year 2001-2002 to date based on the base year (2000-01 = 100)

USEFUL ADDRESSES

Board of Investment
Ataturk Avenue
G-5/1 Islamabad
Pakistan
Tel: (92-51) 922 4103, 922 4101, 920 4339
Fax: (92-51) 921 5554, 920 6160
E-mail: minister@pakboi.gov.pk
Website: www.pakboi.pk

Company Registration Office (Islamabad)
1st floor, State Life Building No. 7
Blue Area, Jinnah Evenue.
Islamabad
Pakistan
Tel: (051) 921 8528, 920 6219
Fax: (051) 920 8740
E-mail: crois@isb.paknet.com.pk

Company Registration Office (Karachi)
3rd & 4th Floor, SLIC Building No. 2
Wallanc road,
Karachi
Pakistan
Tel: (021) 921 3271 & 72
Fax: (021) 921 3278, 921 3424
E-mail: crokhi@khi.paknet.com.pk

Federal Bureau of Statistics
5-SLIC Building
F-6/4, Blue Area
Islamabad
Pakistan
Fax: (92-51) 920 3233
E-mail: statpak@statpak.gov.pk
Website: www.statpak.gov.pk

Pakistan Council of Architects & Town Planners
Suite 111, First Floor
RSM Square
E-1Shaheed-e-Millat Road
Karachi
Pakistan
Tel: (021) 452 3129
Fax: (021) 454 1099
E-mail: tech@pcatp.org.pk
Website: www.pcatp.org.pk

Pakistan Engineering Council (PEC)
Ataturk Avenue (East), G-5/2
P.O. Box 1296
Islamabad
Pakistan
Tel: (92-51) 920 6974, 921 9500, 282 9348, 282 9296, 282 9311
Fax: (92-51) 227 6224
Website: www.pec.org.pk

Securities and Exchange Commission of Pakistan
NIC Building
Jinnah Avenue
Islamabad 44000
Pakistan
Tel: (051) 920 7091
Fax: (051) 920 4915
E-mail: enquiries@secp.gov.pk
Website: www.secp.gov.pk

PHILIPPINES

All data relate to 2008 unless otherwise indicated.

Population	
Population	88.6 mn
Urban population	30.14%
Population under 15	31.97%
Population 65 and over	3.3%
Average annual growth rate (2000 to 2008)	2.04%
Geography	
Land area	300,000 km^2
Agricultural area	41%
Capital city	Manila
Population of capital city	1.35 mn
Economy	
Monetary unit	Philippines Peso (Php)
Exchange rate (average fourth quarter 2008) to:	
the pound sterling	Php 76.26
the US dollar	Php 48.44
the euro	Php 63.65
the yen x 100	Php 50.49
Average annual inflation (2000 to 2008)	5.6%
Inflation rate	9.3%
Gross Domestic Product (GDP)	Php 7,497.5 bn
GDP per capita	Php 84,622
Average annual real change in GDP (1998 to 2008)	6.1%
Private consumption as a proportion of GDP	70.4%
Public consumption as a proportion of GDP	9.5%
Investment as a proportion of GDP	5.3%
Construction	
Gross value of construction output	Php 609.3 bn
Net value of construction output	Php 256.5 bn
Net value of construction output as a proportion of GDP	3.4%

THE CONSTRUCTION INDUSTRY

Construction Output

2007 Construction output, as measured by the real Gross Value Added amounted to Php58.8 billion or a double digit growth of 19.5% that was spurred by government infrastructure spending. On the other hand, construction investments, as measured by the Gross Value in Construction (GVC) in 2007, reached Php109.0 billion in real terms, a remarkable 18.0% improvement from 5.5% growth in 2006.

Public construction activities valued at Php46.1 billion in real terms rose by 30.8%. Meanwhile, private construction activities (valued at Php62.8 billion in real terms) made a rebound of 10.2% from a 3.7% dip in 2006. This could be attributed largely to the increasing demand for condominiums, commercial establishments and tourism facilities.

Despite the global crisis that persisted in the fourth quarter and that has spilled into the Year of 2009, the local economy has not atrophied as fourth quarter GDP grew by a respectable 4.5% compared to 6.4% the previous year. Major growth drivers were Trade, Manufacturing, Agriculture and Fishery, Construction and Private Services. Major contribution on the demand side came from increased household spending aided by the growths in investment, construction and government consumption. Meanwhile, the continued double-digit growth in Net Factor Income from Abroad revved Gross National Product up to 6.4% from previous year's 6.0%.

The gross value of output of the construction industry in 2008 was Php609.3 billion or 8.1% of GDP. The net output was Php256.5 billion or 3.4% of GDP.

Construction continued to post gains in 2008 with 13.1% from 13.4% registered in the previous year benefiting from the projects of the private sector.

A breakdown of new building starts is shown in the table below for the year 2006 to 2007. The table shows an average increase of 5% in commercial buildings.

VALUE OF NEW BUILDING STARTS, 2006 AND 2007

	2006		*2007*	
Type of work	*Php billion*	*% of total*	*Php billion*	*% of total*
Residential	51.2	50.3	53.6	50.2
Commercial	34.4	33.8	38.0	35.6
Industrial	5.8	5.7	5.2	4.9
Institutional	8.6	8.4	8.2	7.6
Agricultural	0.6	0.6	0.4	0.4
Others	1.2	1.2	1.4	1.3
Total	101.8	100.0	106.8	100.0

Source: Construction Division, National Statistics Office
(As compiled by the Construction Industry Authority of the Philippines)

The regional distribution of net value of construction in 2007 in relation to the distribution of population is shown below.

REGIONAL DISTRIBUTION OF CONSTRUCTION COMPARED TO POPULATION

Region		*Population (2007 %)*	*Construction (2007 %)*
National Capital Region (NCR Metro Manila)		13.05	49.41
CAR	Cordillera Administrative Region	1.72	1.13
I	Ilocos Region	5.13	3.96
II	Cagayan Valley	3.45	0.68
III	Central Luzon	10.98	8.37
IVA	Calabarzon	13.26	13.27
IVB	Mimaropa	2.89	0.77
V	Bicol Region	5.77	1.34
VI	Western Visayas	7.73	4.26
VII	Central Visayas	7.23	8.21
VIII	Eastern Visayas	4.42	0.91
IX	Zamboanga Peninsula	3.65	0.40
X	Northern Mindanao	4.46	1.90
XI	Davao Region	4.69	3.62
XII	Soccsksargen	4.32	1.16
XIII	Caraga	2.59	0.61
ARMM	Autonomous Region in Muslim Mindanao	4.65	0.00
Total		100	100

Source: Construction Division, National Statistics Office

The regions with the greatest construction activity – the National Capital Region (or Metro Manila), Central Luzon and Calabarzon indicate that almost 70% of the national construction activity comes from the regions where the concentration of population is relatively high.

Characteristics and Structure of the Industry

All construction companies are required to obtain a licence from the Philippine Contractors Accreditation Board (PCAB) before they are allowed to undertake any work. There are two types of PCAB Contractor's Licence.

- Regular Licence: this is issued to construction firms (sole proprietorships, partnerships or corporations) with at least 60% Filipino equity participation and incorporated under Philippine laws.

- Special Licence: this is issued to a joint venture, consortium, foreign contractor or a project owner for the construction of a specific project.

Licensed contractors are broadly classified under General Building, General Engineering and Specialty Contractors. They are also categorized by their financial worth. The PCAB categories are as follows:

NUMBER OF LICENSED CONTRACTORS BY CATEGORY, 2009*

Category	*Minimum capital*	*Number of licensed contractors*
AAA	Php90 million	174
AA	Php45 million	88
A	Php9 million	534
B	Php4.5 million	954
C	Php3.0 million	571
D	Php0.9 million	1,819
Trade	Php0.045 million	203
Total		4,343

Source: Philippine Contractors Accreditation Board (PCAB)
** As of January 2009*

Small contractors (categories C, D and Trade) form the largest group making up 60% of the total licensed contractors. The number of large contractors (categories AAA and AA) has been steadily increasing and as of January 2009, they comprise 6% of the total in 2009.

The larger general companies include:

MAJOR CONSTRUCTION FIRMS IN THE PHILIPPINES

Major contractors	*Sales (in Php'000)*	*Main work/types*
Hanjin Heavy Industries and Construction Co. Ltd.	21,139,806	General Building and Engineering
Makati Development Corporation	6,170,911	General Building and Engineering
EEI Corporation	5,638,579	General Building and Engineering
D.M. Consunji Inc	4,078,817	General Building and Engineering
Taisei Philippine Construction Inc	3,996,511	General Building and Engineering
Shimizu Corporation	3,115,491	General Building and Engineering
F.F. Cruz & Company, Inc.	2,387,425	General Building and Engineering
Kajima Philippines Incorporated	2,353,914	General Building and Engineering
Monolith Construction & Devt. Corp.	2,334,899	General Building and Engineering
Hillmarcs Construction Corp.	2,327,633	General Building and Engineering
Taisei Corporation – Philippine Branch	2,122,326	General Building and Engineering
CCT Construction Corp.	1,923,085	General Building and Engineering
Obayashi Corporation Philippine Branch	1,721,227	No available information
Aboitiz Construction Group, Inc.	1,448,511	General Building and Engineering

Source: The Top 8000 Corporations in the Philippines; 2008 Edition Securities and Exchange Commission

Selection of Design Consultants

Generally, professional consultants are appointed directly by the client although sometimes they are engaged by another consultant either through referrals or by competition.

In the private and public sector, price is the most important criterion followed by track record, personal contacts and recommendations. The table below sets out the indicative range of professional fees reflective of the current industry practice. It is expressed as a percentage of the total project cost.

SCALE OF PROFESSIONAL FEES

Professional consultants	*Industry practice* %
Architect	3.0 - 6.0%
Structural Engineer	0.5 - 0.6%
Mechanical Engineer	0.1 - 0.3%
Electrical Engineer	0.1 - 0.3%
Sanitary and Plumbing	0.1 - 0.3%
Fire and Safety Engineer	0.1 - 0.3%
Project Manager	1.0 - 3.0%
Construction Manager	1.0 - 2.0%
Quantity Surveyor	0.5 - 1.0%

Contractual Arrangements

Most clients commission their own design by appointing a separate planning or architectural firm and invite contractors to bid for construction. Design-and-build is seldom used but is popular for small scale projects such as housing projects. In recent years, management contracting has been adopted extensively and is fast gaining popularity in the Philippines, especially for large projects.

In almost all cases, building work is undertaken by general contractors. For private sector projects, contractors are usually selected on the basis of their reputation and competency through negotiation or by competition. Public sector contracts are governed by a special law – *Presidential Decree 1594* and its Rules and Regulations. Advance payments equivalent to 15% of the contract price is given upon submission of an irrevocable letter of credit. Government contractors are also compensated for price fluctuations in materials, labour and equipment if the increases exceed 5% of the original contract price.

The principal contract documents comprise conditions of contract, general agreement, schedule of works and bills of quantities.

The selected contractor normally provides all construction materials, manpower, and other inputs. However, in some cases the owner supplies certain materials. Some major contracting companies nominate subcontractors to undertake specialized works such as prestressed concrete, plumbing, electrical, mechanical and drainage.

Development Control and Standards

The National Housing and Land Use Regulatory Board are responsible for controlling land use and building operations in the industry. The board is responsible for issuing development permits and for ensuring that developers comply with the required standards. Guidelines and procedures for obtaining permits are enumerated in their handbooks, *PD 957* for high cost housing and *BP 225* for low cost housing. The request for a permit is processed only after all requirements are met. It takes about one to three months before a certificate of registration and license to sell is issued.

All new construction work in the Philippines have to comply with the provisions set out in the 2004 edition of the *National Structural Code of the Philippines* or *NSCP,* the *ACO-1989 edition*, the 2007 edition of the *National Building Code of the Philippines* and the 1985 edition of the *AISC Steel Manual.* Deviations from the codes may be allowed by the building officials, provided it is shown and verified by tests that such deviation is within the scope of the code. The *ACI-1989* edition covers the proper design and construction of reinforced concrete buildings, and prescribes rules and regulations governing permits, inspections, specifications, materials, concrete quality, mixing, formwork, embedded pipes, strengths and serviceability, loads, specifications and provisions for seismic design. The quality and testing of materials used in construction are covered by the *American ASTM standard specification* and the welding of reinforcement by the *American AWS standard.*

Liability

The contractor is responsible for making good any defects appearing in the materials and workmanship for a minimum period of one year. This is usually supported by a Guarantee Bond of a value equivalent to the contract price. Main contractors will obtain similar guarantees from their subcontractors. Disputes and claims are settled by reference to and in accordance with the provisions of the Construction Industry Arbitration Commission.

CONSTRUCTION COST DATA

Cost of Labour

The figures on the next page are typical of labour costs in Metro Manila as at the fourth quarter of 2008. The wage rate is the basis of an employee's income, while the cost of labour indicates the cost to a contractor of employing that employee. The difference between the two covers a variety of mandatory and voluntary contributions – a list of items which could be included is given in section 2.

	Wage rate (per day) Php	*Cost of labour (per day) Php*	*Number of hours worked per year*
Site operatives			
Mason/bricklayer	450	585	2,496
Carpenter	460	598	2,496
Plumber	450	585	2,496
Electrician	455	592	2,496
Structural steel erector	475	618	2,496
HVAC installer	480	624	2,496
Semi-skilled worker	430	559	2,496
Unskilled labourer	420	546	2,496
Equipment operator	525	683	2,496
Site supervision			
General foreman	641	834	2,496
Trades foreman	597	776	2,496
Clerk of works	411	535	2,496
Contractors' personnel			
Site manager	1,183	1,538	2,496
Resident engineer	641	834	2,496
Resident surveyor	641	834	2,496
Junior engineer	481	625	2,496
Junior surveyor	481	625	2,496
Planner	500	650	2,496
Consultants' personnel			
Senior architect	1,775	2,308	2,080
Senior engineer	1,775	2,308	2,080
Senior surveyor	1,775	2,308	2,080
Qualified architect	1,109	1,442	2,080
Qualified engineer	1,109	1,442	2,080
Qualified surveyor	1,109	1,442	2,080

Cost of Materials

The figures that follow are the costs of main construction materials, delivered to site in the Metro Manila area, as incurred by contractors in the fourth quarter of 2008. These assume that the materials would be in quantities as required for a medium sized construction project and that the location of the works would be neither constrained nor remote. All the costs in this section exclude expanded value added tax.

	Unit	*Cost Php*
Cement and aggregate		
Ordinary portland cement in 40kg bags	bag	185
Coarse aggregates for concrete	m^3	750
Fine aggregates for concrete	m^3	650
Ready mixed concrete (A: 34 MPa)	m^3	4,500
Ready mixed concrete (B: 21 MPa)	m^3	3,400
Steel		
Mild steel reinforcement	tonne	33,000
High tensile steel reinforcement	tonne	35,000
Structural steel sections	tonne	49,200
Bricks and blocks		
Common bricks (2" x 4" x 8")	1,000	N/A
Hollow concrete blocks (6" x 8" x 16")	1,000	12,000
Timber and insulation		
Softwood sections for carpentry	m^3	20,110
Softwood for joinery	m^3	21,500
Hardwood for joinery	m^3	59,000
Exterior quality plywood (13mm)	m^2	240
Plywood for interior joinery (13mm)	m^2	180
Softwood strip flooring (10mm)	m^2	2,500
Chipboard sheet flooring (25mm)	m^2	850
100mm thick quilt insulation	m^2	650
100mm thick rigid slab insulation	m^2	1,850
Softwood internal door complete with frames and ironmongery	each	10,500
Glass and ceramics		
Float glass (6mm)	m^2	850
Plaster and paint		
Good quality ceramic wall tiles (108 x 108mm)	m^2	650
Plaster in 20 kg bags	tonne	42,250
Plasterboard (13mm thick)	m^2	250
Emulsion paint	gallon	550
Tiles and paviors		
Clay floor tiles (8" x 8" x 1")	m^2	450
Vinyl floor tiles (300 x 300 x 3mm)	m^2	550
Precast concrete paving slabs (400 x 185 x 50mm)	m^2	1,250

	Unit	Cost Php
Drainage		
WC suite complete	each	8,700
Lavatory basin complete	each	7,500
150mm diameter cast iron drain pipes	m	2,800

Unit Rates

The descriptions below are generally shortened versions of standard descriptions listed in full in section 4. Where an item has a two digit reference number (e.g. 05 or 33), this relates to the full description against that number in section 4. Where an item has an alphabetic suffix (e.g. 12A or 34B) this indicates that the standard description has been modified. Where a modification is major the complete modified description is included here and the standard description should be ignored; where a modification is minor (e.g. the insertion of a named hardwood) the shortened description has been modified here but, in general, the full description in section 4 prevails.

The unit rates below are for main work items on a typical construction project in the Metro Manila area in the fourth quarter of 2008. The rates include all necessary labour, materials and equipment. Allowances of 5% - 10% to cover preliminary and general items and 3% - 5% to cover contractors' overheads and profit have been included in the rates. All the rates in this section exclude expanded value added tax.

		Unit	Rate Php
Excavation			
01	Mechanical excavation of foundation trenches	m^3	350
02	Hardcore filling making up levels	m^3	650
03	Earthwork support	m^2	450
Concrete work			
04	Plain in situ concrete in strip foundations in trenches	m^3	3,550
05	Reinforced in situ concrete in beds	m^3	4,000
06	Reinforced in situ concrete in walls	m^3	4,200
07	Reinforced in situ concrete in suspended floors or roof slabs	m^3	4,550
08	Reinforced in situ concrete in columns	m^3	4,800
09	Reinforced in situ concrete in isolated beams	m^3	4,550
10A	Precast concrete slab (1500 x 2000 x 100)	each	6,500
Formwork			
11	Softwood formwork to concrete walls	m^2	650
12	Softwood formwork to concrete columns	m^2	650

		Unit	*Rate Php*
13	Softwood formwork to horizontal soffits of slabs	m²	720
Reinforcement			
14	Reinforcement in concrete walls	kg	40
15	Reinforcement in suspended concrete slabs	kg	42
16	Fabric reinforcement in concrete beds	m²	150
Steelwork			
17	Fabricate, supply and erect steel framed structure	tonne	95,000
18	Framed structural steelwork in universal joist sections	tonne	85,000
19	Structural steelwork lattice roof trusses	tonne	75,000
Brickwork and blockwork			
20	Precast lightweight aggregate hollow concrete block walls	m²	950
21A	Solid (perforated) concrete blocks	m²	900
Roofing			
24	Concrete interlocking roof tiles 430 x 380mm	m²	2,500
25	Plain clay roof tiles 260 x 160mm	m²	1,560
28	Particle board roof coverings	m²	1,210
29	3 layers glass-fibre based bitumen felt roof covering	m²	850
30	Bitumen based mastic asphalt roof covering	m²	750
31	Glass-fibre mat roof insulation 160mm thick	m²	850
32	Rigid sheet loadbearing roof insulation 75mm thick	m²	800
33	Troughed galvanized steel roof cladding	m²	780
Woodwork and metalwork			
34	Preservative treated sawn softwood 50 x 100mm	m	600
35	Preservative treated sawn softwood 50 x 150mm	m	700
36	Single glazed casement window in hardwood, size 650 x 900mm	each	2,500
37	Two panel glazed door in hardwood, size 850 x 2000mm	each	16,050
38A	Solid core half hour fire resisting hardwood internal flush doors, size 650 x 900mm	each	8,500
39	Aluminium double glazed window, size 1200 x 1200mm	each	25,900
40	Aluminium double glazed door, size 850 x 2100mm	each	32,100

		Unit	Rate Php
41	Hardwood skirtings	m	650
Plumbing			
42A	UPVC half round eaves gutter, 12" x 8"	m	1,500
43	UPVC rainwater pipes, 4" diameter	m	650
44A	Light gauge copper cold water tubing; 1"diameter	m	1,512
45A	High pressure plastic pipes for cold water supply; 4" diameter	m	4,480
46A	Low pressure plastic pipes for cold water distribution; 2"dia	m	2,430
47	UPVC soil and vent pipes, 4" diameter	m	650
48	White vitreous china WC suite	each	9,500
Electrical work			
52	PVC insulated and copper sheathed cable	m	2,800
53	13 amp unswitched socket outlet	each	350
54	Flush mounted 20 amp, 1 way light switch	each	360
Finishings			
55	2 coats gypsum based plaster on brick walls	m^2	650
56	White glazed tiles on plaster walls	m^2	850
57	Red clay quarry tiles on concrete floors	m^2	1,800
58	Cement and sand screed to concrete floors	m^2	375
59	Thermoplastic floor tiles on screed	m^2	875
60	Mineral fibre tiles on concealed suspension system	m^2	1,200
Glazing			
61	Glazing to wood	m^2	1,200
Painting			
62	Emulsion on plaster walls	m^2	350
63	Oil paint on timber	m^2	550

Approximate Estimating

The building costs per unit area given overleaf are averages incurred by building clients for typical buildings in the Metro Manila area as at the fourth quarter of 2008. They are based upon the total floor area of all storeys, measured between external walls and without deduction for internal walls.

Approximate estimating costs generally include mechanical and electrical installations but exclude furniture, loose or special equipment, and external works;

they also include professional services. The costs shown are for specifications and standards appropriate to the Philippines and this should be borne in mind when attempting comparisons with similarly described building types in other countries. A discussion in this issue is included in section 2. Comparative data for countries covered in this publication, including construction cost data, are presented in Part Three.

Approximate estimating costs must be treated with caution; they cannot provide more than a rough guide to the probable cost of building. All the rates in this section exclude expanded value added tax.

	Cost m^2 *Php*	*Cost* ft^2 *Php*
Industrial buildings		
Factories for letting	16,340	1,518
Factories for owner occupation (light industrial use)	17,300	1,607
Factories for owner occupation (heavy industrial use)	18,060	1,678
Factory/office (high-tech) for letting (shell and core only)	15,500	1,440
Factory/office (high-tech) for letting (ground floor shell, first floor offices)	16,500	1,533
Factory/office (high tech) for owner occupation (controlled environment, fully furnished)	21,000	1,951
High tech laboratory workshop centres (air-conditioned)	26,100	2,425
Warehouses, low bay (6 to 8m high) for letting (no heating)	14,100	1,310
Warehouses, low bay for owner occupation	17,200	1,598
Warehouses, high bay for owner occupation	18,700	1,737
Cold stores/refrigerated stores	27,300	2,536
Administrative and commercial buildings		
Civic offices, non air-conditioned	25,400	2,360
Civic offices, fully air-conditioned	30,000	2,787
Offices for letting, 5 to 10 storeys, non air-conditioned	22,500	2,090
Offices for letting, 5 to 10 storeys, air-conditioned	28,200	2,620
Offices for letting, high rise, air-conditioned	30,100	2,796
Offices for owner occupation 5 to 10 storeys, non air-conditioned	32,000	2,973
Offices for owner occupation 5 to 10 storeys, air-conditioned	35,000	3,252
Offices for owner occupation high rise, air-conditioned	42,000	3,902
Prestige/headquarters office, 5 to 10 storeys, air-conditioned	45,000	4,181
Prestige/headquarters office, high rise, air-conditioned	48,000	4,459
Health and education buildings		
General hospitals (230 beds)	43,500	4,041
Teaching hospitals (100 beds)	36,800	3,419
Private hospitals (100 beds)	38,500	3,577
Health centres	23,000	2,137

	Cost m² Php	*Cost ft² Php*
Nursery schools	22,000	2,044
Primary/junior schools	24,000	2,230
University (arts) buildings	24,500	2,276
University (science) buildings	23,600	2,192
Management training centres	25,800	2,397
Recreation and arts buildings		
Theatres (over 500 seats) including seating and stage equipment	58,000	5,388
Theatres (less than 500 seats) including seating and stage equipment	55,000	5,110
Concert halls including seating and stage equipment	65,000	6,039
Sports halls including changing and social facilities	42,000	3,902
National museums including full air-conditioning and standby generator	62,000	5,760
Local museums including air-conditioning	47,000	4,366
City centre/central libraries	35,000	3,252
Branch/local libraries	33,500	3,112
Residential buildings		
Social/economic single family housing (multiple units)	12,000	1,115
Private/mass market single family housing 2 storey detached/semi detached (multiple units)	14,000	1,301
Purpose designed single family housing 2 storey detached (single unit)	15,500	1,440
Social/economic apartment housing, low rise (no lifts)	18,500	1,719
Social/economic apartment housing, high rise (with lifts)	19,000	1,765
Private sector apartment building (standard specification)	29,200	2,713
Private sector apartment buildings (luxury)	41,000	3,809
Student/nurses halls of residence	28,000	2,601
Homes for the elderly (shared accommodation)	29,000	2,694
Hotel, 5 star, city centre	66,000	6,132
Hotel, 3 star, city/provincial	50,500	4,692
Motel	40,000	3,716

Expanded Value Added Tax (E-VAT)

The standard rate of expanded value added tax (E-VAT) is currently 12%, chargeable on general building work.

EXCHANGE RATES AND INFLATION

The combined effect of exchange rates and inflation on prices within a country and price comparisons between countries is discussed in section 2.

Exchange Rates

The graph below plots the movement of the Filipino peso against the sterling, the euro, the US dollar and 100 Japanese yen since 1998. The values used for the graph are quarterly and the method of calculating these is described and general guidance on the interpretation of the graph provided in section 2. The average exchange rate in the fourth quarter of 2008 was Php76.26 to pound sterling, Php63.65 to euro, Php48.44 to US dollar and Php50.49 to 100 Japanese yen.

THE FILIPINO PESO AGAINST STERLING, EURO, US DOLLAR AND 100 JAPANESE YEN

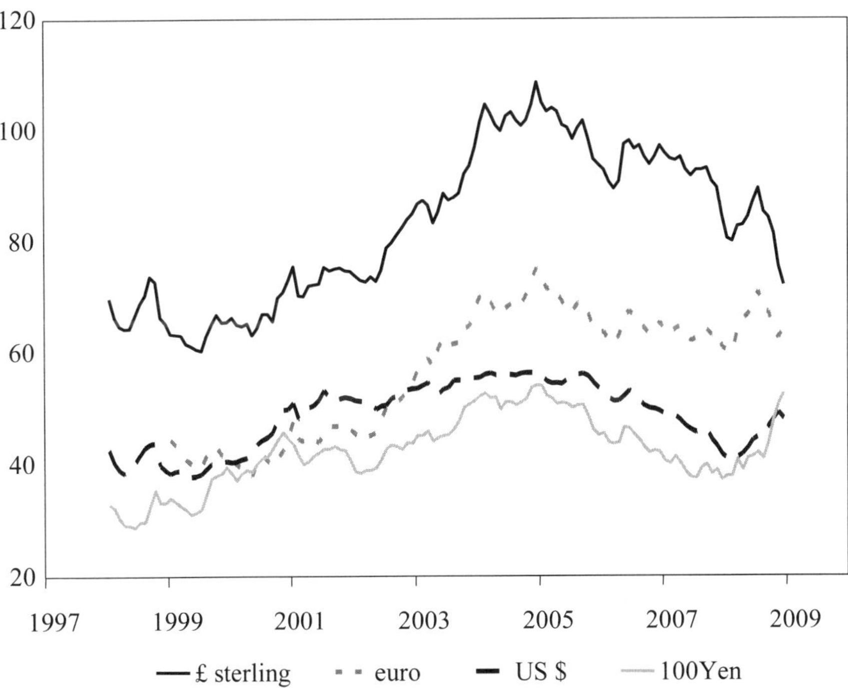

Consumer Price Inflation

The table below presents consumer price inflation in Philippines since 2000.

CONSUMER PRICE INFLATION

Year	*Consumer Price Index*	*Headline Inflation Average %*
2000	100.0	4.0
2001	106.8	6.8
2002	110.0	3.0
2003	113.8	3.5
2004	120.6	6.0
2005	129.8	7.7
2006	137.9	6.2
2007	141.8	2.8
2008	155.0	9.3

Source: National Statistics Office

USEFUL ADDRESSES

Public Organizations

Board of Investments
Industry & Investments Building
385 Sen. Gil Puyat Avenue
Makati City 1200
Tel: (63) 2 897 6682 (trunk line) / 890 1332 / 976 5700 / 896 1166
E-mail: nerbac@boi.gov.ph
Website: www.boi.gov.ph

Construction Industry Arbitration Commission (CIAC)
4F Jupiter I Bldg. 4th Floor Jupiter I Building
56 Jupiter St. Bel-Air Subd
Makati City
Tel: (63) 2 897 0853
Fax: (63) 2 897 9313
E-mail: ciapciac@info.com.ph
Website: www.dti.gov.ph

Construction Industry Authority of the Philippines (CIAP)
Jupiter I Bldg. 4th Floor Jupiter I Building
56 Jupiter St. Bel-Air Subd
Makati City
Tel: (63) 2 895 4424 (Officer-in-Charge) / 895 6826
Fax: (63) 2 897 9336
E-mail: ciap@info.com.ph
Website: www.dti.gov.ph

Department of Public Works and Highways (DPWH)
Bonifacio Drivc, Port Arca
Manila
Tel: (63) 2 304 3000 (trunk line) / 304 3221 (Office of the Secretary)
Fax: (63) 2 304 3455
E-mail: ebdane.jun@dpwh.gov.ph
Website: www.dpwh.gov.ph

Department of Trade and Industry (DTI)
4F Industry and Investment Bldg
385 Sen.Gil Puyat Avenue
Makati City
Tel: (63) 2 751 0384 (trunk line) / 751 3215 (Office of the Secretary)
Fax: (63) 2 895 6487
E-mail: web@dti.dti.gov.ph
Website: www.dti.gov.ph

National Housing Authority (NHA)
Elliptical Road, Quezon City
Tel: (63) 2 9284561 to 66 / 9217828 (Office of the Secretary)
Fax: (63) 2 9222058
Website: www.nha.gov.ph

National Statistics Office (NSO)
Solicarel Building
Ramon Magsaysay Blvd, Sta Mesa
Manila 1008
P.O. Box 779
Tel: (63) 2 716 0807 (Administrator) / 713 7074 / 715 6502
Fax: (63) 2 713 7073 / 715 6503
E-mail: C.Ericta@census.gov.ph
Website: www.census.gov.ph

Philippine Contractors Accreditation Board (PCAB)
Jupiter I Bldg. 4th Floor, Jupiter I Building
56 Jupiter St. Bel-Air Subd.
Makati City
Tel: (63) 2 895 4258
Fax: (63) 2 895 4220
E-mail: pcab@info.com.ph
Website: www.dti.gov.ph

Professional Regulation Commission (PRC)
P. Paredes., Cor. Morayta St.
Sampaloc, Manila
Tel: (63) 2 723 2250 (Office of the Commissioner)
Fax: (63) 2 735 4476
E-mail: webmaster@prc.gov.ph
Website: www.prc.gov.ph

Trade And Professional Associations

Association of Structural Engineers of the Philippines (ASEP)
Unit 713, 7th Floor, Future Point Plaza Condominium
112 Panay Avenue
Quezon City
Tel: (63) 2 410 0483
Fax: (63) 2 411 8606
Website: www.aseponline.org

Philippine Constructors Association (PCA)
3F Padilla Bldg. F. Ortigas Jr. Road
Ortigas Center, Pasig City
Tel: (63) 2 631 2778 / 631 3135
Fax: (63) 2 631 2788
E-mail: email@philconstruct.com
Website: www.philconstruct.com

Philippine Institute of Civil Engineers (PICE)
Unit 701, 703, 705 Future Point Plaza
Condominium I, #112 Panay Avenue
Quezon City 1100
Tel: (63) 2 448 7488 to 90 / 376 4215
Fax: (63) 2 448 7491 / 376 4255
E-mail: picenatl@pice.org.ph
Website: www.pice.org.ph

United Architects of the Philippines (UAP)
UAP National Headquarters
53 Scout Rallos Street, Diliman
Quezon City 1103
Tel: (63) 2 412 6403 / 412 6364 / 412 6374 / 412 3311
Fax: (63) 2 372 1796
E-mail: uap@compass.com.ph / uappdc@compass.com.ph
Website: www.united-architects.org

SINGAPORE

All data relate to 2007 unless otherwise indicated.

Population	
Population	4.6 mn
Urban population	100%
Population under 15	19%
Population 65 and over	9%
Average annual growth rate (2004 to 2007)	2.8%
Geography	
Land area	707 km^2
Agricultural area	2%
Capital city	Singapore
Economy	
Monetary unit	Singapore Dollar (S$)
Exchange rate (average fourth quarter 2008) to:	
the pound sterling	S$ 2.34
the US dollar	S$ 1.49
the euro	S$ 1.96
the yen x 100	S$ 1.55
Average annual inflation (1998 to 2007)	0.74%
Inflation rate	2.1%
Gross Domestic Product (GDP) at 2000 market prices	S$ 229.1 bn
GDP per capita	S$ 52,994
Average annual real change in (GDP) (1998 to 2007)	5.3%
Private consumption as a proportion of GDP	39.6%
Public consumption as a proportion of GDP	10.5%
Construction	
Gross value of construction output	S$ 17.8 bn
Net value of construction output	S$ 8.4 bn
Net value of construction output as a proportion of GDP	3.7%

THE CONSTRUCTION INDUSTRY

Construction Output

Towards the end of 2005, the Singapore construction sector has virtually recovered from the recession that lasted almost 8 years since the Asian Financial Crisis.

The construction demand in 2006 registered an increased of 46% reaching S$16.8 billion from S$11.5 billion in 2005. The upsurge continued and the demand in 2007 was S$24.5 billion, another 46% jump year-on-year, compared with 2006.

According to Building and Construction Authority's (BCA) data, the total construction demand for 2008 (based on actual contracts awarded) was S$34.6 billion, approximately 41% higher than 2007. As of now, the figure for 2008 has set a new record high compared to the last industry peak in 1997 at S$24.0 billion.

Construction demand from the private sector slowed down considerably from the 3rd Quarter 2008 onwards. This reflected the cautious sentiment prevalent in the market since the onset of the global financial crisis.

Public sector works, on the other hand, contributed substantially to the total construction demand for the 3rd and 4th Quarters of 2008. The most prominent being the awards of S$4 billion worth of infrastructure contracts for the construction of Marina Coastal Expressway and Downtown Line from LTA in the 4th Quarter of 2008.

The above represents a brief snapshot of the performance of the construction industry for 2008.

For 2009, BCA forecasts that the total construction demand is approximately between S$18 billion and S$24 billion, which represent a drop of 31% to 48% against the 2008 preliminary figures. Construction demand from the private sector projects is anticipated to decline drastically by almost 58% to 82% (from S$20.1 billion in 2008 to a mere S$3.6 billion – S$8.4 billion in 2009). Total public sector projects, on the other hand, are anticipated to have better growth with estimated volume of work at approximately S$14.4 billion – S$15.6 billion in 2009.

Total construction output in 2007 (in terms of certified progress payment) increased from S$12.9 billion to S$17.8 billion compared to 2006.

In tandem with the increase in construction demand, the construction output in 2008 (in terms of certified progress payment) rose further to S$22.4 billion (preliminary figure).

The table on the next page shows the value of construction contracts awarded in the public and private sectors.

VALUE OF CONSTRUCTION CONTRACTS AWARDED, 2005-2009*

Type of work	*2005*	*2006*	*2007*	*2008*	*2009**
					(S$ bn)
Private					
Residential	2,589.22	4,134.97	5,551.21	6,396.66	1.5 - 2.5
Commercial	902.20	2,304.90	5,125.69	8,356.91	0.9 - 1.7
Industrial	2,748.24	5,374.16	6,775.56	3,679.43	0.8 - 3.0
Institutional and others	511.34	456.30	403.91	916.04	0.2 - 0.6
Civil engineering	720.39	783.89	903.60	727.06	0.2 - 0.5
Total	7,471.39	13,054.22	18,759.97	20,076.10	3.6 - 8.4
Public					
Residential	1,134.80	1,163.41	1,809.83	4,200.75	2.8 - 3.0
Commercial	106.81	67.81	104.56	115.79	0.1 - 0.2
Industrial	370.83	136.47	191.96	44.73	0.1 - 0.1
Institutional and others	1,399.06	1,239.84	1,491.07	2,975.22	2.8 - 3.4
Civil engineering	973.26	1,134.94	2,102.49	7,206.17	8.7 - 8.9
Total	3,984.76	3,742.47	5,699.91	14,542.66	14.4 - 15.6
Total	11,456.15	16,796.69	24,459.88	34,618.76	18.0 - 24.0

Source: Building and Construction Authority as at 1 July 2009
** forecast*

As seen above, construction demand from the private sector projects for 2009 is anticipated to decline drastically by almost 58% to 82% (from S$20.1 billion in 2008 to S$3.6 billion – S$8.4 billion in 2009). The drop is almost identical to the decline of private sector work demand experienced in 1998 during the Asian Financial Crisis and it could be the worst in the last 10 years.

Total public sector projects, on the other hand, are anticipated with an estimated volume of work at approximately S$14.4 billion to S$15.6 billion in 2009. Majority of the public sector works are likely to come from infrastructure works as well as Downtown Line projects. The Government's commitment to inject more public sector projects in 2009/2010 amid the current global economic crisis would certainly aid to stabilise and boost the local construction industry.

Characteristics and Structure of the Industry

The construction industry in Singapore is supported by the Building and Construction Authority (BCA), a statutory board under the auspices of the Ministry of National Development (MND). BCA was established on 1 April 1999 following a merger between the Construction Industry Development Board and the Building Control Division of the Public Works Department. It has the primary role of developing and regulating the building and construction industry.

The Contractors Registry System (CRS)

The Contractors Registry System (CRS) was established in 1984 to register contractors who provide construction-related goods and services to the public sector. Only registered contractors are permitted to tender for public sector construction projects. Contractors applying for registration under the Contractors Registry must have the relevant experience, and financial, technical and management capability. This includes the employment of sufficient number of full time qualified technical personnel in the relevant disciplines. There are six (6) major groups of registration heads, namely Construction Workheads (CW), Construction Related Workheads (CR), Mechanical & Electrical Workheads (ME), Maintenance Workheads (MW), Supply Workheads (SY) and Regulatory Workheads (RW).

From 1 July 1999, ISO 9000 certification is a pre-requisite for contractors and consulting firms undertaking public sector construction projects valued at more than S$30 million. The ISO 9000 certificate must be awarded by a certification body accredited with the Singapore Accreditation Council (SAC).

In June 2006, BCA adopted a credit rating system to indicate the financial standing of larger construction firms in its Contractors Registry. The adopted credit rating system is similar to one developed by credit and business information bureau DP Information Group to assess the financial health of companies. However, the BCA system applies only to the larger construction companies (i.e. those in the top categories of A1, A2 and B1, which may tender for public sector construction contracts valued at S$30 million or more). Government agencies will use the DP credit rating as an additional reference on the financial standing of the firms when evaluating public tenders.

New Tendering Limits for BCA Registered Contractors

In 2002, BCA launched a Tender Limit Variable Component (TLVC) to the tender limits of all registration grades in the Contractors Registry System (CRS). TLVC is determined using the Tender Price Index (TPI) to reflect the impact of tender price movements on project value. Over the years the TPI has moved up significantly, hence resulting in a need to adjust the tender limits of the various CRS registration grades to better reflect the fluctuations in the construction costs in the market.

In November 2007, BCA announced that the tendering limits will be adjusted once a year on the first of July. The current new tendering limits shown on the next page are based on the latest TLVC updated on 19 January 2009.

Construction Workheads (CW01 & 02)	*A1*	*A2*	*B1*	*B2*	*C1*	*C2*	*C3*
Tendering limit (S$m)	unlimited	105.0	50.0	15.0	5.0	1.5	0.75
Specialist Workheads (CR, ME, MW & SY)	*L6*	*L5*	*L4*	*L3*	*L2*	*L1*	
Tendering limit (S$m)	unlimited	15.0	7.5	5.0	1.5	0.75	

Source: Building and Construction Authority

Construction Quality Assessment System (CONQUAS)

In 1989, BCA launched the Construction Quality Assessment System (CONQUAS) to provide a yardstick for the measure of the workmanship and quality achieved in a completed building project. CONQUAS has been widely used in both the public sector and the private sector. The system sets out the standards for the various categories of work. Points are awarded for work which falls within the acceptance standards or tolerances to derive the total CONQUAS score for the building. Since 1989, more than 2,240 public and private sector projects worth S$82 billion have been assessed by BCA.

As a de facto national quality yardstick for the industry, CONQUAS has been periodically fine-tuned to keep pace with changes in technology and quality demands of more sophisticated Singaporeans.

CONQUAS covers three main aspects of the general building works:

(1) Structural Works – relates to the structural integrity and safeguard on safety
(2) Architectural Works – deals with the aesthetic of the building such as finishes and other architectural components. This is the part where the quality and standard of workmanship are most visible
(3) Mechanical & Electrical (M&E) Works – concerns with the performance of selected mechanical and electrical services and installations to ensure the comfort of the building occupants.

A Bonus Scheme for Construction Quality (BSCQ) was promulgated in June 1998. Contractors who achieve a CONQUAS score exceeding the stipulated standard for the relevant building category will be paid a bonus by the Government. Similarly, they will be penalised if their quality of workmanship is poor.

Design work is undertaken mainly by architects and professional engineers. It is a pre-requisite to be registered with the Board of Architects, Singapore (a statutory board governing the practice of architects) before being allowed to practise in Singapore as an architect or use the designation ‘architect’. The Singapore Institute of Architects (SIA) is the only body representing professional architects in Singapore. There are currently about 855 registered architects who are SIA members. It serves as a link between the profession and the government and technical authorities on matters affecting the profession.

The designation ‘professional engineer’ is registered and protected by the Professional Engineers Board (a statutory board governing the practice of engineers) and no one is allowed to practise in Singapore as a professional engineer unless registered with the Board. There are two bodies representing professional engineers: the Institution of Engineers Singapore (IES) which is the national society for engineers in Singapore, and the Association of Consulting Engineers Singapore (ACES) which is the national association for consulting engineers practices in Singapore. As of 2008, the IES has about 5,200 members all of whom are registered engineers of various disciplines. Membership to ACES is a privilege only open to practising consulting firms, and the current number of consulting firms who are members of ACES is 134. Altogether, there are about 500 licensed Professional Engineers working in these consulting firms.

The Singapore Institute of Surveyors and Valuers (SISV) is the only body representing the surveying and valuation profession in Singapore. There are three divisions in the institute: quantity surveying, land surveying and valuation and general practice. As of 2008, there are about 1,207 members.

Minimum Buildability Score

The legislation of buildable design came into effect on 1 January 2001. Projects submitted for planning after 1 January 2001 will be affected by the legislation and are required to comply with the minimum buildability score as stipulated in the Code of Practice for Buildable Design.

Over the years, the minimum buildability scores have been progressively raised.

In September 2005, all new building projects with gross floor area equal or greater than 2,000 m^2 are required to comply with the minimum buildability score.

The minimum buildability score requirement shall also apply to addition and alteration works to an existing building if the building works involve the construction of new floor and/or reconstruction of existing floor for which their total gross floor area is 2,000 m^2 or more.

The Code of Practice for Buildable Design (September 2005 edition) has stipulated the minimum buildability scores for building works with applications for planning permission made on or after 1 September 2005, 1 January 2007 and 1 August 2008 for different building types as follows:

Category of Building Work / Development	Minimum Buildability Score								
	1 September 2005			1 January 2007			1 August 2008		
	$2,000m^2 \leq$ GFA < $5,000m^2$	$5,000m^2 \leq$ GFA < $25,000m^2$	GFA ≥ $25,000m^2$	$2,000m^2 \leq$ GFA < $5,000m^2$	$5,000m^2 \leq$ GFA < $25,000m^2$	GFA ≥ $25,000m^2$	$2,000m^2 \leq$ GFA < $5,000m^2$	$5,000m^2 \leq$ GFA < $25,000m^2$	GFA ≥ $25,000m^2$
Residential (landed)	57	59	62	60	62	65	60	65	68
	57 (A&A work with existing building)								
Residential (non-landed)	63	65	68	66	68	71	67	72	75
	60 (A&A work with existing building)								
Commercial	65	72	75	67	74	77	69	74	77
	62 (A&A work with existing building)								
Industrial	67	74	77	69	74	77	69	74	77
	62 (A&A work with existing building)								
School	64	69	72	64	69	72	64	69	72
	60 (A&A work with existing building)								
Institutional and others	60	66	69	60	66	69	60	66	69
	60 (A&A work with existing building)								

Building and Construction Industry Security of Payment Act 2004

The Building and Construction Industry Security of Payment Act ("BCISOP Act") 2004 came into force in Singapore on 1 April 2005.

The BCISOP Act was enacted to facilitate payments for construction work done or for related goods and services supplied, under a contract in the building and construction industry relating to construction work which includes professional consultancy services.

The underlining objectives of the BCISOP Act are to:

- improve cash flow by expediting payment;
- provide a statutory entitlement to progress payments to contractors, subcontractors and suppliers for work carried out, even if no such entitlement is provided in their contract;
- provide a procedure of adjudication to claim payment; which is intended to be a more cost and time efficient way of resolving disputes on payment claims between the parties; and
- provide remedies when adjudicated amount not paid.

The BCISOP Act provides a new regime of claim, adjudication and enforcement procedures which include the right to suspend work for non-payment. It also renders unenforceable "pay when paid" provisions in contracts. This benefits the subcontractors and suppliers.

The BCISOP Act is supplemented by the BCISOP Regulations 2005 where the Act confers power on the MND to set out the regulations to facilitate the implementation of the Act.

However, the BCISOP Act is not applicable to construction work and goods and services relating to residential property (defined under the Residential Property Act) not requiring approval under the BCA Building Control Act, construction work carried outside Singapore, goods and services supplied to construction work outside Singapore and employment contracts.

Legislation on Environmental Sustainability for Buildings

The BCA Green Mark Scheme was launched in 2005 to promote the development of environmentally sustainable buildings.

With the objective to push for a wider adoption of green building technologies, BCA has enhanced the Building Control Act to include a minimum environmental sustainability standard that is equivalent to the Green Mark Certified Level for new buildings and existing ones that undergo major retrofitting.

The new Building Control (Environmental Sustainability) Regulations 2008 stipulates a minimum Green Mark score of 50 for relevant building works since 15 April 2008. It applies to:

All new building works with Gross Floor Area of 2,000 m^2 or more;

- Additions or extensions to existing buildings which involve increasing Gross Floor Area of the existing buildings by 2,000 m^2 or more;
- Building works which involve major retrofitting to existing buildings with existing Gross Floor Area of 2,000 m^2 or more.

The requirements on environmental sustainability of buildings is be integrated with the Building Plan process. The Qualified Person (QP) who submits the Building Plan and the other appropriate practitioners will be responsible for assessing and scoring the building works under their charge using the criteria and scoring methodology spelled out in the Code for Environmental Sustainability of Buildings.

Under the Legislation, Green Mark assessments are no longer required to be conducted as an independent third party certification. Compliance to the regulations will be based on QP's declaration and random audit and site checks prior or during Temporary Occupation Permit (TOP). However, third party assessment by BCA will be conducted for projects targeting Green Mark Gold rating and above.

In line with the above new regulations, the BCA Green Mark Assessment Criteria for new buildings has been revised and took effect from 31 January 2008 onwards. It is sub-divided into 2 categories namely:

BCA Green Mark for Non-Residential Buildings
BCA Green Mark for Residential Buildings

The Green Mark rates the environmental friendliness of a building based on a point scoring approach. Depending on the score, the rating is categorized in four levels – Platinum, Gold^{Plus}, Gold and Certified. The salient difference from Version 2 to Version 3 is summarised below:

Green Mark Rating	*Green Mark Points (Version 2) 2006*	*Green Mark Points (Version 3) With effect from 31 Jan 2008*
Green Mark Platinum	85 and above	90 and above
Green Mark Gold^{Plus}	80 to <85	85 to <90
Green Mark Gold	70 to <80	75 to <85
Green Mark Certified	50 to <70	50 to <75

Clients and Finance

Private sector projects are usually undertaken by private developers and institutions and financed in a number of ways. This includes loans from banks and financial institutions and the developer's own funds. Public sector projects are mainly undertaken by the Housing and Development Board (HDB) and the MND.

HDB was established as a statutory board in 1960. Its main activity is the provision of suitable housing for lower and middle income groups. As of 2007, the HDB stock of flats housed about 84% of the population in Singapore. The sources of finance for the HDB for capital expenditure are mainly government financed loans.

Selection of Design Consultants

There are no prescribed criteria or published guidelines for the selection of design consultants in Singapore. In the public sector, a prequalification exercise through quality/fee methodology requiring the submission of credentials, including relevant experience, followed by interviews is usually adopted. A design competition may sometimes be held. In the private sector, the design competition method of selection is adopted on some occasions. Clients usually select the design consultants known to them or those who have a reputation for a specific type of building.

Most of the professional bodies publish fee scales though these are not mandatory and are rarely used.

Contractual Arrangements

For all public sector construction projects, there are two forms of contract that are currently in use:

- Public Sector Standard Conditions of Contract for Construction Works
- Public Sector Standard Conditions of Contract for Design and Build

The latest edition for both the above forms was published in December 2008.

Public sector contracts are generally awarded on the basis of the lowest compliant tenders submitted by contractors registered with BCA although price is sometimes not the sole criterion.

For contracts exceeding S$3.0 million in value, evaluation of tenders would be based on the Price Quality Method (PQM). The weightage between price and quality will range from 60:40 to 80:20, depending on the complexity of the project. Under Quality, various criteria like track record, financial capacity and safety performance would be included.

Most private sector projects use the SIA standard forms of building contract which comprise the Measurement Contract (for use with bills of quantities); the Lump Sum Contract (where quantities does not form part of the contract); the Minor Works Contract and the Conditions of Subcontract.

With an increasing inclination of projects adopting the design and build approach, the Real Estate Developers' Association of Singapore (REDAS) first launched a design and build form in August 2001 which was specially drafted to meet the industry's demand. This form of contract is currently in its 2nd Edition which was published in October 2007.

In the private sector, contracts are awarded either through competition or by negotiation (especially during period of high pricing and lack of contractors in the construction industry) or a combination of both.

Liability and Insurance

Professional indemnity insurance is compulsory for architectural and engineering firms practising as limited companies. In the case of partnerships, it is not compulsory, although most of the large practices do hold professional indemnity insurance.

Development Control and Standards

The Urban Redevelopment Authority (URA) is the National Planning and Conservation Authority regulating and facilitating the physical development of Singapore. Most types of development require written planning permission but certain types are not considered material or are specifically exempted and thus do not require planning permission. The URA has published a series of development control handbooks to guide and inform applicants of the procedures to be observed in submitting development applications.

BCA is responsible for setting and monitoring building regulations which cover, for example, structural integrity, lighting, ventilation and thermal transmission. The Building Control Act requires a developer to appoint an Accredited Checker (AC) to check and endorse all structural designs and calculations prior to submission to BCA for approval. ACs are independent registered professional engineers approved by BCA.

The Fire Safety & Shelter Bureau (FSSB) of the Singapore Civil Defence Force is responsible for all building inspections prior to the issuance of the TOP or Certificate of Statutory Completion (CSC) by the Building Control Division

(BCD). In 1994, FSSB introduced the Registered Inspectors Scheme to speed up such building inspections. Under this scheme, the Registered Inspectors (RI) inspect the buildings on behalf of the authority and issue Inspection Certificates. This certificate is required before the architect can apply for building occupation.

The Land Transport Authority (LTA) is responsible for planning and implementation of road and other transport systems such as Mass Rapid Transit (MRT) and Light Rail Transit (LRT) in Singapore. LTA works closely with URA and HDB to ensure that roads and other transport systems are well planned and properly integrated with the urban development.

The Singapore Productivity and Standards Board (SPRING) draws up and promulgates the Singapore Standards (SS), the standard specifications for products, and it is usual for manufacturers to comply with these. Architects and other building professionals generally follow the recommendations of the SS when specifying building products.

CONSTRUCTION COST DATA

Cost of Labour

The figures below are typical of labour costs in Singapore as at June 2007. The wage rate is the basis of an employee's income, while the cost of labour indicates the cost to a contractor of employing that employee. The difference between the two covers a variety of mandatory and voluntary contributions – a list of items which could be included is given in section 2.

	Wage rate (per day) S$	*Cost of labour (per day) S$*	*Number of hours worked per year*
Site operatives			
Bricklayer	42.00	53.00	2,288
Carpenter	52.00	65.00	2,288
Plumber	55.00	69.00	2,288
Electrician	68.00	85.00	2,288
Structural steel erector	60.00	75.00	2,288
Welder	60.00	75.00	2,288
Labourer	38.00	48.00	2,288
Equipment operator	70.00	87.00	2,288
Site supervision	*(per month)*	*(per month)*	
General foreman	2,270	3,310	2,288
Trades foreman	2,650	3,870	2,288
Clerk of works	2,540	3,710	2,288
Resident engineer	4,000	5,840	2,280
Contractors' personnel			
Site manager	4,700	6,860	2,288
Site engineer	3,270	4,770	2,288

	Wage rate (per month) S$	*Cost of labour (per month) S$*	*Number of hours worked per year*
Site quantity surveyor	3,000	4,380	2,288
Consultants' personnel			
Senior architect	5,750	8,400	2,080
Senior engineer	4,800	7,010	2,080
Senior surveyor	4,500	6,570	2,080
Qualified architect	3,750	5,480	2,080
Qualified engineer	3,300	4,820	2,080
Qualified surveyor	3,000	4,380	2,080

Cost of Materials

The figures that follow are the costs of main construction materials, delivered to site in Singapore, as incurred by contractors in the fourth quarter of 2008. These assume that the materials would be in quantities as required for a medium sized construction project and that the location of the works would be neither constrained nor remote.

All the costs in this section exclude goods and services tax (GST).

	Unit	*Cost S$*
Cement and aggregate		
Ordinary portland cement in 50kg bags	tonne	123.00
Coarse aggregates for concrete	tonne	26.00
Fine aggregates for concrete	tonne	33.00
Ready mixed concrete (Grade 30)	m^3	121.00
Ready mixed concrete (Grade 20)	m^3	119.00
Steel		
Mild steel reinforcement	tonne	1,091.00
High tensile steel reinforcement	tonne	1,091.00
Structural steel sections	tonne	1,590.00
Bricks and blocks		
Common bricks (215 x 102.5 x 65mm)	1,000	220.00
Good quality facing bricks (215 x 102.5 x 65mm)	1,000	500.00
Hollow concrete blocks (390 x 190 x 100mm)	1,000	700.00
Timber and insulation		
Hardwood for joinery	m^3	1,040.00
Exterior quality plywood (12mm)	m^2	8.00
Plywood for interior joinery (12mm)	m^2	8.00
50mm thick quilt insulation (16kg/m^3)	m^2	3.50
50mm thick rigid slab insulation (60kg/m^3)	m^2	12.00

	Unit	*Cost S$*
Hardwood internal door complete with frame and ironmongery	each	750.00
Glass and ceramics		
Float glass (10mm)	m^2	40.00
Sealed double glazing units (6/12/6) including frame	m^2	400.00
Plaster and paint		
Good quality ceramic wall tiles (300 x 300 x 8mm)	m^2	30.00
Plasterboard (13mm thick) – gypsum	m^2	5.00
Emulsion paint in 5 litre tins	litre	3.50
Gloss oil paint in 5 litre tins	litre	7.00
Tiles and paviors		
Clay floor tiles (100 x 200 x 8mm)	m^2	25.00
Vinyl floor tiles (300 x 300 x 2mm)	m^2	20.00
Clay roof tiles	1,000	2,500.00
Precast concrete roof tiles	1,000	N.A
Drainage		
WC suite complete	each	325.00
Wash hand basin complete	each	130.00
100mm diameter clay drain pipes	m	20.00
150mm diameter cast iron drain pipes (medium grade)	m	N.A

Unit Rates

The descriptions on the next page are generally shortened versions of standard descriptions listed in full in section 4. Where an item has a two digit reference number (e.g. 05 or 3), this relates to the full description against that number in section 4. Where an item has an alphabetic suffix (e.g.12A or 34B) this indicates that the standard description has been modified. Where a modification is major the complete modified description is included here and the standard description should be ignored; where a modification is minor (e.g. the insertion of a named hardwood) the shortened description has been modified here but, in general, the full description in section 4 prevails.

The unit rates below are for main work items on a typical construction project in the city area in the fourth quarter of 2008. The rates include all necessary labour, materials and equipment.

All the rates in this section exclude goods and services tax (GST).

		Unit	Rate S$
Excavation			
01	Mechanical excavation of foundation trenches	m^3	23.00
02	Hardcore filling making up levels	m^3	50.00
Concrete work			
04	Plain in situ concrete in strip foundations in trenches (Grade 20)	m^3	191.00
05	Reinforced in situ concrete in beds (Grade 30)	m^3	195.00
06	Reinforced in situ concrete in walls (Grade 30)	m^3	195.00
07	Reinforced in situ concrete in suspended floors or roof slabs (Grade 30)	m^3	195.00
08	Reinforced in situ concrete in columns (Grade 30)	m^3	195.00
09	Reinforced in situ concrete in isolated beams (Grade 30)	m^3	195.00
Formwork			
11A	Waterproof plywood formwork to concrete walls	m^2	39.00 - 44.00
12A	Waterproof plywood formwork to concrete columns	m^2	39.00 - 44.00
13A	Waterproof plywood formwork to horizontal soffits of slabs	m^2	39.00 - 44.00
Reinforcement			
14	Reinforcement in concrete walls	tonne	1,800.00 - 2,000.00
15	Reinforcement in suspended concrete slabs	tonne	1,800.00 - 2,000.00
16	Fabric reinforcement in concrete beds	m^2	9.00
Steelwork			
17	Fabricate, supply and erect steel framed structure	tonne	4,500.00 - 5,500.00
18	Framed structural steelwork in universal joist sections	tonne	4,500.00 - 5,500.00
19	Structural steelwork lattice roof trusses	tonne	4,500.00 - 5,500.00
Brickwork and blockwork			
20	Precast lightweight aggregate hollow concrete block walls	m^2	28.00 - 30.00
21A	Solid (perforated) common brick	m^2	36.00 - 42.00
23	Facing bricks (half brick thick)	m^2	48.00 - 54.00
Roofing			
24	Concrete interlocking roof tiles 430 x 380mm	m^2	N.A

		Unit	*Rate S$*
25	Plain clay roof tiles 260 x 160mm	m²	60.50
29	3 layers glass-fibre based bitumen felt roof covering	m²	36.00
33	Troughed galvanized steel roof cladding	m²	42.00
Woodwork and metalwork			
34	Preservative treated sawn hardwood 50 x 100mm	m	12.00
35	Preservative treated sawn hardwood 50 x 150mm	m	18.00
37	Two panel glazed door in Kapur hardwood, size 850 x 2100mm	each	800.00
38	Solid core half hour fire resisting hardwood internal flush door, size 900 x 2100mm	each	1,000.00
39	Aluminium double glazed window, size 1200 x 1200mm	each	750.00
41	Hardwood skirtings	m	8.50
Plumbing			
42	Galvanished half round eaves gutter	m	32.00
43	UPVC rainwater pipes	m	65.00
44	Light gauge copper cold water tubing	m	19.00
45	High pressure plastic pipes for cold water supply	m	20.00
47	UPVC soil and vent pipes	m	65.00
48	White vitreous china WC suite	each	495.00
49	White vitreous china wash hand basin	each	275.00
51A	Stainless steel double bowl sink and double drainer	each	440.00
Electrical work			
52	PVC insulated and copper sheathed cable	m	3.80
53	13 amp unswitched socket outlet	each	22.00
54	Flush mounted 20 amp, 1 way light switch	each	95.00
Finishings			
55A	2 coats cement and sand (1:4) plaster on brick walls	m²	20.00
56	White glazed tiles on plaster walls	m²	40.00
57	Red clay quarry tiles on concrete floors	m²	70.00
58	Cement and sand screed to concrete floors	m²	22.50
59	Thermoplastic floor tiles on screed	m²	25.00
60	Mineral fibre tiles on concealed suspension system	m²	38.00
Glazing			
61	Glazing to wood	m²	42.00

		Unit	Rate S$
Painting			
62	Emulsion on plaster walls	m^2	3.50 - 4.00
63	Oil paint on timber	m^2	9.40

Approximate Estimating

The building costs per unit area that follow are averages incurred by building clients for typical buildings in Singapore as at the fourth quarter of 2008. They are based upon the total floor area of all storeys, measured between external walls and without deduction for internal walls.

Approximate estimating costs generally include mechanical and electrical installations but exclude furniture, loose or special equipment, and external works; they also exclude fees for professional services. The costs shown are for specifications and standards appropriate to Singapore and this should be borne in mind when attempting comparisons with similarly described building types in other countries. A discussion of this issue is included in section 2. Comparative data for countries covered in this publication, including construction cost data, are presented in Part Three.

Approximate estimating costs must be treated with caution; they cannot provide more than a rough guide to the probable cost of building. All the rates in this section exclude goods and services tax (GST).

	Cost m^2 S$	Cost ft^2 S$
Industrial buildings		
Factories for letting	1,450	135
Factories for owner occupation (light industrial use)	1,600	149
Factories for owner occupation (heavy industrial use)	1,950	181
Factory/office (high tech) for letting (shell and core only)	1,800	167
Factory/office (high tech) for letting (ground floor shell, first floor offices)	2,100	195
Factory/office (high tech) for owner occupation (controlled environment, fully finished)	2,250	209
High tech laboratory workshop centres (air-conditioned)	1,950	181
Warehouses, low bay (6 to 8m high) for letting	1,300	121
Warehouses, low bay for owner occupation	1,400	130
Warehouses, high bay for owner occupation	2,100	195
Administrative and commercial buildings		
Offices for letting, 5 to 10 storeys, non air-conditioned	2,100	195
Offices for letting, 5 to 10 storeys, air-conditioned	2,300	214
Offices for letting, high rise, air-conditioned	2,500	232
Offices for owner occupation high rise, air-conditioned	2,650	246

	Cost m² S$	*Cost ft² S$*
Prestige/headquarters office, 5 to 10 storeys, air-conditioned	2,850	265
Prestige/headquarters office, high rise, air-conditioned	3,050	283
Health and education buildings		
General hospitals (100 beds)	3,550	330
Private hospitals (100 beds)	3,750	348
Health centres	2,100	195
Primary/junior schools	1,500	139
Secondary/middle schools	1,700	158
University	2,500	232
Recreation and arts buildings		
Theatres (less than 500 seats)	4,000	372
Sports halls including changing and social facilities	2,200	204
Swimming pools (international standard) (Olympic size)	each	2,100,000
Swimming pools (schools standard) including changing facilities	each	1,900,000
City centre/central libraries	2,300	214
Branch/local libraries	2,200	204
Residential buildings		
Private/mass market single family housing 2 storey detached/semi detached (multiple units)	3,600	334
Purpose designed single family housing 2 storey detached (single unit)	4,650	432
Social/economic apartment housing, high rise (with lifts)	1,650	153
Private sector apartment building (standard specification)	3,250	302
Private sector apartment buildings (luxury)	4,950	460
Student/nurses halls of residence	2,200	204
Homes for the elderly (shared accommodation)	2,150	200
Hotel, 5 star, city centre	4,550	423
Hotel, 3 star, suburbs	3,350	311

Goods and Services Tax (GST)

The standard rate of goods and services tax (GST) is currently 7%, chargeable on general building work.

EXCHANGE RATES AND INFLATION

The combined effect of exchange rates and inflation on prices within a country and price comparisons between countries is discussed in section 2.

Exchange Rates

The graph below plots the movement of the Singapore dollar against sterling, the euro, the US dollar and 100 Japanese yen since 1998. The values used for the graph are quarterly and the method of calculating this is described and general guidance on the interpretation of the graph provided in section 2. The average exchange rate in the fourth quarter of 2008 was S$2.34 to pound sterling, S$1.96 to euro, S$1.49 to US dollar and S$1.55 to 100 Japanese yen.

THE SINGAPORE DOLLAR AGAINST STERLING, EURO, US DOLLAR AND 100 JAPANESE YEN

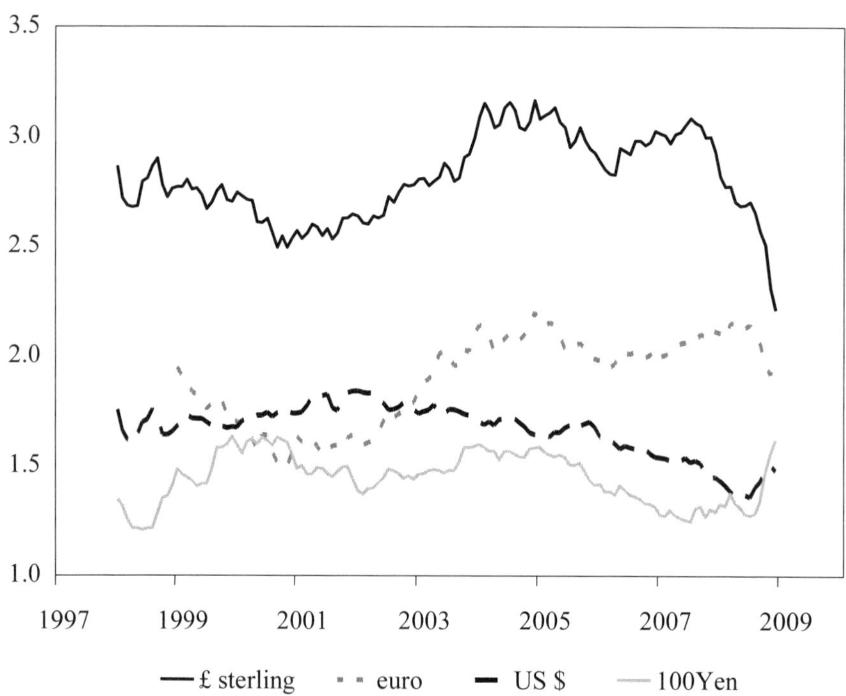

Price Inflation

The table below presents tender price index in Singapore since 1998.

TENDER PRICE INDEX (BASE YEAR AT 1998)

Year	*Tender price index*	*Year*	*Tender price index*
1998	100	2004	94.7
1999	89.7	2005	97.7
2000	88.4	2006	105.5
2001	89.1	2007	128.5
2002	88.5	2008^p	146.5
2003	89.3		

p= preliminary

USEFUL ADDRESSES

Public Organizations

Board of Architects Singapore
5 Maxwell Road
1st Storey Tower Block
MND Complex
Singapore 069110
Tel: (65) 6222 5295
Fax: (65) 6222 4452
E-mail: boarch@singnet.com.sg
Website: www.boa.gov.sg

Building and Construction Authority
5 Maxwell Road, #16-00
Tower Block MND Complex
Singapore 069110
Tel: (65) 6325 7720
Fax: (65) 6325 4800
E-mail: bca_enquiry@bca.gov.sg
Website: www.bca.gov.sg

Land Transport Authority
No. 1 Hampshire Road
Singapore 219428
Tel: (65) 1800 6225 5582
Website: www.lta.gov.sg

Ministry of the Environment and Water Resources
Environment Building
40 Scotts Road, #24-00
Singapore 228231
Tel: (65) 6731 9000
Fax: (65) 6731 9456
E-mail: mewr_feedback@mewr.gov.sg
Website: www.env.gov.sg

Ministry of Manpower
18 Havelock Road
Singapore 059764
Tel: (65) 6438 5122
Fax: (65) 6534 4840
Website: www.mom.gov.sg

Ministry of National Development
5 Maxwell Road, #21 / 22-00
Tower Block MND Complex
Singapore 069110
Tel: (65) 6222 1211
Fax: (65) 6325 7254
E-mail: mnd_hq@mnd.gov.sg
Website: www.mnd.gov.sg

Professional Engineers Board
5 Maxwell Road
1st Storey Tower Block, MND Complex
MND Complex
Singapore 069110
Tel: (65) 6222 9293
Fax: (65) 6222 9471
E-mail: registrar@peb.gov.sg
Website: www.peb.gov.sg

Singapore Department of Statistics
100 High Street, #05-01
The Treasury
Singapore 179434
Tel: (65) 6332 7686
Fax: (65) 6332 7689
E-mail: info@singstat.gov.sg
Website: www.singstat.gov.sg

Singapore Productivity and Standards Board (SPRING Singapore)
2 Bukit Merah Central
Singapore 159835
Tel: (65) 6278 6666
Fax: (65) 6278 6667
Website: www.spring.gov.sg

Urban Redevelopment Authority
The URA Centre
45 Maxwell Road
Singapore 069118
Tel: (65) 6221 6666
Fax: (65) 6227 5069
Website: www.ura.gov.sg

Trade And Professional Associations

Association of Consulting Engineers Singapore
70 Palmer Road, #04-06
Palmer House
Singapore 079427
Tel: (65) 6324 2682
Fax: (65) 6324 2581
E-mail: acesing@starhub.net.sg
Website: www.aces.org.sg

The Institution of Engineers Singapore
70 Bukit Tinggi Road
Singapore 289758
Tel: (65) 6469 5000
Fax: (65) 6467 1108
Website: www.ies.org.sg

Real Estate Developers' Association of Singapore (REDAS)
190 Clemenceau Avenue, #07-01
Singapore Shopping Centre
Singapore 239924
Tel: (65) 6336 6655
Fax: (65) 6337 2217
E-mail: enquiry@redas.com
Website: www.redas.com

Singapore Business Federation
10 Hoe Chiang Road, #22-01
Keppel Towers
Singapore 089315
Tel: (65) 6827 6828
Fax: (65) 6827 6807
Website: www.sbf.org.sg

Singapore Contractors Association Ltd
Construction House
1 Bukit Merah Lane 2
Singapore 159760
Tel: (65) 6278 9577
Fax: (65) 6273 3977
E-mail: enquiry@scal.com.sg
Website: www.scal.com.sg

Singapore Institute of Arbitrators
3 St. Andrew's Road
City Hall Building Level 3
Singapore 178958
Tel: (65) 6332 5132
Fax: (65) 6338 2245
E-mail: siarb@siarb.org.sg
Website: www.sairb.org.sg

Singapore Institute of Architects
79B Neil Road
Singapore 088904
Tel: (65) 6226 2668
Fax: (65) 6226 2663
E-mail: info@sia.org.sg
Website: www.sia.org.sg

Singapore Institute of Building Limited
70 Palmer Road, #03-09C
Palmer House
Singapore 079427
Tel: (65) 6223 2612
Fax: (65) 6223 2568
E-mail: josephine@sib.com.sg
Website: www.sib.com.sg

Singapore Institute of Surveyors and Valuers
20 Maxwell Road, #10-09B
Maxwell House
Singapore 069113
Tel: (65) 6222 3030
Fax: (65) 6225 2453
E-mail: sisv.info@sisv.org.sg
Website: www.sisv.org.sg

DAVIS LANGDON & SEAH KOREA CO. LTD

Davis Langdon & Seah Korea Co. Ltd manages risk and deliver value to clients contemplating investment in or development of property or construction project.

TYPICAL PROJECT STAGES AND INTEGRATED SERVICES

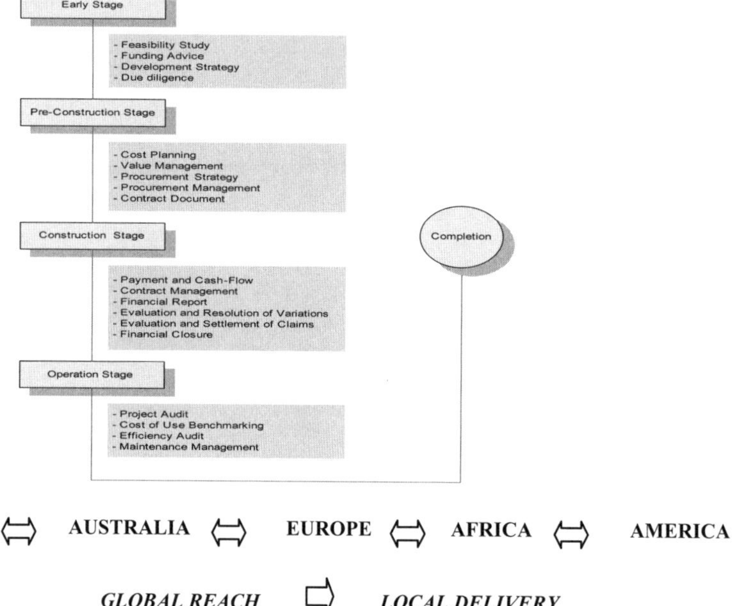

ASIA ⇔ AUSTRALIA ⇔ EUROPE ⇔ AFRICA ⇔ AMERICA

GLOBAL REACH ⇨ LOCAL DELIVERY

SEOUL OFFICE
#429 G-Five Central Plaza
1685-8 Seocho 4-Dong
Seocho-Gu, Seoul
Korea 137-882
Tel : (82-2) 543 3888
Fax: (82-2) 533 0234

SOUTH KOREA

All data relate to 2008 unless otherwise indicated.

Population	
Population	48.6 mn
Urban population (2006)	82%
Population under 15	17.5%
Population 65 and over	10.3%
Average annual growth rate (2000 to 2008)	0.42%
Geography	
Land area	99,678.12 km^2
Agricultural area	22%
Capital city	Seoul
Population of capital city (2006)	9.7 mn
Economy	
Monetary unit	South Korean Won (W)
Exchange rate (average fourth quarter 2008) to:	
the pound sterling	W 2,013
the US dollar	W 1,102
the euro	W 1,606
the yen x 100	W 1,076
Average annual inflation (2002 to 2008)	2.8%
Inflation rate	4.2%
Gross Domestic Product (GDP)	W 883,219 bn
GDP per capita	W 20,112,229
Average annual real change in (GDP) (2000 to 2008)	4.93%
Private consumption as a proportion of GDP (2007)	44.8%
Public consumption as a proportion of GDP (2007)	11.2%
Investment as a proportion of GDP (2007)	25%
Construction	
Gross value of construction output	W 242,012.5 bn
Net value of construction output (2007)	W 161,257.3 bn
Net value of construction output as a proportion of GDP (2007)	18%

THE CONSTRUCTION INDUSTRY

Construction Output

The gross output of construction in 2008 was about W242,012.5 billion, equivalent to US$219 billion, or 27.4% of GDP. The net output of construction in 2007 was about W161,257.3 billion, equivalent to US$146 billion, or 18.2% of GDP.

Construction investment which accounts for 14.2% of Korea's GDP has plunged from 7.9% to -0.4% during these couple of years. In 2007, first half fiscal spending had increased civil engineering works investment of 5.4% which influenced overall construction investment to 3.5%. But in the third quarter, construction investment has fallen to 1.1% due to frontload public budget drying up. In the same vein, according to the Korea National Statistical Office, while public sector recorded 14.4% growth, private sector had only registered 1.7% increase for the first half of 2007 and 3.1% increase for the third quarter of 2007.

TRENDS IN THE INCREASE RATES OF CONSTRUCTION INVESTMENT INDICES (UNIT YEAR-ON-YEAR, %)

Category	*2003*	*2004*	*2005*	*2006*	*2007*	
					First-Half	*Third-Quarter*
Construction Investment	7.9	1.1	-0.2	-0.4	3.5	1.1
Building Construction	11.6	2.0	-1.4	-2.0	2.3	-
Residential Construction	9.0	4.7	2.2	-2.4	0.1	-
Non-Residential Construction	14.2	-0.7	-5.0	-1.6	4.8	-
Civil Engineering Works	3.0	-0.2	1.5	2.0	5.4	-
Construction Works Completed	16.6	11.1	4.1	3.7	5.8	3.6
(by) Public Sector	10.7	4.9	-3.7	1.9	14.4	1.6
(by) Private Sector	20.6	14.8	7.5	4.9	1.7	3.1

Source: Bank of Korea, Korea National Statistical Office

Since 2008, 365 construction companies have gone bankrupt, soaring by 40.4% against the same period of last year (260), according to construction association of Korea.

Overdue rate of construction industry is the highest in industries (0.97%). Most construction companies have been looking for any resources to improve their cash flow by HR restructuring and selling their property, their sub-company and resort.

As Korea ratings announced that it lowered ratings of 25 construction companies, it slashed credit ratings on construction companies en masse.

Domestic construction received orders in both private and public sectors in October 2008 recorded the amount of W9,722 billion, a 21.5% fall compared to the same period of previous year.

Reconstruction and redevelopment received orders in October 2008 fell by 56.8% against the same month of previous year; 39.9% in reconstruction and 70% in redevelopment because of construction market downturns.

Sector-By-Sector Analysis of Construction Investment

A) Residential

The housing boom in South Korea ended in 2005. The market moderated slightly in the beginning of 2006 and began to descend into negative territory in the second quarter of 2006, where it has remained except for the first quarter of 2007.

TREND IN THE INCREASE RATE OF RESIDENTIAL CONSTRUCTION INVESTMENT

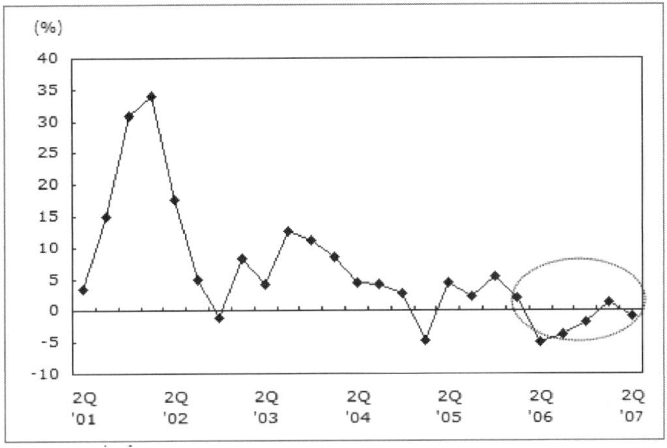

Source : Bank of Korea

Bank of Korea turned from easy monetary policy to monetary tightening in the second half of 2005, raising its target rate 1.5% points from 3.5% in October 2005 to 5.0% in August 2007. The hostile government policies and higher interest rates fractured the housing business cycle. Between 2004 and 2006, an annual average 460,000 to 470,000 units were provided. It was a major drop compared to the average 594,000 units in the years between 2001 and 2003. The reduction was seen largely in the Seoul metropolitan area, where nearly 50% of the national population resides. Seoul metropolitan area was under reconstruction regulations, inevitably decreasing residential construction investment.

HOUSING SUPPLY (BASED ON AUTHORIZATION)

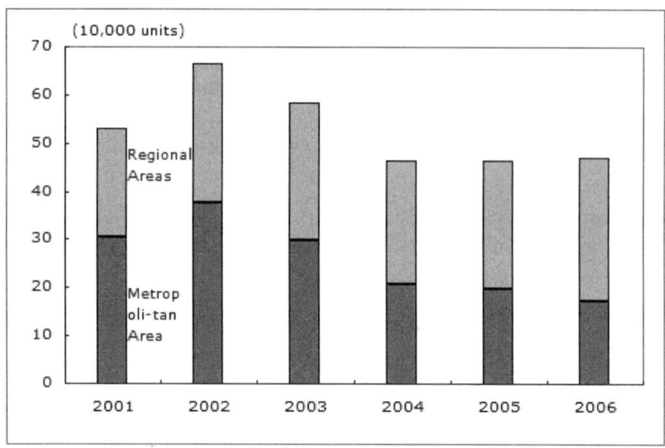

Source : Ministry of Construction and Transportation

The decline in residential construction has led to an oversupply in other areas of Korea. In 2006, the weight of residential construction outside of the Seoul metropolitan area rose to 63.4% of the national total from 43.6% in 2002. Regional area already had more than adequate housing supply. Furthermore, it coupled with further regulations on housing ownership and lowered housing demand in regional areas ahead of the Seoul area.

B) Non-Residential

Non-residential construction has been on a recovery track, thanks to rising consumers' expenditure and firms' facility investment. Non-residential construction comprises mostly shops, offices, factories and warehouses, which is why it responds to domestic demand and trails GDP by two quarters. This is evident if the history of non-residential construction is examined. In 2000 and 2001, consumption soared, boosting non-residential construction up to the first half of 2004. However, from the second half of 2004 to the first half of 2006, domestic demand weakened, hence resulted in the reduction in the construction orders. Private consumption started to recover from the first half of 2005. The increase rate of non-residential construction investment has steadily been growing since the third quarter of 2006.

NON-RESIDENTIAL CONSTRUCTION AND PRIVATE CONSUMPTION

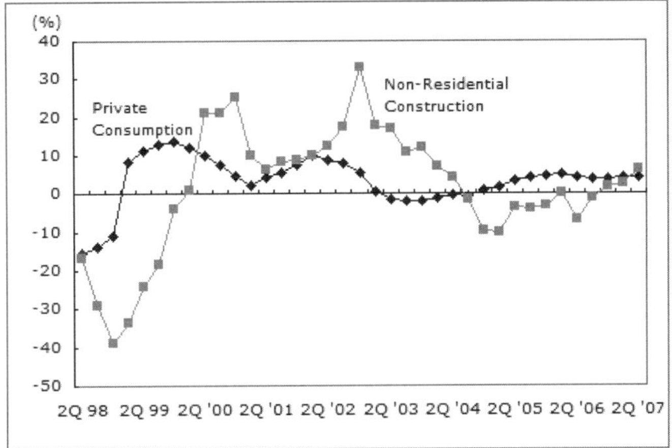

Source : Bank of Korea

C) Civil Engineering

Civil engineering works increased considerably in the first half of 2007. In 2006, the public sector accounted for 59.6% of the total construction orders in this area and the private sector accounted for 26.3%. Private-funded public projects that take up 14.1% have been announced and driven by the government. Civil engineering works tend to be more sensitive to SOC (Social Overhead Capital) construction and fiscal spending than the overall business cycle of the whole economy.

The government frontloaded the budget entering 2007, in an attempt to boost the economy, expanding public works. This followed a recent trend in which the government has responded actively to the business cycle, causing fiscal spending on public works to fluctuate widely throughout a year. For example, in 2005, the budget frontloading increased public works significantly in the first half, but the activity declined sharply in the second half. This led to a base effect in 2006, reducing the growth rate of public works in the first half and increasing greatly in the second half.

TREND IN PUBLIC WORKS INVESTMENT

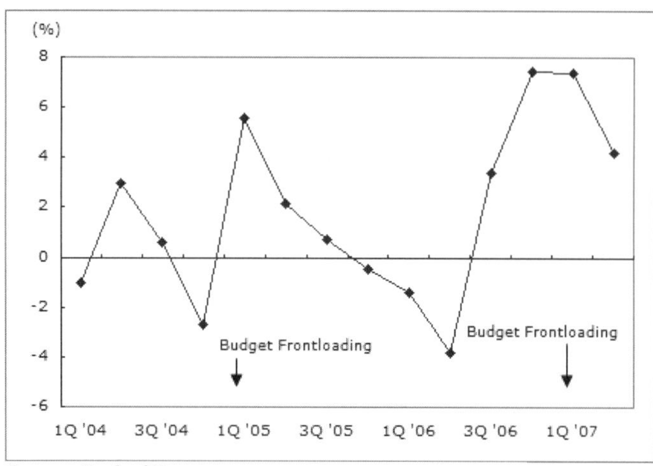

Source : Bank of Korea

Characteristics and Structure of the Industry

The major contractors in terms of revenue are given on the table below.

MAJOR KOREAN CONTRACTORS

Major contractors	*2008 Revenue Won (million)*	*Rank in 2008 Top 10*
Daewoo Engineering & Construction	8,927,241	1
Samsung Corp	7,733,819	2
Hyundai Engineering & Construction	6,907,849	3
GS Engineering & Construction	6,735,770	4
Daerim Corp	6,148,979	5
POSCO Engineering & Construction	5,232,033	6
Hyundai Corp	5,205,436	7
Lotte Engineering & Construction	4,487,681	8
SK Engineering & Construction	2,898,782	9
Doosan Engineering & Construction	2,113,861	10

Source : Contractors Association of Korea

Under the *Construction Industry Promotion Act*, any person who desires to operate a construction business has to be licensed by the Minister of Construction and Transportation (MOCT). The various types of licences are as follows:

Classification	*Type of licence*
General Contractor	• Civil engineering and building • Building • Civil engineering
Specialty Contractor	• 32 types of different trades including decoration, steelwork, etc.

Source: Construction Economic Bureau, MOCT

There are altogether 22,550 licences, of which 2,970 are for general contractors and 19,580 speciality contractors.

The top 100 firms dominate the construction market, accounting for 60% of the total contract value. The 50 larger firms made up about 47% of the total contract value.

The industry is heavily dependent on subcontractors. There are 32 different types of licences as defined and authorised by Presidential Decree.

Engineers are employed as engineering consultants, safety inspectors and research development specialists. The National Technical Qualification Act stipulates that a construction engineer must be assigned to the site. Construction engineers are classified under four levels depending on experience and educational background: Master engineer, Class 1 engineer, Class 2 engineer and entry level engineers. According to the Korea Construction Engineers Association, there were 11,028 Master engineers, 118,000 Class 1 engineers and 108,319 Class 2 engineers.

Only licensed architects are allowed to practise either as a partnership, corporation or individual proprietorship. Their activities are governed by the National Architectural Code and include architectural design, construction management and supervision.

Clients and Finance

The global plant market has been growing at an average rate of 6% every year due to surging crude oil prices, economic expansion in Brazil, Russia, India and China and heightening competition to secure enough natural resources. The plant orders won by Korean companies increased from $6.4 billion in 2003 to $42.2 billion in 2008.

In 2007, the total overseas orders were worth $39.8 billion. That accounts as the largest amount in Korean history and more than double the amount of 2006. Driving the growth are Persian Gulf nations that are building more refineries and power plants for sustaining economic growth, according to the Ministry of Land, Transportation and Maritime Affairs. Today, the Middle East remains the largest customer for Korean firms. According to the association, 152 companies are involved in 246 projects valued at $57.2 billion, as of June 1997.

In 2008 alone, Korean builders have won 23.4 billion won ($22.5 million) worth of construction, plant and civil engineering orders from around the world, including $11.5 billion in orders from the Middle East and $8.1 billion from Asia, according to the Korea Plant Industries Association. Overseas orders won by local firms have been growing quickly during the recent years, especially in the Middle East and Asia as oil rich nations have been earning more oil dollars with rising crude oil prices.

In addition, domestic funds account for a majority of transactions in the local real estate market, competing with foreign funds that have dominated the market since the 1997-98 Asian Financial Crisis.

According to one of the local real estate consulting firm's survey, of more than 1,000 buildings in Seoul that are taller than 10 stories or whose gross floor area is more than 6,611 square meters (71,162 square feet), 28 building sales or rental deals, worth 2.3 trillion won ($2.4 billion), were made during the first half of 2004. Among all the deals, approximately 72% of the total gross areas, worth 940 billion won, were done by domestic funds. Domestic funds have been making a comeback in the building market in Seoul since 2007, when the value of their transactions in the local market exceeded foreign firms' investments, accounting for 52% of the total value of all building transactions.

One of the main reasons for the recovery of domestic funds is that an increasing number of local asset management firms have bought more real estate properties to enhance their REIT (Real Estate Investment Trust) operations, and local educational foundations running colleges and other educational institutions are snapping up more real estate properties, which are to be used for school buildings.

Selection of Design Consultants

In the public sector, consultants are selected in competition. The private sector client usually chooses the designer but for large projects, a competition may be held on an international basis to include foreign consultants. Design is generally separated from construction but for certain projects, design and build or turnkey method of procurement is adopted.

Contractual Arrangements

The recently revamped evaluation system for selection of suitable contractors focuses on qualitative performance. It measures technical ability, quality achievement, environment and safety observance and is computed by the authority responsible for construction licensing of all general contractors and some speciality contractors. The results are promulgated to the public and used as guidelines in the selection process.

There are four methods of selecting contractors: open, limited, selective and negotiation. Open tendering is used for small projects while for large projects, limited tendering is preferred. Where there are less than ten eligible contractors because of the nature of the project, or the project is too small for open tendering, or preferential treatment is to be given to outstanding small and medium sized firms, selective tendering is used. Negotiation is only pursued under exceptional circumstances such as for projects relating to natural disasters and national security, contracts between public sectors, ongoing projects where contracting with the current contractor is very favourable. Competitive tendering is used extensively in the public sector and for large projects, limited and selective tendering is also adopted. The private sector prefers selective tendering and even negotiation.

Firm price contract is the most common form of contractual arrangement. However, during periods of price fluctuations, contracts usually contain a 'rise-and-fall clause' which allows the contractor to charge the client any increase in the actual labour and material prices over those prevailing at the time of contracting. At tendering stage, alternative design bids are sometimes allowed with the cost savings accruing to the tenderer. During the construction stage, contractors are encouraged to implement cost saving methods through financial incentives.

The contract documents include specifications, tax certificates, written guarantees, etc. and the contract itself covers the normal contractual arrangements including provisions for delay and defects liability.

Public sector projects are covered by the *Government Procurement Agreement (GPA)* under the World Trade Organization (WTO), which Korea is a signatory since 1996. Tendering and award procedures follow strictly the GPA rules in the case of central government projects valued at more than W5 billion, and W15 billion for local government projects.

Liability and Insurance

Various insurances and guarantees are compulsory; those relating to performance in bidding, carrying out the work, maintenance, and also for protection against dumping. It is also obligatory to insure against workers' compensation and fire. Some contractors have other additional insurance arrangement.

Development Control and Standards

In each city and town, there is an area zoning for overall and regional development planning. While alterations to the plans are possible, they are very unusual and difficult to comply.

Each project is assessed in terms of its traffic impact and then in terms of the building itself its energy usage, structure and aesthetics. The speed of the approval process depends on the region, the size of the project and its purpose but the whole process normally takes six to eight months. Sophisticated buildings such as hotels, condominiums, sports centres or fire stations may take longer. There are laws on requirements for building structures and facility standards which must be rigidly adhered to. *Korean Standards* (KS) also exist for building materials such as bricks, glass, steel and aggregates.

DEVELOPMENT PROCEDURES

General Building Construction	Redevelopment Project
	Assignment of Redevelopment Zone
	↓
Project Plan Preparation	Project Plan Preparation
	↓
	Request for Re-zoning (if necessary)
	↓
	Urban Planning Review Board of District Office
↓	↓
	Urban-Architectural Design Review Board
	↓
	Approval for Re-zoning (if necessary)
	↓
Traffic & Environmental Impact Assessment	Traffic & Environmental Impact Assessment
↓	↓
Design Review Board	Design Review Board
↓	↓
Design Development and Permit Drawing Preparation	Design Development and Permit Drawing Preparation
↓	↓
Application for Building Permit	Application for Building Permit
↓	↓
Building Permit	Building Permit (Project Development Permit)
↓	↓
Construction Drawing Preparation	Construction Drawing Preparation
	↓
	Evacuation
↓	↓
	Demolition
	↓
Application for Construction Commencement	Application for Construction Commencement
↓	↓
Construction Commencement	Construction Commencement
↓	↓
Construction Period	Construction Period
↓	↓
Construction Completion	Construction Completion

CONSTRUCTION COST DATA

Cost of Labour

The figures that follow are typical of labour costs in South Korea as at the fourth quarter of 2008. The wage rate is the basis of an employee's income, while the cost of labour indicates the cost to a contractor of employing that employee. The difference between the two covers a variety of mandatory and voluntary contributions – a list of items which could be included is given in section 2.

	Wage rate (per day) Won	*Cost of labour (per day) Won*	*Number of hours worked per year*
Site operatives			
Mason/bricklayer	86,508	96,100	3,100
Carpenter	102,164	113,500	3,100
Plumber	83,392	92,600	3,100
Electrician	88,317	98,100	3,100
Structural steel erector	100,401	111,500	3,100
HVAC installer	82,269	91,400	3,100
Semi-skilled worker	70,889	78,700	3,100
Unskilled labourer	60,547	67,300	3,100
Equipment operator	100,234	111,300	3,100
Watchman/security	65,500	72,800	3,100
Site supervision			
General foreman	80,830	89,800	3,100
Trades foreman	70,200	78,000	3,100
Clerk of works	60,547	67,300	3,100
	(per month)	*(per month)*	
Contractors' personnel			
Site manager	6,670,000	7,400,000	2,730
Resident engineer	5,000,000	5,550,000	2,730
Resident surveyor	5,000,000	5,550,000	2,730
Junior engineer	2,920,000	3,240,000	2,730
Junior surveyor	2,920,000	3,240,000	2,730
Planner	2,500,000	2,770,000	2,730
Consultants' personnel			
Senior architect	4,580,000	5,080,000	1,920
Senior engineer	4,580,000	5,080,000	1,920
Senior surveyor	4,580,000	5,080,000	1,920
Qualified architect	5,830,000	6,470,000	1,920
Qualified engineer	5,830,000	6,470,000	1,920
Qualified surveyor	5,830,000	6,470,000	1,920

Cost of Materials

The figures that follow are the costs of main construction materials, delivered to site in the Seoul area, as incurred by contractors in the fourth quarter of 2008. These assume that the materials would be in quantities as required for a medium sized construction project and that the location of the works would be neither constrained nor remote. All the costs in this section exclude value added tax.

	Unit	*Cost Won*
Cement and aggregate		
Ordinary portland cement in 40kg bags	bag	3,370
Coarse aggregates for concrete	m^3	17,500
Fine aggregates for concrete	m^3	20,000
Ready mixed concrete (40-135-8)	m^3	41,530
Ready mixed concrete (25-210-12)	m^3	49,080
Precast concrete piles D400 x 65 x 10m (class B)	each	246,400
Steel		
Mild steel reinforcement (over D16)	tonne	593,000
High tensile steel reinforcement (over D16)	tonne	598,000
Structural steel sections (rolled H beam)	tonne	700,000
Bricks and blocks		
Common bricks (190 x 90 x 57mm)	1,000	45,000
Good quality facing bricks (190 x 90 x 57mm)	1,000	400,000
Hollow concrete blocks (150 x 190 x 390mm)	each	600
Autoclaved lightweight concrete blocks (600 x 400 x 100mm)	each	10,500
Precast concrete cladding units with plain surface finish	m^2	40,000
Timber and insulation		
Softwood sections for carpentry	m^3	305,390
Softwood for joinery	m^3	1,047,900
Exterior quality plywood (15mm)	m^2	4,780
Plywood for interior joinery (12mm)	m^2	7,550
Softwood strip flooring (22 x 129 x 3700mm)	m^2	92,000
Chipboard sheet flooring (18 x 200 x 2424mm)	m^2	37,000
100mm thick quilt insulation (Mineral Wool, 100kg/m^3)	m^2	8,680
100mm thick rigid slab insulation (expanded polystyrene)	m^2	4,940
Softwood internal door complete with frames and ironmongery	each	231,400
Glass and ceramics		
Float glass (3mm)	m^2	4,770
Sealed double glazing units (24mm green)	m^2	34,800

	Unit	*Cost Won*
Plaster and paint		
Good quality ceramic wall tiles (200 x 200mm)	m^2	8,000
Plaster in 40kg bags	bag	3,300
Plasterboard (9.5mm thick)	m^2	1,920
Emulsion paint in 5 litre tins (white)	litre	4,300
Gloss oil paint in 5 litre tins (white)	litre	5,200
Tiles and paviors		
Clay floor tiles (200 x 200 x 7mm)	m^2	8,000
Vinyl floor tiles (300 x 300 x 3mm)	m^2	6,400
Precast concrete paving slabs (300 x 300 x 60mm)	m^2	6,210
Clay roof tiles (360 x 345 x 13mm)	1,000	1,300,000
Precast concrete roof tiles (400 x 350 x 12mm)	1,000	650,000
Granite 20 - 24mm thick polished finish medium quality	m^2	29,000
Granite 20 - 24mm thick polished finish high quality	m^2	55,000
Drainage		
WC suite complete	each	170,000
Lavatory basin complete	each	165,000
100mm diameter PVC drain pipes	m	25,000
150mm diameter cast iron drain pipes	m	94,000

Unit Rates

The descriptions below are generally shortened versions of standard descriptions listed in full in section 4. Where an item has a two digit reference number (e.g. 05 or 33), this relates to the full description against that number in section 4. Where an item has an alphabetic suffix (e.g. 12A or 34B) this indicates that the standard description has been modified. Where a modification is major the complete modified description is included here and the standard description should be ignored; where a modification is minor (e.g. the insertion of a named hardwood) the shortened description has been modified here but, in general, the full description in section 4 prevails.

The unit rates below are for main work items on a typical construction project in the Seoul and adjoining areas in the fourth quarter of 2008. The rates include all necessary labour, materials and equipment. Allowances to cover preliminary and general items and contractor's overheads and profit have been added to the rates. All the rates in this section exclude value added tax.

		Unit	*Rate Won*
Excavation			
01	Mechanical excavation of foundation trenches	m^3	4,000
02	Hardcore filling making up levels	m^3	11,000
03	Earthwork support	m^2	90,000

		Unit	Rate Won
Concrete work			
04	Plain in situ concrete in strip foundations in trenches	m^3	70,000
05	Reinforced in situ concrete in beds	m^3	70,000
06	Reinforced in situ concrete in walls	m^3	70,000
07	Reinforced in situ concrete in suspended floors or roof slabs	m^3	70,000
08	Reinforced in situ concrete in columns	m^3	70,000
09	Reinforced in situ concrete in isolated beams	m^3	70,000
10	Precast concrete slabs	m^3	60,000
Formwork			
11	Softwood formwork to concrete walls	m^2	30,000
12	Softwood or metal formwork to concrete columns	m^2	35,000
13	Softwood or metal formwork to horizontal soffits of slabs	m^2	30,000
Reinforcement			
14	Reinforcement in concrete walls	tonne	900,000
15	Reinforcement in suspended concrete slabs	tonne	900,000
16	Fabric reinforcement in concrete beds	m^2	2,000
Steelwork			
17	Fabricate, supply and erect steel framed structure	tonne	1,650,000
18	Framed structural steelwork in universal joist sections	tonne	1,900,000
19	Structural steelwork lattice roof trusses	tonne	2,300,000
Brickwork and blockwork			
20	Precast lightweight aggregate hollow concrete block walls	m^2	38,000
21A	Solid (perforated) concrete blocks	m^2	28,000
23	Facing bricks	m^2	47,000
Roofing			
24	Concrete interlocking roof tiles 430 x 380mm	m^2	35,000
25	Plain clay roof tiles 260 x 160mm	m^2	38,000
26	Fibre cement roof slates 600 x 300mm	m^2	32,000
27	Sawn softwood roof boarding	m^2	85,000
28	Particle board roof coverings	m^2	12,000
29	3 layers glass-fibre based bitumen felt roof covering	m^2	30,000
30	Bitumen based mastic asphalt roof covering	m^2	33,000
31A	Glass-fibre mat roof insulation 100mm thick	m^2	20,000
32	Rigid sheet loadbearing roof insulation 75mm thick	m^2	15,000

		Unit	*Rate Won*
33	Troughed galvanized steel roof cladding	m^2	32,000
Woodwork and metalwork			
34	Preservative treated sawn softwood 50 x 100mm	m	15,000
35	Preservative treated sawn softwood 50 x 150mm	m	23,000
36	Single glazed casement window in Lanan hardwood, size 650 x 900mm	each	150,000
37	Two panel glazed door in Lanan hardwood, size 850 x 2000mm	each	420,000
38A	Solid core half hour fire resisting aluminium internal flush doors, size 800 x 2000mm	each	1,300,000
39	Aluminium double glazed window, size 1200 x 1200mm	each	380,000
40	Aluminium double glazed door, size 850 x 2100mm	each	280,000
41A	Hardwood skirtings (Lanan)	m	15,000
Plumbing			
42	UPVC half round eaves gutter	m	45,000
43	UPVC rainwater pipes	m	25,000
44	Light gauge copper cold water tubing	m	58,000
45	High pressure plastic pipes for cold water supply	m	14,000
46	Low pressure plastic pipes for cold water distribution	m	14,000
47	UPVC soil and vent pipes	m	13,000
48	White vitreous china WC suite	each	150,000
49	White vitreous china lavatory basin	each	130,000
50	Glazed fireclay shower tray	each	220,000
51	Stainless steel single bowl sink and double drainer	each	140,000
Finishings			
55	2 coats gypsum based plaster on brick walls	m^2	18,000
56	White glazed tiles on plaster walls	m^2	35,000
56A	Granite veneer 20mm thick for walls, fixed with cement mortar	m^2	135,000
57	Red clay quarry tiles on concrete floors	m^2	35,000
58	Cement and sand screed to concrete floors	m^2	7,000
59	Thermoplastic floor tiles on screed	m^2	48,000
60	Mineral fibre tiles on concealed suspension system	m^2	45,000
Glazing			
61A	6mm clear float glass; glazing to wood	m^2	17,000

		Unit	Rate Won
Painting			
62	Emulsion on plaster walls	m²	2,000
63	Oil paint on timber	m²	3,000

Approximate Estimating

The building costs per unit area below are averages incurred by building clients for typical buildings in the Seoul and adjoining areas as at the fourth quarter of 2008. They are based upon the total floor area of all storeys, measured between external walls and without deduction for internal walls.

Approximate estimating costs generally include mechanical and electrical installations but exclude furniture, loose or special equipment, and external works; they also exclude fees for professional services. The costs shown are for specifications and standards appropriate to South Korea and this should be borne in mind when attempting comparisons with similarly described building types in other countries. A discussion of this issue is included in section 2. Comparative data for countries covered in this publication, including construction cost data, are presented in Part Three.

Approximate estimating costs must be treated with caution; they cannot provide more than a rough guide to the probable cost of building. All the rates in this section exclude value added tax.

	Cost m² Won	Cost ft² Won
Industrial buildings		
Factories for letting	600,000	56,200
Factories for owner occupation (light industrial use)	760,000	70,200
Factories for owner occupation (heavy industrial use)	910,000	84,300
Factory/office (high tech) for letting (shell and core only)	1,060,000	98,300
Factory/office (high tech) for letting (ground floor shell, first floor offices)	1,100,000	102,200
Factory/office (high tech) for owner occupation (controlled environment, fully finished)	1,150,000	106,800
High tech laboratory workshop centres (air-conditioned)	1,510,000	140,500
Warehouses, low bay (6 to 8m high) for letting (no heating)	610,000	57,000
Warehouses, low bay for owner occupation (including heating)	700,000	64,800
Warehouses, high bay for owner occupation (including heating)	820,000	76,200
Cold stores/refrigerated stores	1,460,000	135,200

	Cost m^2 Won	Cost ft^2 Won
Administrative and commercial buildings		
Civic offices, non air-conditioned	1,060,000	98,800
Civic offices, fully air-conditioned	1,210,000	112,300
Offices for letting, 5 to 10 storeys, non air-conditioned	1,060,000	98,800
Offices for letting, 5 to 10 storeys, air-conditioned	1,210,000	112,300
Offices for letting, high rise, air-conditioned	1,490,000	138,200
Offices for owner occupation 5 to 10 storeys, non air-conditioned	1,180,000	109,500
Offices for owner occupation 5 to 10 storeys, air-conditioned	1,340,000	124,500
Offices for owner occupation high rise, air-conditioned	1,650,000	153,700
Prestige/headquarters office, 5 to 10 storeys, air-conditioned	1,600,000	148,800
Prestige/headquarters office, high rise, air-conditioned	1,820,000	169,100
Health and education buildings		
General hospitals (100 beds)	1,750,000	163,000
Teaching hospitals (100 beds)	1,540,000	142,700
Private hospitals (100 beds)	1,320,000	122,300
Health centres	1,320,000	122,300
Nursery schools	1,100,000	101,900
Primary/junior schools	970,000	89,700
Secondary/middle schools	970,000	89,700
University (arts) buildings	1,100,000	101,900
University (science) buildings	1,320,000	122,300
Management training centres	1,320,000	122,300
Recreation and arts buildings		
Theatres (over 500 seats) including seating and stage equipment	2,150,000	199,700
Theatres (less than 500 seats) including seating and stage equipment	2,430,000	225,800
Concert halls including seating and stage equipment	2,730,000	253,300
Sports halls including changing and social facilities	1,900,000	176,500
Swimming pools (international standard) including changing and social facilities	1,800,000	167,200
Swimming pools (schools standard) including changing facilities	1,800,000	167,200
National museums including full air-conditioning and standby generator	1,650,000	153,300
Local museums including air-conditioning	1,600,000	148,600
Branch/local libraries	1,350,000	125,400

	Cost m² Won	Cost ft² Won
Residential buildings		
Social/economic single family housing (multiple units)	1,050,000	97,500
Private/mass market single family housing 2 storey detached / semi detached (multiple units)	1,200,000	111,500
Purpose designed single family housing 2 storey detached (single unit)	1,670,000	155,500
Social/economic apartment housing, low rise (no lifts)	970,000	90,100
Social/economic apartment housing, high rise (with lifts)	1,150,000	107,000
Private sector apartment building (standard specification)	1,490,000	138,200
Private sector apartment buildings (luxury)	1,600,000	148,600
Student/nurses halls of residence	1,060,000	98,600
Homes for the elderly (shared accommodation)	1,220,000	113,300
Homes for the elderly (self contained with shared communal facilities)	1,450,000	134,700
Hotel, 5 star, city centre	2,600,000	241,900
Hotel, 3 star, city/provincial	2,000,000	185,800
Motel	1,670,000	154,900

Regional Variations

The approximate estimating costs are based on projects in Seoul and other big cities. For other parts of country, add 5% to these costs.

Value Added Tax (VAT)

The standard rate of value added tax (VAT) is currently 10%, chargeable on general building work.

EXCHANGE RATES AND INFLATION

The combined effect of exchange rates and inflation on prices within a country and price comparisons between countries is discussed in section 2.

Exchange Rates

The graph below plots the movement of the South Korean won against the sterling, the euro, the US dollar and 100 Japanese yen since 1998. The values used for the graph are quarterly and the method of calculating these is described and general guidance on the interpretation of the graph provided in section 2. The average exchange rate in the fourth quarter of 2008 was W2,013 to pound sterling, W1,606 to euro, W1,102 to US dollar and W1,076 to 100 Japanese yen.

THE SOUTH KOREAN WON AGAINST STERLING, EURO, US DOLLAR AND 100 JAPANESE YEN

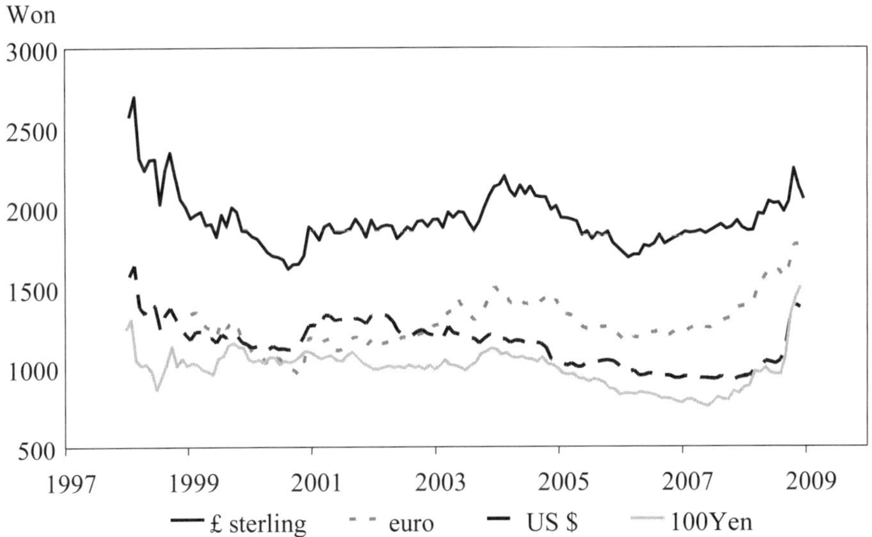

Price Inflation

The table below presents consumer prices in South Korea since 2003.

CONSUMER PRICE INFLATION

	Consumer price index	
Year	*Average Index*	*Average change %*
2003	93.95	3.5
2004	97.32	3.6
2005	100.00	2.8
2006	102.2	2.2
2007	104.8	2.5
2008	109.7	4.7

Source: Ministry of Strategy and Finance Useful Address

USEFUL ADDRESSES

Public Organizations

Korea Institute of Construction Technology
2311 Daehwa-Dong
Ilsan-Gu, Goyang-City
Gyeonggi-Do 411-712
Tel: (82) 31 910 0113
Fax: (82) 31 910 0111
E-mail: Webmaster@kict.re.kr
Website: www.kict.re.kr/eng/

Korea Land Corporation
217 Jungja-Dong
Bundang-Gu, Sungnam-Si
Kyungki Do 463-755
Tel: (82) 31 738 7114 / 8114
Website:world.lplus.org.kr

Korea National Housing Corporation
175 Gumi-Dong
Bundang-Gu, Seongnam-Si
Gyungki Do 463-500
Tel: (82) 2 3414 3522-9
Website: www.jugong.co.kr/ ->English

Korea National Statistical Office
139 Seonsa-ro Seo-Gu
Daejeon 302-701
Tel: (82) 42 481 4114
Website: www.nso.go.kr/eng2006/emain/index.html

Korean Standards Association
Korea Technology Centre
701-7 Yeoksam-Dong, Gangnam-Gu
Seoul 135-513
Tel: (82) 2 6009 4630 – 2
Fax: (82) 2 6009 4639
Website: www.ksa.or.kr/eng

Ministry of Construction and Transportation
1 Jungang-Dong
Gwacheoun-City
Gyeonggi-Do
Tel: DAY: (82) 2 504 9114
NIGHT: (82) 2 503 7400
Fax: (82) 2 2150 1000
Website: www.moct.go.kr/EngHome/

Ministry of Strategy and Finance
Government Complex II, 88 Gwanmoonro
Gwacheon City
Gyeonggi Province 427-725
Tel: (82) 2 2150 2451
Fax: (82) 2 504 1335
E-mail: fppr@mosf.go.kr
Website: english.mosf.go.kr

Trade And Professional Organizations

Architectural Institute of Korea
1044-34 Sadang-Dong
Dongjak-Gu, Seoul 156-827
Tel: (82) 2 525 1841-4
Fax: (82) 2 525 1845
Website: www.aik.or.kr/english/index.htm

Construction Association of Korea
7-8F The Hall of Construction B / D
71-2 Nonhyon-Dong, Kangnam-Gu
Seoul 135-701
Tel: (82) 2 3485 8200
E-mail: webmaster@cak.or.kr
Website: www.cak.or.kr ->English

The International Contractors Association of Korea
13F Booyoung Bldg
120-23 Seosomun-Dong
Joong-Gu, Seoul 100-764
Tel: (82) 2 3406 1044
Fax: (82) 2 3406 1123
Website: www.icak.or.kr/eng

Korea Construction Engineers Association
238-5 Nonhyon-Dong, Kangnam-Ku
Seoul 135-830
Tel: (82) 2 3416 9511
Fax: (82) 2 3416 9090
Website: www.kocea.or.kr ->English

Korea Engineering & Consulting Association
1049-1 Sadang-Dong, Dongjak-Gu
Seoul 156-090
Tel: (82) 2 3019 3200
Fax: (82) 2 3019 3300
E-mail: engineering@kenca.or.kr
Website: www.kenca.org/index_eng.jsp

Korea Housing Builders Association
4F Korea Housing Center, 45-11
Yeouido-Dong, Youngdeungpo-Gu
Seoul 150-010
Tel: (82) 2 785 0990
Fax: (82) 2 785 3915
Website: www.khba.or.kr/en/index-en.htm

Korea Institute of Registered Architects
1603-55 Seocho-1Dong
Seocho-Gu
Seoul 137-877
Tel: (82) 2 581 5711~4
Fax: (82) 2 525 8379
Website: www.kira.or.kr/eng/main.asp

Korea Research Institute for Human Settlements
1591-6 Gwangyang-dong
Dongan-gu, Anyang-si
Gyeonggi-do 431-712
Tel: (82) 31 380 0114
Fax: (82) 31 380 0470
E-mail: jhkim@krihs.re.kr
Website: www.krihs.re.kr/->English

Korean Professional Engineer Association
635-4 Yeaksam-Dong
Gangnam-Gu
Seoul 135-703
Tel: (82) 2 538 3159
Fax: (82) 2 557 7048
E-mail: kpea@kpea.or.kr
Website: www.kpea.or.kr/english/

Others

Korea Development Bank
16-3, Yeouido-Dong
Yeongdeungpo-Gu
Seoul 150-973
Tel: (82) 2 787 6450 / 6479
Fax: (82) 2 787 6496 / 6498
Website: www.kdb.co.kr/-> English

Korea Housing Association
4F Kunsul Bldg
71-2 Nonhyun-Dong
Gangnam-Gu
Seoul
Tel: (82) 2 514 3167
Fax: (82) 2 511 6974
Website: www.housing.or.kr/eng/index.asp

SRI LANKA

All data relate to 2007 unless otherwise indicated.

Population	
Population	20.0 mn
Urban population	20.5%
Population under 15	28%
Population 65 and over	2.45%
Average annual growth rate	1.1%
Geography	
Land area	65,610 km^2
Agricultural area	36%
Capital city	Colombo
Population of capital city	2.2 mn
Economy	
Monetary unit	Sri Lanka Rupee (SLR)
Exchange rate (average fourth quarter 2008)	
the pound sterling	SLR 172.70
the US dollar	SLR 109.82
the euro	SLR 144.87
the yen x 100	SLR 114.50
Average annual inflation	10.16.%
Inflation rate	15.9%
Gross Domestic Product (GDP)	SLR 3,578 bn
GDP per capita	SLR 179,000
Average annual real change in (GDP)	4%
Private consumption as a proportion of GDP	67.0%
Public consumption as a proportion of GDP	15.0%
Investment as a proportion of GDP	29.0%
Construction	
Gross value of construction output	SLR 264.1 bn
Net value of construction output	SLR 143 bn
Net value of construction output as a proportion of GDP	4%

THE CONSTRUCTION INDUSTRY

Construction Output

Analysis reveals that growth in the construction sector averaged 21.6% per year in real terms for the period between 2003 and 2007. Output growth was at its peak at 31.6% per year in the year 2005, but the subsequent decline in public and private investment from 2005 to 2007 led to a deceleration in construction growth to 29.1% in 2006 and to 21.8% in 2007. Unsettled security situation and adversity of the oil price hike together with the general economic slowdown has affected the construction industry.

The gross value of construction output in 2007 was SLR 264.10 billion equivalents to US$ 2.4 billion or 7.38% of GDP. The net value of construction output was SLR 143 billion equivalents to US$ 1.3 billion or 4% of GDP. The chart below clearly indicates the drop in growth rate in construction output (as a proportion of GDP) in the years 2006 and 2007.

COMPOSITION OF GDP AT CURRENT PRICES 2002-2007

Element/Year	*2002*	*2003*	*2004*	*2005*	*2006*	*2007*
GDP	1,636,037	1,822,468	2,090,841	2,452,782	2,938,656	3,577,438
GDP Growth Rate(%)	-	11.4	14.7	17.3	19.8	21.7
Construction Output	100,404	110,111	127,692	167,999	216,833	264,104
Growth Rate (Construction)	-	9.7	16.0	31.6	29.1	21.8

Source: Central Bank Of Sri Lanka

Up-to-date (as at end 2008) accurate statistics on the breakdown of the workload in the construction sector are not available. However, the Institute for Construction Training and Development (ICTAD) estimated that in 2007 building work contributed about 60% of total construction output with the remaining 40% from infrastructure. Building work consists of housing (about 35%) and non-residential building (about 25%). Infrastructure work includes roads, tunnels, bridges, water supply, power plants, airports and irrigation.

In recent years the overall level of public sector work has generally declined while private sector work has increased. In fact the Sri Lankan government is encouraging private investment in infrastructure projects – water supply, waste water disposal, power generation, roads, industrial estates, car parks and buildings. Both local and foreign investors are encouraged to participate in infrastructure and property development through various government schemes and policies.

Government has implemented schemes to generate more work for the industry locally initiated or foreign funded. It is determined to ensure that the projects funded by the World Bank, IMF, ADB and others are channelled properly, performed and brought in to the local economy. In the past, many such projects have not materialised due to inadequate arrangements for public private partnerships, lack of organised local counterpart funding and without doubt the over archaic desire to employ the foreign architects and contractors.

The main priority of the government is the provision of social housing and infrastructure development. The Urban Development Authority and National Housing Development Authority (NHDA) with Ministry of Construction and Engineering Services are tasked with the implementation of housing programmes using public resources. Apart from the involvement of NHDA, several low cost housing programmes have also been implemented by private developers under BOI (Board of Investment) duty free concessions to meet the housing needs in the country.

The value added of the construction sector is estimated to have grown considerably in the past. The growth impetus in this sector came from both government sector spending on public infrastructure facilities and private sector involvement in construction activities, particularly in condominium projects and private housing units. Also, during the first half of the year, there has been an increase in the usage of cement and other building materials as indicated by domestic production and imports of the same. The growing trend observed in the first half of 2008 is expected to continue in the second half as well. However, the increasing cost of construction in terms of materials and wages as well as deterioration in the security situation in the island could pose some down-side risks.

Donor countries have agreed to provide concessionary loans for development of construction industry in Sir Lanka. The objective of the development is to strengthen the capacity and improve the ability of the Construction Industry to facilitate its participation in rehabilitation of tsunami damaged infrastructure expeditiously. This facility will be used for purchase of construction equipment and also for working capital requirements of the small and medium entrepreneurs in the sector. Section of this facility will be utilised to strengthen the management competence and technical capacity of the employees in the construction sector by contributing to the establishment Advanced Construction Training Academy (ACTA) by the National Construction Association.

Characteristics and Structure of the Industry

There are two sectors in the industry – the formal sector and the informal sector. The informal sector's real output is hardly quantified but it is estimated to be approximately 75% of house building. Much of the social building and infrastructure work is carried out by the people themselves.

The formal sector of the construction industry was previously dominated by two state sector contractors until the later part of the 1970s. The promulgation of open economic policies by the government in the early 1980s led to an explosive growth in the number of contractors as well as in the volume of work. With the establishment of free economy and influx of foreign investors and property developers there had been deployment of overseas contractors in the recent past to handle major projects in Sri Lanka. Currently the bulk of the building work is carried out by general contractors who employ labour-only subcontractors. Specialist subcontractors may be nominated by the client's consultant team or employed by the general contractor.

A Construction Industry Development Act presented by newly formed Chamber of Construction Industry Sri Lanka is to be adopted soon to tackle problems faced by the domestic construction industry. This new act has been formulated as an alternative to the Construction Industry Authority Act, currently in force, which only dealt with regulations. The proposed legislation contains provisions for the establishment of a construction industry authority as a regulating body. This will also address providing uniform treatment related to unsolicited developments involving construction contracts and strict adherence to environmental standards.

Most of the contractors in the industry are privately owned companies and they are registered by ICTAD under ten grades (other than specialist contractors) based on financial limits. As of March 2008 there are 5,454 registered companies as given in the table below.

NUMBER OF REGISTERED COMPANIES BY GRADE, 2007

Building Construction

Registration grade	*Limit of project size*	*Number of registered companies*
Grade M1	SLR300 million and above	29
Grade M2	SLR150 - 300 million	25
Grade M3	SLR50 - 150 million	60
Grade M4	SLR20 - 50 million	116
Grade M5	SLR10 - 20 million	145
Grade M6	SLR5 - 10 million	333
Grade M7	SLR2 - 5 million	1,150
Grade M8	SLR 1 - 2 million	192
Grade M9	< SLR1 million	138
Grade M10		18
Total		2,206

Source: Institute for Construction Training and Development

Highway Construction

Registration grade	*Limit of project size*	*Number of registered companies*
Grade M1	SLR300 million and above	19
Grade M2	SLR150 - 300 million	7
Grade M3	SLR50 - 150 million	20
Grade M4	SLR20 - 50 million	52
Grade M5	SLR10 - 20 million	76
Grade M6	SLR5 - 10 million	203
Grade M7	SLR2 - 5 million	1,507
Grade M8	SLR 1 - 2 million	226
Grade M9	< SLR1 million	159
Grade M10		18
Total		2,287

Source: Institute for Construction Training and Development

Dredging & Reclamation

Registration grade	*Limit of project size*	*Number of registered companies*
Grade M1	SLR300 million and above	6
Grade M2	SLR150 - 300 million	1
Grade M3	SLR50 - 150 million	4
Grade M4	SLR20 - 50 million	3
Grade M5	SLR10 - 20 million	44
Grade M6	SLR5 - 10 million	127
Total		185

Source: Institute for Construction Training and Development

Irrigation & Land Drainage

Registration grade	*Limit of project size*	*Number of registered companies*
Grade M1	SLR300 million and above	5
Grade M2	SLR150 - 300 million	3
Grade M3	SLR50 - 150 million	12
Grade M4	SLR20 - 50 million	31
Grade M5	SLR10 - 20 million	79
Grade M6	SLR5 - 10 million	160
Total		290

Source: Institute for Construction Training and Development

Water Supply & Drainage

Registration grade	*Limit of project size*	*Number of registered companies*
Grade M1	SLR300 million and above	14
Grade M2	SLR150 - 300 million	8
Grade M3	SLR50 - 150 million	10
Grade M4	SLR20 - 50 million	21
Grade M5	SLR10 - 20 million	37
Grade M6	SLR5 - 10 million	123
Total		213

Source: Institute for Construction Training and Development

Bridge Construction

Registration grade	*Limit of project size*	*Number of registered companies*
Grade M1	SLR300 million and above	5
Grade M2	SLR150 - 300 million	6
Grade M3	SLR50 - 150 million	11
Grade M4	SLR20 - 50 million	11
Grade M5	SLR10 - 20 million	29
Grade M6	SLR5 - 10 million	81
Total		143

Source: Institute for Construction Training and Development

Specialised Contractors

Registration grade	*Limit of project size*	*Number of registered companies*
Grade EM1	SLR300 million and above	52
Grade EM2	SLR150 - 300 million	7
Grade EM3	SLR50 - 150 million	1
Grade EM4	SLR20 - 50 million	7
Grade EM5	SLR5 - 10 million	45
Grade F-1		10
Grade F-2		1
Grade F-4		1
Grade P-1		5
Grade P-2		1
Total		130

Source: Institute for Construction Training and Development

The fields of specialization for main contractors comprise of building, highways, bridges, water supply & drainage, irrigation & land drainage, dredging & reclamation and specialist contractors (building services & finishes).

Registration grade	*Limit of project size*	*Number of registered companies*
Grade M1	SLR300 million and above	130
Grade M2	SLR150 - 300 million	57
Grade M3	SLR50 - 150 million	118
Grade M4	SLR20 - 50 million	241
Grade M5	SLR10 - 20 million	410
Grade M6	SLR5 - 10 million	1,072
Grade M7	SLR2 - 5 million	2,657
Grade M8	SLR 1 - 2 million	418
Grade M9	< SLR1 million	297
Grade M10		36
Grade F1		10
Grade F2		1
Grade F4		1
Grade P1		5
Grade P2		1
Total		5,454

Registered main contractors who have done major contracts in Sri Lanka are:

- State Engineering Corporation
- State Development and Construction Corporation
- Access International Projects
- Road Development Authority
- International Construction Consortium
- Link Engineering (Pvt) Ltd
- Maga Engineering (Pvt) Ltd
- Tudawe Brothers Ltd
- Nawaloka Construction Company
- Sierra Construction Ltd
- Sanken Lanka Ltd

Building work has been administered by professional consultants appointed by the building client. The consultancy field in the construction industry was largely non existent prior to 1970 except for a few architectural practices which catered to the private sector. The consultants are responsible for the design and contract administration and the supervision of the project. Of late, design-and-build contractors offering a single point responsibility have also made inroads to the industry, especially in the industrial building sector. Further, other non-traditional forms of arrangements are beginning to emerge as the industry. Design work is undertaken mainly by architects and engineers. Majority of the architects are members of the Sri Lanka Institute of Architects (a professional body governing the practice of architecture). A large number of the Institute's members are working in the private sector as partners or hold senior positions while the remainder work in the government, education or semi governmental bodies.

Civil engineers are members of the Institution of Engineers Sri Lanka (a professional body governing the practice of engineering).

The quantity surveyor has now established its prominence in Sri Lanka with the Chartered status incorporated in the parliament of Sri Lanka with the introduction of undergraduate programmes in the universities in the mid 1980s, the services offered by a quantity surveyor are more appreciated by the industry so much so that demand surpasses supply. The profession has already gained its statutory recognition and approval.

Clients and Finance

Until the early 1980s the Sri Lankan construction industry has been dominated by the public sector. However, since the mid 80s, with the free economy there is a marked shift towards the private sector and a gradual decline of the importance of the state sector.

Public investment in key areas of infrastructure development is only around 5.5% of GDP and therefore reliance is placed almost entirely on foreign aid. The participation of the private sector in infrastructure projects on BOO/BOT basis presents attractive opportunities to meet this challenge. The establishment of the Private Sector Infrastructure Development Company (PSIDC) which offers long-term subordinate loans is once again a driving force to boost private sector involvement in economic infrastructure development.

The Sri Lankan economy, which recorded the highest growth of 7.7 per cent in 2006, was projected to expand further by around 6.7 per cent in 2007. This indicates the continuing resilience of the economy; its ability to grow amidst the adversity of the oil price hike, and the unsettled security situation. The challenge ahead is to transform this salutary feature to a sustainable, long-term growth through a conductive environment for continued investment in both infrastructure and new economic activities. To maintain this growth momentum, the country has, on a priority basis, addressed the infrastructure deficiencies and problems in resource allocations and attain macro-economic stability.

As projected, the growth momentum witnessed in the past few years has continued during first half of 2008 with a growth of 7% in the second quarter and this growth broad based, with the three sectors; i.e. Agriculture, Industry and Services. Despite the 7% economic growth recorded in the second quarter of 2008, the construction industry and in particular the private sector was faced with severe hardships during latter part of the year in view of high interest rates and rising material prices.

Year 2008 can be considered as one of the most difficult periods for business in the recent times. The problems stemmed from a rapidly deteriorating economic outlook both locally and globally.

Developments in the commercial and residential real estate sector too slowed down or had halted their constructions during the latter part of the year 2008 due to financial crisis. Demand for houses and apartments had fallen drastically in the last quarter of 2008, due to the uncertainty in the market, declining resale values and rental guarantees.

Selection of Design Consultants

ICTAD is in the process of revising its Consultancy documents and this revision of the Consultancy documents was made essentially to denote more precisely the activities pertaining to the different disciplines, and to reduce overlapping of functions as far as possible. It is expected that the revised documents would guide both the Client and the Consultants in the choice of the Consultant and the Services assigned to them.

The Services provided by each profession was categorized under the following three headings:

a) Basic Services relevant to the profession.
b) Extra Services pertaining to the profession.
c) Other Services which are not pertaining to the particular profession that could be provided on special assignments.

In the public sector, a pre-qualification exercise and through the submission of credentials and relevant experience followed by interviews is the norm. In some instances a design competition is held unlike in the private sector, where this is seldom adopted. Private clients may select the design consultants known to them or from those who have a reputation for a specific type of building. The fee is based on a scale stipulated by respective professional organisation.

Contractual Arrangements

The *ICTAD Standard Form of Contract* is the standard document widely accepted by the industry and its use is mandatory for all public sectors for different types of contracts. The most common type of contract assumes the use of measured bills of quantities that are normally prepared by the quantity surveyor. The tender documentation structure under this form comprises the following:

- Form of tender
- Conditions of tender
- Drawings
- Specifications
- Bills of quantities, schedules of works or schedules of rates

Procurement of all public works and services are now governed by the *Guidelines on Government Tender Procedure (Revised edition August 1997)* published by the Ministry of Finance and Planning. These guidelines also provide a comprehensive procedure to be followed when dealing with BOO/BOT projects.

In addition to these forms of contracts, independent consultants have documented their own bespoke forms based on common standard forms of contract such as JCT, FIDIC, etc. and most overseas clients investing for property development here used amended versions of these documents to suit their requirements.

Contracts are usually let on a measure and pay basis. Fluctuations in labour, materials and plant costs are reimbursed. Such increases in costs are paid according to a formula, which is based on input percentages and indices published by ICTAD.

There are a number of alternative contractual arrangements – most notably management contracting, design-and-build, construction management and guaranteed maximum price. With respect to design-and-build procurement, a publication of the conditions of contract is available from ICTAD. Under management contracting, the client enters into a separate contract with a designer and a management contractor who in turn enters into subcontracts with individual works contractors. For construction management, the client enters into separate contracts with a designer, a construction manager and the works contractors. This form of contractual arrangement is increasingly being used in preference to management contracting.

Liability and Insurance

Almost all professional practices do not carry professional indemnity insurance, probably for the reason that no major legal proceedings have been pursued against such firms.

Dispute resolution in construction has been made faster with the newly introduced Arbitration Act in Sri Lanka. The Sri Lanka National Arbitration Centre is the oldest institution in the country for resolution of construction of commercial disputes where as Institution of Commercial Law & Practice is a fairly a new facility for conducting dispute resolution.

Development Control and Standards

The Urban Development Authority (UDA) is the national planning authority regulating and facilitating the physical development of Sri Lanka. All types of development require written planning permission and this covers obtaining clearance for reports on Traffic Impact Assessment and Environmental Impact Assessment.

The industry-specific standards which apply to the whole industry are promulgated by ICTAD, Sri Lankan Standards Institution (SLSI) and the Road Development Authority (RDA). ICTAD standards are applicable to both building and civil engineering works.

The Sri Lanka Standards Institution draws up and promulgates the *Sri Lanka Standards (SLS*), the standard specification for products, and it is usual for manufacturers to comply with these.

Architects and other building professionals generally follow the recommendations of the *SLS* or in its absence the *BS* specifications when specifying building products. In addition to these standards other codes such as Fire Code and Energy Efficient Code are in place for controlling building standards.

A Construction Industry Authority Act is now in force to give effect to state policies on construction which dealt with regulation and are aimed at increasing the efficiency of the construction industry, and make it more responsible for the national development efforts and economic needs of the country. This has been implemented by ICTAD who will be given a much greater responsibility and authority to establish itself as the authority for the construction industry. There will soon be an act called Construction Industry Development Act presented by the Chamber of Construction Industry Sri Lanka which addresses security of payments to contractors and consultants, occupation health and safety standards to be adhered to by those undertaking construction contracts.

Research and Development

The main organizations engaged in construction research are ICTAD, the Building Economics Research Unit (BERU) of the University of Moratuwa, the University of Peradeniya and National Building Research Organization (NBRO). The addresses are given under the Useful Addresses section.

Prospects for the Future

By 2008/2009, the construction sector is expected to grow by around 10 per cent as in previous year with active participation of the private sector as well as the public sector. Public sector activities are expected to increase with the currently ongoing and proposed infrastructure projects in the areas of electricity generation and development of transport infrastructure. The private sector is also expected to focus on housing construction including condominiums and apartment complexes.

CONSTRUCTION COST DATA

Cost of Labour

The figures are typical of labour costs in the Western Province (Colombo and its suburbs) as at the fourth quarter of 2008. The wage rate is on the basis of an employee's income, while the cost of labour (all-in rates) indicates the cost to a contractor of employing that employee. The difference between the two covers a variety of contributions – among them are EPF (Employees Provident Fund), ETF (Employees Trust Fund), holidays, bonus, insurance, welfare, training, uniforms and any other fringe benefits.

	Wage rate (per day) SLR	*Cost of labour (per day) SLR*	*Number of worked hours per year*
Site operatives			
Mason/bricklayer	850	1,300	2,038
Carpenter	850	1,300	2,038
Plumber	850	1,300	2,038
Electrician	850	1,300	2,038
Structural steel erector	1,000	1,500	2,038
HVAC installer	1,000	1,500	2,038
Semi-skilled worker	800	1,200	2,038
Unskilled labourer	650	975	2,038
Equipment operator	750	1,100	2,038
Watchman/security	700	1,050	2,038
	(per month)	*(per month)*	
Site supervision			
General foreman	25,000	50,000	2,160
Trades foreman	30,000	55,000	2,160
Clerk of works	35,000	60,000	2,160
Contractor's personnel			
Site manager	90,000	140,000	2,160
Resident engineer	90,000	140,000	2,160
Resident surveyor	70,000	120,000	2,160
Junior engineer	40,000	65,000	2,160
Junior surveyor	40,000	65,000	2,160
Planner	45,000	70,000	2,160
Consultants' personnel			
Senior architect	110,000	200,000	1,920
Senior engineer	110,000	200,000	1,920
Senior surveyor	100,000	180,000	1,920
Qualified architect	50,000	75,000	1,920
Qualified engineer	50,000	75,000	1,920
Qualified surveyor	50,000	75,000	1,920

Wages for site supervision / contractor's personnel / consultants personnel are based on private sector salary structure (top range)

Cost of Materials

The figures that follow are the costs of main construction materials, delivered to site in Colombo city area, as incurred by contractors in the fourth quarter of 2008. These assume that the materials would be in quantities as required for a medium sized construction project and that the location of the works would be neither constrained nor remote.

All the costs in this section excludes Goods & Services Tax (GST).

	Unit	*Cost SLR*
Cement and aggregate		
Ordinary Portland cement in 50kg bags	tonne	15,000
Coarse aggregate for concrete (20mm)	m^3	19,000
Fine aggregates for concrete	m^3	2,475
Ready mixed concrete(15N/mm^2)	m^3	10,300
Ready mixed concrete(20N/mm^2)	m^3	11,000
Ready mixed concrete(25N/mm^2)	m^3	11,300
Ready mixed concrete(30N/mm^2)	m^3	12,300
Ready mixed concrete(35N/mm^2)	m^3	12,600
Ready mixed concrete(40N/mm^2)	m^3	13,000
Steel		
Mild steel reinforcement	tonne	100,000
High tensile steel reinforcement	tonne	110,000
Bricks and blocks		
Common bricks (215 x 102.5 x 65mm) – hand cut	1,000	7,500
Good quality facing bricks (215 x 102.5 x 65mm)	1,000	14,700
Hollow cement blocks (400 x 200 x 100mm)	1,000	35,700
Rubble (150mm - 25mm)	m^3	1,100
Timber and Insulation		
Formwork timber class III(3/4" thick)	m^2	280
Timber class I (25 x 100mm)	m	850
Plywood sheets (8' x 4') – imported (15mm thick)	each	2,500
Plywood doors (2'9" x 6'9")	each	6,300
Glass and ceramics		
Plain glass (3mm)	m^2	550
Good quality ceramic wall tiles (108 x 108mm)	m^2	1,350
Plaster and paint		
Lime plaster in 25kg bags	tonne	8,500
Emulsion paint in 4 litre bucket	litre	575
Gloss enamel paint in 4 litre bucket	litre	710
Coloured pigment (Red)	kg	575

	Unit	Cost SLR
Tiles and paviors		
Ceramic floor tiles (300 x 300mm) – white	m²	1,400
In situ terrazzo	m²	4,000
Granite tiles (300 x 300mm)	m²	13,000
Drainage		
Sanitary ware-imported	3pcs	116,000
110mm diameter PVC pipes	m	1,800
Roof covering		
Calicut roof tiles	1000	35,350
Corrugated asbestos cement sheet	m²	540
Precast items		
Bent type fence posts intermediate 7'7"+1'6"(5" x 5") - (3" x 3")	nos	1,120
Concrete slabs 2'0" x 2'0" x 3"	nos	435
Kerbs concrete grade 20 150 x 300 915 type A	nos	500
Pre-stressed up rights – height 1050mm	nos	890
Pre-stresses hand rails 2000mm long 100mm diameter	nos	1,100

Unit Rates

The descriptions below are generally shortened versions of standard descriptions listed in section 4. Where an item has a two digit reference number (e.g. 05 or 33), this relates to the full descriptions against that number in section 4. Where an item has an alphabetic suffix (e.g. 12A or 34B) this indicates that the standard has been modified. Where a modification is a major one the complete modified description is included here and the standard description should be ignored, where a modification is minor (e.g. the insertion of a named hardwood) the shortened description has been modified here but, in general, the full description in section 4 prevails.

The unit rates below are for main work items on a typical construction project in the Colombo area in the fourth quarter of 2008. The rates include all necessary labour, materials and equipment. Allowances to cover preliminary and general items and contractor's overheads and profit have been included in the rates. All the rates in this section exclude goods and services tax (GST).

		Unit	Rate SLR
Excavation			
01A	Mechanical excavation of foundation trenches (not exceeding 1m depth)	m³	300
02	Hardcore filling making up levels	m³	1,700

		Unit	*Rate SLR*
Concrete work			
04A	Plain in situ concrete in strip foundation in trenches (15N/m^2)	m^3	8,300
05	Reinforced in situ concrete in beds (20N/m^2)	m^3	11,300
06	Reinforced in situ concrete in walls (20N/m^2)	m^3	11,500
07A	Reinforced in situ concrete suspended floors or roof slabs (25N/m^2)	m^3	11,850
08A	Reinforced in situ concrete in columns (30N/m^2)	m^3	12,590
09A	Reinforced in situ concrete in isolated beams (30N/m^2)	m^3	12,590
Formwork			
11A	Plywood formwork to concrete walls	m^2	1,600
12A	Plywood or metal formwork to concrete columns	m^2	1,530
13A	Plywood or metal formwork to horizontal soffits of slabs	m^2	1,450
Reinforcement			
14A	Reinforcement in concrete walls (10mm)	tonne	160,000
15A	Reinforcement in suspended concrete slabs (10mm)	tonne	157,000
16A	Fabric reinforcement in concrete (A142 steel mesh)	m^2	871
Brickwork and block work			
20	Precast lightweight aggregate hollow concrete block walls (100mm thick)	m^2	1,700
21A	Brickwork in common bricks bedded in 1:5 cement sand mortar (1 brick thick)	m^2	3,200
21B	Brickwork in common bricks bedded in 1:5 cement sand mortar (1/2 brick thick)	m^2	1,650
23A	Facing bricks bedded in 1:5 cement sand mortar (1 brick thick)	m^2	4,800
Roofing			
24A	Calicut roof tiles 400 x 250mm	m^2	640
25A	Plain clay roof tiles 200 x 150mm	m^2	1,350
26A	Half round roof tiles	m^2	1,750
27A	Lunumidella roof boarding	m^2	1,400
31A	Double sided reflective aluminium foil with wool blanket for thermostatic insulation (including sound insulation)	m^2	800
33A	Zinc/Aluminium steel roof sheeting	m^2	3,100

		Unit	*Rate SLR*
Woodwork and metalwork			
34A	Preservative treated sawn timber 75 x 100mm	m	1,100
35	Preservative treated sawn softwood 50 x 150mm	m	1,375
36A	Single glazed casement window in class 1 timber size 630 x 900mm	m^2	10,500
38A	Solid core half hour fire resisting hardwood internal flush door, size 838 x 1981mm	m^2	15,600
39A	Aluminium glazed window, size 1200 x 1200mm	m^2	16,800
40A	Aluminium glazed door, size 8050 x 2100mm	m^2	19,150
41A	Timber skirtings (class 1 timber) 25 x 100mm roof trusses	m	660
Plumbing			
42A	UPVC half round eaves gutter (112mm)	m	580
43A	UPVC rainwater pipes (110mm)	m	850
45A	High pressure plastic pipes for cold water supply (50mm)	m	550
46A	Low pressure plastic pipes for cold water distribution	m	250
47A	UPVC soil and vent pipes (110mm/type 600)	m	1,630
48	White vitreous china WC suite	each	7,830
49	White vitreous china lavatory basin	each	6,600
50	White glazed fireclay shower tray	each	7,850
51	Stainless steel single bowl sink and double drainer	each	8,000
Electrical work			
52	PVC insulated and PVC sheathed copper cable core	m	250
53	13 amp unswitched socket outlet	each	750
54A	Flush mounted 5amp, 1 way light switch	each	520
Finishings			
55A	2 coats cement based plaster on brick walls (rough finish)	m^2	350
56	White glazed tiles on plaster walls	m^2	2,700
57A	Non slip ceramic floor tiles	m^2	2,700
58A	Cement and sand screed to concrete floors (12mm thick)	m^2	415
59A	PVC floor tiles on screed	m^2	2,200
Glazing			
61	Glazing to wood	m^2	1,860
Painting			
62	Emulsion on plaster walls	m^2	360
63	Oil paint on timber	m^2	410

Approximate Estimating

The building costs per unit area given below are averages incurred by building clients for typical buildings in Sri Lanka as at the fourth quarter of 2008. They are based upon the total floor area of all storeys, measured between external walls and without deduction for internal walls.

Approximate estimating costs generally include mechanical and electrical installations but exclude furniture, loose or special equipment, and external works; they also exclude fees for professional services. The costs shown are for specifications and standards appropriate to Sri Lanka and this should be borne in mind when attempting comparisons with similarly described building types in other countries. A discussion on this issue is included in Section 2. Comparative data for countries covered in this publication, including construction cost data, are presented in Part Three.

Approximate estimating costs must be treated with caution; they cannot provide more than a rough guide to the probable cost of building. All the rates in this section exclude Goods and Services Tax (GST).

	Cost m^2 *SLR*	*Cost* ft^2 *SLR*
Industrial buildings		
Factories for owner occupation (light industrial use)	24,300	2,258
Factories for owner occupation (heavy industrial use)	59,000	5,481
Factory/office (high tech) for owner occupation (controlled environment, fully finished)	51,500	4,784
Warehouse, low bay for owner occupation	40,000	3,716
Warehouse, high bay for owner occupation	31,000	2,880
Administrative and commercial buildings		
Civic offices, non air-conditioned	49,725	4,620
Civic offices, fully air-conditioned	53,400	4,961
Offices for letting/owner occupation high rise, air-conditioned 10 to 15 storeys.	77,200	7,172
Headquarters office, 5 to 10 storeys, air-conditioned	93,700	8,705
Prestige office, high rise with air-conditioning and parking (intelligent buildings)	145,300	13,499
Health and education buildings		
General hospitals	41,200	3,828
Private hospitals	82,600	7,674
Health centres	53,600	4,980
Nursery schools	22,500	2,090
University buildings	28,500	2,648
Management training centres	26,000	2,415

	Cost m² SLR	Cost ft² SLR
Recreation and arts buildings		
Concert halls including seating and stage equipment	14,800	1,375
Swimming pools (international standard) including changing and social facilities (surface tension)	19,300	1,793
Swimming pools (school standard) including changing facilities	-	-
Local museums	21,200	1,970
City centre/shopping complex including parking	88,400	8,213
Book shops/libraries	33,000	3,066
Town development/shopping/bus stands	30,500	2,834
Studio/engineering buildings for television network	41,400	3,846
Shopping arcades	16,400	1,524
Stadia	19,300	1,793
Residential buildings		
Social/economic single family housing (single units)	27,800	2,583
Private/Private single family housing 2 storey detached	30,300	2,815
Purpose designed single family housing 2 storey detached (single unit)	37,950	3,526
Local/economic apartment housing, low rise (no lifts) – low cost	28,500	2,648
Social/economics apartment housing, low rise (with lifts)	38,400	3,567
Private sector apartment building (standard specification)	43,000	3,995
Private sector apartment building (luxury)	56,000	5,203
Students/nurses hall of residence low cost	26,500	2,462
Hotel, 5 star, city centre	142,000	13,192
Hotel, 3 star, city	101,800	9,457
Resorts	46,300	4,301
Resorts -cottage type	67,600	6,280
Motel	39,300	3,651

Construction Cost Index

The table below presents construction cost index in Sri Lanka since 1999.

Year	*1999*	*2000*	*2001*	*2002*	*2003*	*2004*	*2005*	*2006*	*2007*
Construction Cost Index (1990 = 100)	167.8	175.5	196.6	208.6	220.6	245.2	290.6	343.5	387.6
Change (%)	1.2	5.4	12.0	5.9	5.7	11.1	18.5	18.4	12.7

Source: Central Bank of Sri Lanka

EXCHANGE RATES

The graph below plots the movement of the Sri Lankan rupee against the sterling, the euro, the US dollar and 100 Japanese yen since 1998. The values used for the graph are quarterly and the method of calculating these is described and generally guidance on the interpretation of the graph provided in section 2. The average exchange rate in the fourth quarter of 2008 was SLR172.70 to pound sterling, SLR144.87 to euro, SLR 109.82 to US dollar and SLR114.50 to 100 Japanese yen.

THE SRI LANKAN RUPEE AGAINST STERLING, EURO, US DOLLAR AND 100 JAPANESE YEN

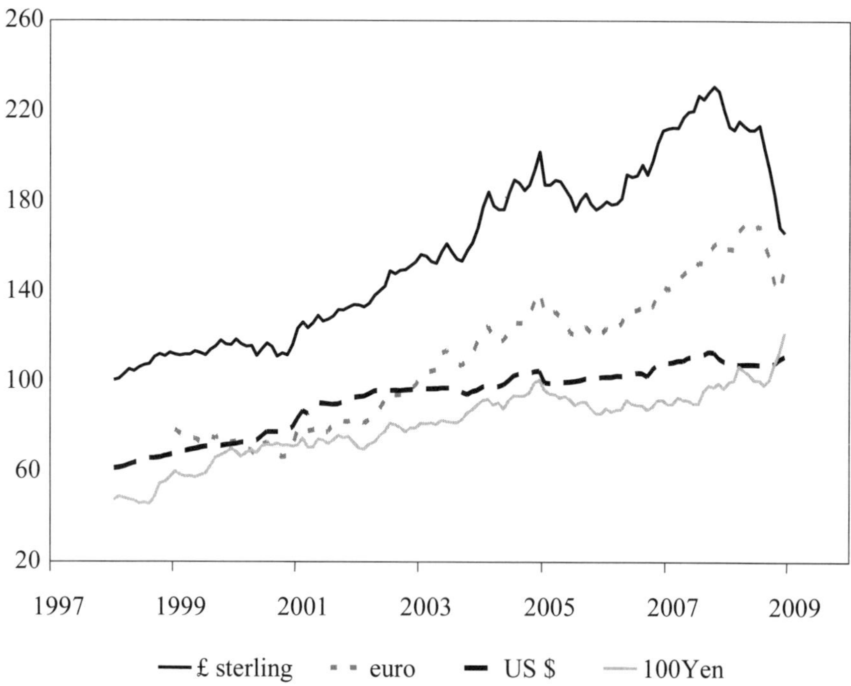

USEFUL ADDRESSES

Public Organizations

Board of Investment of Sri Lanka
Level 26, West Tower
World Trade Centre
Echelon Square
Colombo 01
Tel: (94) 112 434403-5 / 112 435027 /
112 447531 / 112 386953-4
Fax: (94) 112 447994-5 / 112 422407
E-mail: infoboi@boi.lk
Website: www.boi.lk

Building Economics Research Unit (BERU)
Department of Building Economics
Faculty of Architecture
University of Moratuwa
Tel: (94) 112 645301 / 112 645401 / 112 645671 – Ext 263

The Central Bank of Sri Lanka
P.O. Box 590
30, Janadhipathi Mawatha
Colombo 01
Tel: (94) 112 477000 / 112 440330 / 112 330220
E-mail: cbslgen@cbls.lk
Website: www.cbsl.gov.lk

Department of Buildings
2nd Floor
Sethsiripaya
Sri Jawardenapura
Battaramulla
Tel: (94) 112 861489 / 112 861494 /
112 862921 / 112 862922 / 112 889456
Fax: (94) 112 864771
Website: www.buildings.gov.lk

The Department of Census and Statistics
P.O. Box 563
Colombo
Tel: (94) 112 675297
Fax: (94) 112 697594
E-mail: information@statistics.gov.lk
Website: www.statistics.gov.lk

Institute for Construction Training and Development (ICTAD)
123 Wijerama Mawatha
Colombo 07
Tel: (94) 112 686092
Fax: (94) 112 699738
E-mail: ictad@sltnet.lk
Website: www.ictad.lk

National Housing Development Authority (NHDA)
Sir Chittampalam A. Gardiner Mawatha
P.O. Box 1826, Colombo 02
Tel: (94) 112 431932 / 112 431707
Fax: (94) 112 449622
E-mail: nhdadgmseve@sltnet.lk
Website: www.nhda.lk

Road Development Authority
Head Office
Sethsiripaya
Battaramulla
Tel: (94) 112 862721 / 112 862722
Website: www.rda.gov.lk

Sri Lanka Standards Institution
17 Victoria Place
Elvitigala Mawatha
Colombo 08
Tel: (94) 112 671567-72
Fax: (94) 112 671579
E-mail: slsi@slsi.slt.lk
Website: www.slsi.lk

The Urban Development Authority
7th Floor, Sethsiripaya
Battaramulla
Tel: (94) 112 346091
Fax: (94) 112 877472
E-mail: charman@uda.lk
Website: www.urbanlanka.lk

Trade And Professional Associations

Chamber of Construction Industry
No. 120/7, Widya Mawatha
Off Wijerama Mawatha
Colombo 7
Tel: (94) 112 697109 / 112 682472 / 112 691710
Fax (94) 112 682757
E-mail: cci@sltnet.lk
Website: www.buildsrilanka.com\ccisl

Institute of Project Managers of Sri Lanka
189/1B, Nawala Road
Nugegoda
Tel: (94) 112 815391-2 / 114 406893
Fax: (94) 112 815390
E-mail: rsvpgsm@yahoo.com, lanmds@slt.lk
Website: www.ipmsl.org

Institute of Quantity Surveyors (IQS)
2nd Floor
275/5, Prof Stanley Wijesundara Mawatha
Colombo 07
Tel (94) 112 595570
E-mail: iqssl@sltnet.lk
Website: www.iqssl.org

National Building Research Organisation (NBRO)
99/1 Jawatta Road
Colombo 05
Tel: (94) 112 588946
Fax: (94) 112 502611
E-mail: nbro@sltnet.lk
Website: www.nbro.gov.lk

National Construction Contractors Association of Sri Lanka (NCCASL)
No. 350 A, Pannipitiya Road
Pelawatthe
Battaramulla
Tel: (94) 112 786325 / 114 541743
Fax: (94) 112 784355
Website: www.ncasrilanka.com

Organisation of Professional Association (OPA)
275/5 Baudhaloka Mawatha
Colombo 07
Tel: (94) 112 697109

Sri Lanka Institute of Architects (SLIA)
120/7, Vidya Mawatha
Colombo 07
Tel: (94) 112 697101
E-mail: slialib1@sltnet.lk
Website: www.slia.lk

Sri Lanka Institute of Engineers
120/15 Wijerama Mawatha
Colombo 07
Tel: (94) 112 698426 / 112 685490
Fax: (94) 112 699202
E-mail: iesl@slt.lk
Website: www.iesl.lk

TAIWAN

All data relate to 2008 unless otherwise indicated.

Population	
Population (2008)	23.04 mn
Urban population (2007)	55%
Population under 15 (2008)	16.95%
Population 65 and over (2008)	10.42%
Average annual growth ratc (2005 to 2008)	3.81%
Geography	
Land area	36,189 km^2
Agricultural area	23%
Capital city	Taipei
Population of capital city	2.62 mn
Economy	
Monetary unit	New Taiwan Dollar (NT$)
Exchange rate (average fourth quarter 2008) to:	
the pound sterling	NT$ 50.06
the US dollar	NT$ 33.05
the euro	NT$ 42.97
the yen x 100	NT$ 35.56
Average annual inflation (2005 to 2008)	2.06%
Inflation rate (2008)	3.53%
Gross Domestic Product (GDP)	NT$ 12,365 bn
GDP per capita	NT$ 536,675
Average annual real change in (GDP) (2005 to 2008)	3.7%
Private consumption as a proportion of GDP (2008)	54.4%
Public consumption as a proportion of GDP (2008)	11.2%
Investment as a proportion of GDP	16.6%
Construction	
Gross value of construction output	NT$ 286.7 bn
Net value of construction output (2007)	NT$ 269.8 bn
Net value of construction output as a proportion of GDP (2007)	2.18%

THE CONSTRUCTION INDUSTRY

Construction Output

The gross output of construction in 2008 was NT$286.7 billion, equivalent to US$8.7 billion, or 2.3% of GDP. The net output of construction in 2007 was approximately NT$269.8 billion, equivalent to US$8.2 billion, or 2.2% of GDP.

The output of construction for the years 2005-2008 is illustrated in the table below.

OUTPUT OF CONSTRUCTION, 2005-2008 (MN NT$)

	NT$	*Real Growth Rate (%)*
2005	227,548	-
2006	257,757	13.27
2007	273,326	6.04
2008	286,696	4.89

Source: Directorate–General of Budget, Accounting and Statistics

Since 1997 the output of the construction industry had declined in consecutive years and had regained a positive growth rate till 2005. The situation was affected by mixed effects of overall economy sluggishness caused by the Asian Financial Crisis, a slack housing market, and especially the reduced government fixed investment. The Financial tsunami triggered by the sub-prime housing mortgage in US in mid-2008 has further impacted the economy of Taiwan. The Taiwan Government has tried to boost the public expenditure trying to compensate for the shrinkage in the private investment.

The new government elected in March 2008 has committed to launch a series of major projects trying to re-boost the momentum of the economy as well as upgrade the infrastructure for the island in order to realize the election campaign promises. The major projects along with the estimated budget which relates to construction are Convenient Transportation Network Island Wide (NT$1,450bn), Kaohsiung Free Trade and Ecological Harbor (NT$57.7bn), Taichung Asia-Pacific Sea Air Freight Planning Center (NT$50bn), Taoyuan International Airport City (NT$67bn), Anti-flooding System (NT$18.6bn), Construction of Sewer System (NT$240bn), Industry Renovation Alley (NT$11.5bn). Among the above budgets, 2/3 will be invested by the government and the other 1/3 will be private investment.

The residential market was booming in 2006 and 2007 after sluggishness over a long period of time. However, due to the financial crisis along with a large number of residential vacancies in the market, the market is slowing in 2008. By lifting the ban on mainland China's capital investment in Taiwan's real estate (mainly in office building and resort development), the market may regain its prominence in the future.

Characteristics and Structure of the Industry

It is a legal requirement in Taiwan for construction firms to be registered under a government administered licensing system. There are five types of licences: Class A, B, C, specialist contractor and small contractor. The licences are awarded according to a company's capital, technical ability and experience. The highest category, Class A, enables the contractor to tender for projects of a reasonable size. Class B enables the contractor to work on projects up to a value of NT$75 million and Class C on projects up to a value of NT$22.5 million and 10 times of its capital for specialist contractor and NT$6 million for small contractor.

A contractor's licence can be upgraded after a set period of time and subject to satisfactory performance at the class in which it is currently operating. Licences can also be transferred whereby a contractor may purchase a Class A licence from another company which no longer has the technical expertise to handle large projects.

In the past the licensing system has served to exclude foreign contractors from entering the marketplace. However, based on the newly revised contractor law, foreign contractors can apply for the contractor licence as long as the applicant meets all the requirement set forth under the said law. As of end 2008, there were 1,814 contractors in Class A, 1,276 in Class B, 6,108 in Class C, 243 in specialist contractor and 5,115 in small contractor.

Typically the local construction companies are small family run businesses, short on expertise and modern technology. Commitment to research and development is limited with new technologies being obtained by means of technology transfer from foreign joint venture partners or simply purchased as necessary. The largest local construction company is Ret-Ser Engineering Agency (RSEA), a previously quasi-government organisation which had, prior to April 1997, received preferential consideration when tendering for public works. Amid allegations of corruption and in line with a general restructuring of government procurement practices aimed at minimising inefficiencies, these special privileges were withdrawn. The company has been privatized and changed its name to RSEA Engineering Corporation on 1st July 1998.

When applying for admission into the World Trade Organisation (WTO), Taiwan was strongly requested by most of the members that an Agreement on Government Procurement be signed. It was therefore for the government to conduct fundamental reforms to its government procurement systems and review and revise all relevant law and code in order to meet the spirit of international collaboration. Normally, for projects in excess of NT$50 million, the procurement will be administered by the Public Construction Commission who are authorised to solicit foreign tenders. Tenders are announced through the *Government Procurement Gazette* and through the *Government Procurement Information System (GPIS)* bulletin board, accessed on the Internet through gpis.pcc.gov.tw. The tendering and procurement process is monitored by the Ministry of Audit (MOA) in order to ensure transparency, non-discrimination, efficiency and accountability. A complaints and appeal procedure has also been established to investigate and rule on any complaints received about the tendering process. The tenderer also has the right to file an action in a court against the procuring entity for any breach of a contract awarded.

Whilst Taiwan recorded an unemployment rate of 5.03% in 2008 due to the global financial crisis, the unemployment rate has been on the downward trend. In June 2008 the construction industry employed an average of 855,000 people, of which approximately 12,300 (or 1.44%) were foreigners.

Clients and Finance

After long sluggishness in the construction sector, land prices remained in its low level as well as labour rates. The enforcement of Government Procurement Law, the reform of the country procurement system and the recent ease in the relationship with mainland China have improved the overall investment environment. The poor performance of the construction industry in 2008 was mainly caused by the global depression, the skyrocketed raw material prices as well as the US housing sub-prime crisis. The situation deteriorates further in the second half year of 2008. Currently, signs of recovery of the market are not anticipated to occur in year 2009.

In order to provide a boost for the sector, the government has accelerated programmes for infrastructure and public works projects and reviewed the legal codes governing the industry. However, faced with the difficulty of balancing the budget and looking to achieve effective management and cost control on public works projects, the government need more participation from the private sector.

The housing market is the single largest sector of the construction industry. Up to October 2008, approval was given for the commencement of 22,456,767m^2 of building projects. Funding comes from both private and public sectors. The government also offers a low interest loan scheme to assist low-income families in meeting their housing needs.

It is anticipated that foreign investment especially for those capital from mainland China in the construction market will increase as a result of both the BOT projects and the relaxation of foreign investment restrictions. Overall, it is however expected that domestic investment and demand will continue to dominate for the foreseeable future.

APPROVED FOREIGN INVESTMENT IN CONSTRUCTION PROJECTS

Year	*Amount US$ (Thousand)*
1997	4,266,629
1998	3,738,758
1999	4,231,404
2000	7,607,755
2001	5,128,518
2002	3,271,749
2003	3,575,674
2004	3,952,148
2005	4,228,068
2006	13,969,247
2007	15,361,173
2008	8,232,059

Source: Investment Commission Ministry of Economic Affairs Republic of China (Jan 2009)

Selection of Design Consultants

In the case of architects, the normal basis of selection will be on their experience. Personal contacts and recommendations also play a part. The profession of architect in Taiwan is split into two categories: Class A and Class B. These classes are used as a basis for establishing their suitability for different types of work. As of 2007, there were 3,283 Class A Architects and 28 Class B Architects.

As with the architects, the structural engineer will normally be selected according to track record and general suitability for the project in question.

The title of quantity surveyor or cost consultant is not formally recognized in Taiwan. The preparation of estimates, tender documents, interim payments and advising on contractual matters all fall within the architect's purview or be part of the professional construction management service.

Contractual Arrangements

The procedures of the tendering of public sector works is regulated in the Government Procurement Law and is briefly described as follows (assuming a procurement authority is acting on end user's behalf in carrying out the tender process):

(1) Preparation of Invitation Documentation
Upon receipt of a procurement authorization from a government agency or enterprise (the end-user), procurement authority will first review the end-user's specifications with special terms and conditions to ensure the suitability for a tender from the points of view of government regulations and international commercial practice.
(2) Public Notice of an Invitation to Tender
For open tendering procedures or selective tendering procedures, procurement authority will publish a notice of invitation to tender or of qualification evaluation on the *Government Procurement Gazette*. The time-limit for submission of tenders from the date of publishing a notice to receiving documents varies with the value and contents of the procurement.
(3) Submission of Tenders
All prospective tenders are required to use the standard Invitation, Tender and Contract form that are available at the procurement authority office. The tender shall be submitted to procurement authority before the deadline for tendering. Any tender that is received at the procurement authority after the deadline will not be considered unless a tender from an overseas company is airmailed to procurement authority before the tendering deadline and it is stated in the company's fax reaching procurement authority prior to the tendering deadline.
(4) Requirement for a Bid Bond
Unless otherwise specified, a bid bond is required. The bid bond shall be deposited by cash, bank's promissory note, bank's check, certified check, bearer's government bond of ROC, a certificate of deposit pledged to the procuring entity, irrevocable stand-by letter of credit issued or confirmed by a bank, or bank

guarantee or insurance policy under which the bank or insurer shares the liability with the tender jointly and severally.

Bid bonds provided by unsuccessful tenderers will be returned without interest after announcement of the award of a contract or contracts.

(5) Tender Opening & Tender Evaluation

The end-user shall set a government estimate before the opening of the tenders. Procurement authority official will publicly open all the tenders received and read out the essential points of each tender, such as names of the tenderer and the manufacturer, source of supply, price and shipment date, etc. Tenderers and their suppliers or manufacturers are always welcome to attend the opening of tenders. After that, the end-user will be responsible for the evaluation of specifications contained in the tenders. Tenders of simple contents may be evaluated on the spot and then a contract will be awarded to the winning tenderer. Sometimes clarification is sought to ensure the acceptability of the tenders before any decision is made. Procurement authority will record all the opening processes of each tender and provide the end-user and supervision personnel with a copy of the record.

(6) Award of a Contract

The award of a contract will, in principle, be made to the tenderer whose tender meets the requirements set forth in the tender documentation and is the lowest tenderer within the government estimate. That is to say, the lowest acceptable tenderer may not win the award if his price exceeds the government estimate, but he will usually have a chance to reduce his price to the extent equal to or lower than the government estimate.

(7) Signing of a Contract

After the award is made, a Notice of Award will be issued to the winning tenderer and a contract is signed.

Generally, public contracts will be let under a government standard form of contract. In cases where international bids are being invited, *FIDIC* (an international form of contract), the *UK Joint Contracts Tribunal (JCT)*, or the *American Institute of Architects Form of Contract* may be used.

(8) Requirement for a Performance Bond

Usually the contractor is required to deposit a performance bond within certain days after the date of Notice of Award.

The performance bond shall also be posted in one of the forms enumerated for the bid bond except that the validity and contents of the stand-by L/C are somewhat modified to meet the requirements of contract.

Tender for Private Project:

In the private sector, the market is very commercial. The employer or client will himself take responsibility for the issue of the tender. He will also conduct any subsequent negotiations. In international terms, the tendering process could be viewed as relatively unsophisticated with little emphasis on tendering procedures and fairplay.

Liability and Insurance

Most contracts for public works are let under the government standard forms of contract. These forms set out the risks, rights and obligations of the various parties to the contract. Generally, the contracts place a significant amount of the risk on the contractor.

For government projects, disputes are referred to the Public Construction Procurement Appeal Review Committee. The decree of the said committee was first published by Public Construction Commission (PCC) in April 1999 and revised in September 2002. The committee was separated into two levels, i.e. central government level and local government level who deal with these disputes arise in central and local government procurement respectively. The role of the Procurement Appeal Committee is as follows:

- To settle disputes relating to tender invitation procedures for any public construction project.
- To settle contract disputes relating to public construction.
- To settle disputes of any other nature relating to public construction projects.

The period for which a contractor remains liable under a contract varies and must be written into the contract. In addition, under the standard form of contract, the contractor is responsible for providing insurances covering the works themselves, workmen engaged on the works and third party liability. It is also common for the contract to call for a performance bond (bank bond), often involving a substantial sum of money.

Development Control and Standards

The Construction and Planning Agency (CPA) was established in March 1981 and is the governmental department in charge of regional planning, city planning and building regulations among others. The following departments fall under its auspices.

The Department of Regional Planning is responsible for the overall planning of national land use. This will include development and ratification of regional plans, supervision, promotion and coordination of affairs related to regional development. On a more local level, the Department of City Planning is responsible for urban development policies. City planning laws, new towns planning, urban renewal and coordination of metropolitan development also fall under their jurisdiction. All planning applications and licences for construction are processed through this department.

Land usage in Taiwan is strictly regulated under the Urban Planning, Area Planning and Construction Laws. The purpose for which a site may be used is defined in terms of Zones. Re-zoning is possible but can be difficult, protracted and ultimately dependent upon government discretion. The Statute for Upgrading Industry (SUI) also sets forth important provisions specifying land use and land rights.

In common with most developed countries, there are increasing concerns over the environment and the environmental impact of any construction project. As people become more affluent, they are demanding a higher quality of life and are less willing to sacrifice their environment for apparent economic progress. Thus there was a need for comprehensive environmental legislation and enforcement. Towards that end, in 1987 Taiwan's Environmental Protection Agency (EPA) was formed. The primary function of the EPA is to articulate and develop environmental policy and draft regulations implementing existing legislation. The task of enforcing environmental laws and regulations has been delegated to the Environmental Protection Bureau.

The Construction and Planning Administration also regulates the industry. This regulation includes approval of applications to establish new construction companies and new subcontracting firms and approval of advancement of construction companies to higher classifications.

CONSTRUCTION COST DATA

Cost of Labour

The figures below are typical of labour costs in the Taipei area in the fourth quarter of 2008. The wage rate is the basis of an employee's income, while the cost of labour indicates the cost to a contractor of employing that employee. The difference between the two covers a variety of mandatory and voluntary contributions – a list of items which could be included is given in section 2.

	Wage rate (per day) NT$	*Cost of labour (per day) NT$*	*Number of hours worked per year*
Site operatives			
Bricklayer	2,500	2,900	2,200
Carpenter	2,500	2,900	2,200
Plumber	2,300	2,700	2,200
Electrician	2,300	2,900	2,200
Structural steel erector	2,600	4,400	2,200
Semi-skilled worker	2,300	3,000	2,200
Unskilled labourer	1,800	2,300	2,200
Steel bender	2,480	3,640	2,200
Scaffolder	2,430	3,550	2,200
Plasterer	2,500	2,900	2,200
	(per month)	*(per month)*	
Site supervision			
General foreman	66,000	92,400	2,200
Trades foreman	60,000	84,000	2,200
Clerk of works	72,000	100,000	2,200

	Wage rate (per month) NT$	*Cost of labour (per month) NT$*	*Number of hours worked per year*
Contractors' personnel			
Site manager	70,000	105,000	2,200
Resident engineer	50,000	70,000	2,200
Junior engineer	35,000	50,000	2,200
Planner	45,000	65,000	2,200
Consultants' personnel			
Senior architect	160,000	320,000	2,080
Senior engineer	120,000	240,000	2,080
Qualified architect	100,000	200,000	2,080
Qualified engineer	90,000	180,000	2,080

Cost of Materials

The figures that follow are the costs of main construction materials, delivered to site in the Taipei area, as incurred by contractors in the fourth quarter of 2008. These assume that the materials would be in quantities as required for a medium sized construction project and that the location of the works would be neither constrained nor remote.

All the costs in this section exclude value added tax (VAT).

	Unit	*Cost NT$*
Cement and aggregate		
Ordinary portland cement in 50kg bag	bag	175
Coarse aggregates for concrete	m^3	700
Fine aggregates for concrete	m^3	750
Ready mixed concrete (4000 Psi)	m^3	2,700
Ready mixed concrete (2000 Psi)	m^3	2,100
Steel		
Reinforcement ($fy \geq 4200kg/cm^2$)	tonne	17,500
Reinforcement ($fy \geq 2800kg/cm^2$)	tonne	16,800
H section (below 700mm)	kg	25
H section (above 700mm)	kg	26
Channel	kg	21
Angle	kg	21
Bricks and Blocks		
Common bricks (210 x 100 x 50mm)	1,000	2,400
Good quality facing bricks (210 x 100 x 50mm)	1,000	4,000
Hollow concrete blocks (390 x 190 x 190mm)	1,000	4,700
Timber and insulation		
Exterior quality plywood (12mm)	m^2	200

	Unit	*Cost NT$*
Plywood for interior quality (12mm)	m²	260
4mm thick quilt insulation	m²	400
100mm thick rigid slab insulation	m²	240
Hardwood internal door complete with frame and ironmongery	each	9,000
Glass and ceramics		
Semi-reflective glass (6mm)	m²	750
Plaster and paint		
Good quality ceramic wall tiles (200 x 300mm)	m²	630
Plasterboard (12mm thick)	m²	450
Emulsion paint	litre	120
Tiles and paviors		
Clay floor tiles (250 x 250 x 9mm)	m²	450
Vinyl floor tiles (300 x 300 x 2.3mm)	m²	345
Precast paving slabs (250 x 250 x 25mm)	m²	550
Drainage		
WC suite complete	each	15,000
Lavatory basin complete	each	5,000
150mm cast iron drain pipes	m	700

Unit Rates

The descriptions below are generally shortened versions of standard descriptions listed in full in section 4. Where an item has a two digit reference number (e.g. 05 or 33), this relates to the full description against that number in section 4. Where an item has an alphabetic suffix (e.g. 12A or 34B) this indicates that the standard description has been modified. Where a modification is major the complete modified description is included here and the standard description should be ignored; where a modification is minor (e.g. the insertion of a named hardwood) the shortened description has been modified here but, in general, the full description in section 4 prevails.

The unit rates below are for main work items on a typical construction project in Taipei in the fourth quarter of 2008. The rates include all necessary labour, materials and equipment. An allowance of 5% - 10% to cover preliminary and general items has been added to the rates. All the rates in this section exclude value added tax (VAT).

		Unit	*Cost NT$*
Excavation			
01	Mechanical excavation of foundation trenches including earthwork support	m³	725

		Unit	*Cost NT$*
02	Hardcore filling in bed; 150mm thick	m^2	300
Concrete work			
04A	Plain in situ concrete (2000 psi) in beds	m^3	2,150
05A	Reinforced in situ concrete (4000 psi) in beds	m^3	2,850
06A	Reinforced in situ concrete (4000 psi) in walls	m^3	2,850
07A	Reinforced in situ concrete (4000 psi) in suspended floors	m^3	2,850
08A	Reinforced in situ concrete (4000 psi) in columns	m^3	2,850
09A	Reinforced in situ concrete (4000 psi) in suspended beams	m^3	2,850
Formwork			
11	Formwork to sides of wall	m^2	520
12	Formwork to sides of columns	m^2	520
13	Formwork to soffit of suspended slabs	m^2	580
Reinforcement			
14	Reinforcement in concrete walls	kg	23
15	Reinforcement in suspended concrete slabs	kg	23
Steelwork			
17	Fabricate, supply and erect steel frame structure	tonne	50,000
Brickwork and blockwork			
22A	Solid (perforated) sand lime bricks (half brick thick)	m^2	692
Roofing			
24A	300 x 300 x 20mm thick concrete tiles	m^2	424
30A	Waterproof sheet membrane	m^2	368
30B	Waterproof cement and sand screed; average 90mm thick	m^2	480
32A	Polystyrene board insulation on roof slabs	m^2	312
Woodwork and metalwork			
37A	Proprietary plastic laminated door; size 900 x 2100mm (excluding ironmongery)	each	10,606
38A	One hour fire rated proprietary plastic laminated door; size 900 x 2100mm (excluding ironmongery)	each	31,262
39A	Double glazed aluminium window; size 900 x 2100mm	each	8,708
40A	Proprietary steel door; size 2000 x 1400mm	each	45,000
41A	38mm diameter stainless steel tubular rails	m	3,200
41B	50mm diameter stainless steel tubular rails	m	3,500

		Unit	*Cost NT$*
Plumbing			
44A	50mm diameter galvanised steel pipes; fixed to wall	m	350
44B	75mm diameter galvanised steel pipes; fixed to wall	m	600
44C	100mm diameter galvanised steel pipes; fixed to wall	m	800
47A	300mm wide x 600mm average depth surface channels	m	2,791
47B	Precast concrete channel covers	m	1,071
47C	600 x 400 x 30mm thick cast iron gratings	no	1,339
Finishings			
55A	20mm thick cement and sand plaster to wall	m^2	357
56A	200 x 200 x 5mm white glazed tiles	m^2	1,116
56B	Metallic lustre ceramic facing tiles to external wall	m^2	2,010
56C	100 x 100 x 9mm unglazed porcelain tiles	m^2	1,060
56D	Paperhanging; vinyl sheet covering to walls	m^2	712
58A	150mm thick lightweight concrete to floors	m^2	279
58B	50mm thick cement and sand paving; steel trowelled smooth	m^2	446
60A	Mineral fibreboard suspended ceiling system (2' x 2' x 5/8")	m^2	669
60B	Aluminium suspended ceiling system	m^2	2,679
Glazing			
61A	6mm thick clear float glass	m^2	1,172
61B	Reflective double glazing to metal	m^2	2,679
Painting			
62A	Emulsion paint with acrylic alkali resisting primer to ceilings	m^2	196
62B	Cement paint in two coats to plastered ceilings	m^2	100
62C	Spraying polyurethane paint to walls	m^2	257

Approximate Estimating

The building costs per unit area given on the next page are averages incurred by building clients for typical buildings in the Taipei area in the fourth quarter of 2008. They are based upon the total floor area of all storeys, measured between external walls and without deduction for internal walls.

Approximate estimating costs generally include mechanical and electrical installations but exclude furniture, loose or special equipment, and external works; they also exclude fees for professional services. The costs shown are for specifications and standards appropriate to Taiwan and this should be borne in mind when attempting comparisons with similarly described building types in other countries. A discussion of this issue is included in section 2. Comparative data for countries covered in this publication, including construction cost data, are presented in Part Three.

Approximate estimating costs must be treated with caution; they cannot provide more than a rough guide to the probable cost of building.

	Cost *m² NT$*	*Cost* *ft² NT$*
Industrial buildings		
Factories for letting	25,700	2,388
Factories for owner occupation (light industrial use)	27,200	2,527
Factories for owner occupation (heavy industrial use)	33,300	3,094
Factory/office (high-tech) for letting (shell and core only)	30,250	2,810
Warehouses, low bay for owner occupation	27,200	2,527
Administrative and commercial buildings		
Civic offices, fully air-conditioned	47,400	4,404
Offices for letting, 5 to 10 storeys, air-conditioned	36,300	3,372
Offices for letting, high rise, air-conditioned	40,800	3,790
Prestige/headquarters office, high rise, air-conditioned	54,450	5,059
Health and education buildings		
General hospitals (100 beds)	75,625	7,026
Health centres	60,500	5,621
Primary/junior schools	30,250	2,810
Secondary/middle schools	36,300	3,372
Recreation and arts buildings		
Theatres (over 500 seats) including seating and stage equipment	56,332	5,233
City centre/central libraries	46,893	4,356
Residential buildings		
Social/economic apartment housing, low rise	25,700	2,388
Social/economic apartment housing, high rise (with lifts)	30,250	2,810
Private sector apartment building (standard specification)	36,300	3,372
Private sector apartment buildings (luxury)	45,400	4,218
Hotel, 5 star, city centre	65,800	6,113

Regional Variations

The approximate estimating costs are based on projects in Taipei. For Kaohsiung and other areas, the costs should be reduced by approximately 6%.

Value Added Tax (VAT)

The standard rate of value added tax (VAT) is currently 5%, chargeable on general building work.

EXCHANGE RATES AND INFLATION

The combined effect of exchange rates and inflation on prices within a country and price comparisons between countries is discussed in section 2.

Exchange Rates

The graph below plots the movement of the New Taiwan Dollar against the sterling, the euro, the US dollar and 100 Japanese yen since 1998. The values used for the graph are quarterly and the method of calculating these is described and general guidance on the interpretation of the graph provided in section 2. The average exchange rate in the fourth quarter of 2008 was NT$50.06 to pound sterling, NT$42.97 to euro, NT$33.05 to US dollar and NT$35.56 to 100 Japanese yen.

THE NEW TAIWAN DOLLAR AGAINST STERLING, EURO, US DOLLAR AND 100 JAPANESE YEN

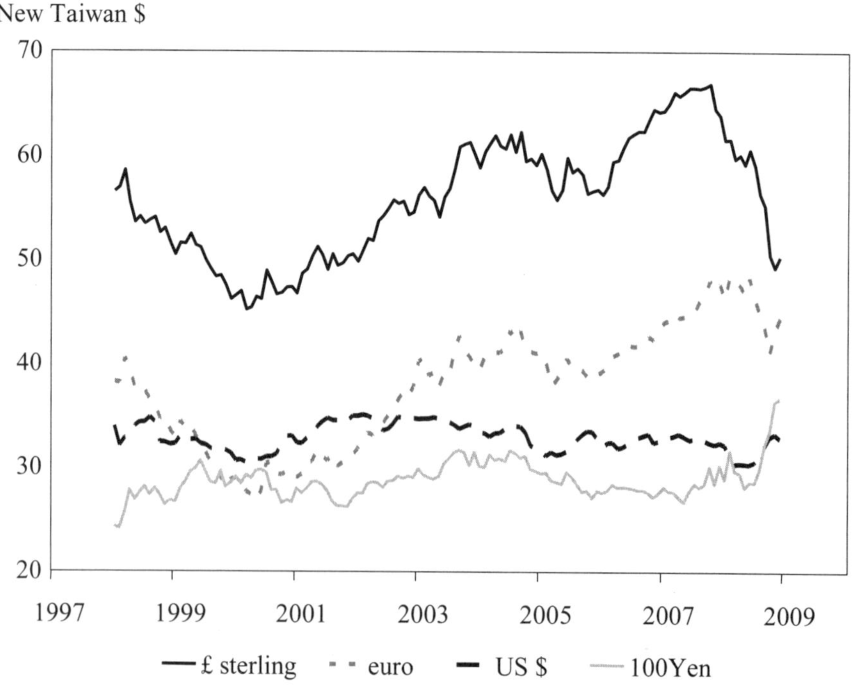

Price Inflation

The table below presents consumer price inflation in Taiwan since 1997.

CONSUMER PRICE INDICES

Year	Consumer Price Index		Construction Cost Index	
	average index	average change %	average index	average change %
1997	93.17	0.90	75.62	2.07
1998	94.73	1.68	77.50	2.49
1999	94.90	0.18	77.06	-0.57
2000	96.09	1.25	76.69	-0.48
2001	96.08	-0.01	75.92	-1.00
2002	95.89	-0.20	77.52	2.11
2003	95.62	-0.28	81.14	4.67
2004	97.17	1.61	92.60	14.12
2005	99.41	2.31	93.24	0.69
2006	100.00	0.60	100.00	7.25
2007	101.80	1.80	109.00	9.00
2008	105.39	3.52	124.25	13.99

Base: 2006 = 100

Source: Commodity – Price Statistics Monthly in the Taiwan Area of the Republic of China by the Directorate–General of Budget, Accounting & Statistics and Central Bank of China Annual Report

Construction costs rose from 2003 to 2008 mostly as a result of a strong residential market and the strong demand of construction material worldwide. Typically over the period from 1997 to 2007, labour costs have increased slightly whilst over the same period; especially from 2004 to 2007 material prices have rose significantly according to market demand. This trend remained unchanged in the early half year of 2008. However, due to the global financial crisis, construction costs began to decline from August 2008 onwards.

USEFUL ADDRESSES

Public Organizations

Building Administration Office of Taipei City Government
1 Shih Fu Road
Taipei
Tel: (886) 2 2720 8889
Fax: (886) 2 2720 3988

Bureau of Foreign Trade
Ministry of Economic Affairs
1 Hu-kou Street
Taipei
Tel: (886) 2 2351 0271
Fax: (886) 2 2351 7080

Construction and Planning Agency
Ministry of Interior
342 Section 2
Bade Road Songshan District
Taipei City
Tel: (886) 2 8771 2345
Fax: (886) 2 8771 2929
Website: www.cpami.gov.tw

Council for Economic Planning and Development
Executive Yuan
3 Baocing Road Jhongjheng District
Taipei City
Tel: (886) 2 2316 5300
Fax: (886) 2 2370 0415

Department of Rapid Transit Systems of Taipei City
7 Lane 48
Chungshan North Road
Section 2
Taipei City
Tel: (886) 2 2521 5550
Fax: (886) 2 2521 7639

Department of Urban Development of Taipei City Government
1 Shih Fu Road
Taipei
Tel: (886) 2 2720 8889
Fax: (886) 2 2759 3321

Government Information Office
Executive Yuan
2 Tianjin Street
Taipei City
Tel: (886) 2 3356 8888
Fax: (886) 2 2356 8733
E-mail: service@mail.gio.gov.tw
Website: www.gio.gov.tw

Ministry of Audit
1 Han Chou North Road
Taipei City
Tel: (886) 2 2397 1366
Fax: (886) 2 2397 7889
Website: www.audit.gov.tw

Ministry of Economic Affairs
15 Fujhou Street
Taipei City
Tel: (886) 2 2321 2200
Fax: (886) 2 2391 9398
E-mail: service@moea.gov.tw
Website: www.moea.gov.tw

Ministry of the Interior
5 Syujhou Road
Jhongjheng District
Taipei City
Tel: (886) 2 2356 5000
Fax: (886) 2 2356 6201
E-mail: gethis@mail.moi.gov.tw
Website: www.moi.gov.tw

Ministry of Transportation and Communication
50 Ren Ai Road
Section 1
Taipei City
Tel: (886) 2 2349 2900
Fax: (886) 2 2349 2491
Website: www.motc.gov.tw

Public Construction Commission
Executive Yuan
9/F, 3 Songren Road
Taipei City
Tel: (886) 2 8789 7500
Fax: (886) 2 8789 7800
Website: www.pcc.gov.tw

Public Works Department of Taipei City Government
1 Shih Fu Road
Taipei
Tel: (886) 2 2720 8889
Fax: (886) 2 2720 5817

Urban Department Office of Taipei City Government
9/F, 8 Roosevelt Road
Section 1
Taipei City
Tel: (886) 2 2321 5696
Fax: (886) 2 2397 4327

Trade Organizations

Chinese Institute of Civil and Hydraulic Engineering
4/F, 1 Jen Ai Road
Section 2
Taipei City
Tel: (886) 2 2392 6325
Fax: (886) 2 2396 4260

Chinese Institute of Engineers
3/F, 1 Jen Ai Road
Section 2
Taipei City
Tel: (886) 2 2392 5128
Fax: (886) 2 2397 3003

Chinese National Association of General Contractors
2/F, 40 Kaifeng Street
Section 2
Taipei City
Tel: (886) 2 2381 3488
Fax: (886) 2 2381 8366

National Federation of Professional Electrical Engineer
11/F, 69-10 Jhongsiao East Road
Taipei City
Tel: (886) 2 2778 8898
Fax: (886) 2 2778 8900

Taiwan Union Building Materials Association
Room 4
7/F, 374 Bade Road
Section 2
Taipei City
Tel: (886) 2 2751 8834
Fax: (886) 2 2777 2101

DAVIS LANGDON & SEAH (THAILAND) LTD

Davis Langdon & Seah (Thailand) Ltd manages client requirements, controls risk, manages cost and maximises value for money, throughout the course of construction projects, always aiming to be – and to deliver – the best.

TYPICAL PROJECT STAGES, INTEGRATED SERVICES AND THEIR EFFECT

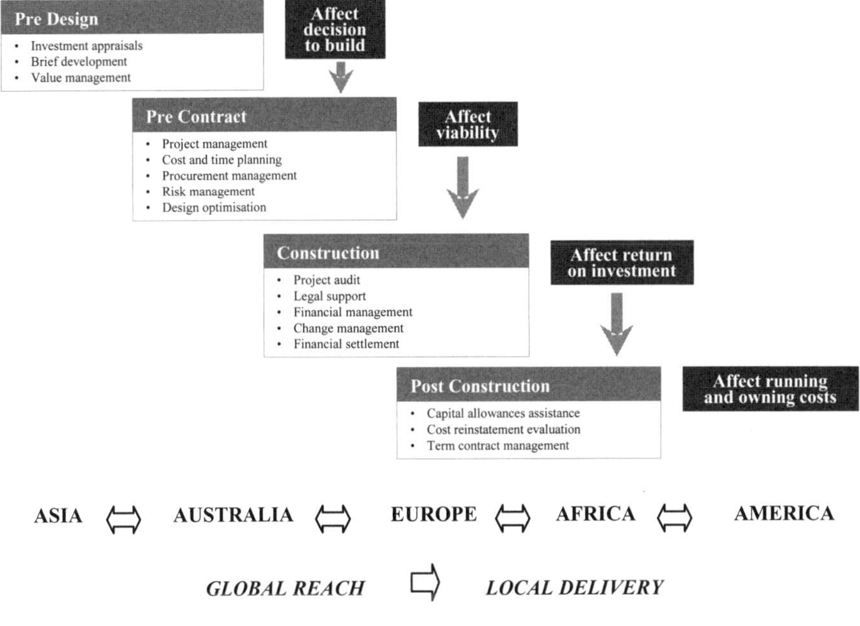

ASIA ⇔ AUSTRALIA ⇔ EUROPE ⇔ AFRICA ⇔ AMERICA

GLOBAL REACH ⇨ LOCAL DELIVERY

THAILAND

All data relate to 2007 unless otherwise indicated.

Population	
Population	66.04 mn
Urban population	33%
Population under 15	23%
Population 65 and over	11%
Average annual growth rate (2004 to 2007)	0.79%
Geography	
Land area	514,000 km^2
Agricultural area	35%
Capital city	Bangkok
Population of capital city	10 mn
Economy	
Monetary unit	Thai Baht (Bt)
Exchange rate (average fourth quarter 2008) to:	
the pound sterling	Bt 52.07
the US dollar	Bt 35.04
the euro	Bt 47.18
the yen x 100	Bt 38.38
Average annual inflation (1998 to 2007)	2.8%
Inflation rate	2.3%
Gross Domestic Product (GDP)	Bt 8,493 bn
GDP per capita	Bt 128,607
Average annual real change in (GDP)(1998 to 2007)	4.48%
Private consumption as a proportion of GDP	53.7%
Public consumption as a proportion of GDP	12.2%
Investment as a proportion of GDP	26.5%
Construction	
Gross value of construction output	Bt 722 bn
Net value of construction output	Bt 247 bn
Net value of construction output as a proportion of GDP	2.9%

Source: *National Economic and Social Development Board*
Bangkok Bank, Research Department

THE CONSTRUCTION INDUSTRY

Construction Output

The net output of construction industry in 2007 was Bt 247 billion, equivalent to US$ 7.04 billion, or 2.9% of GDP.

The total construction area approved in municipal zone for 2007 was 17.35 million m² of which 12.92 million m² were residential development.

The graphs below show the value of land transfer, construction area permitted and the net output of construction industry at current prices for the last 20 years.

LAND TRANSACTION

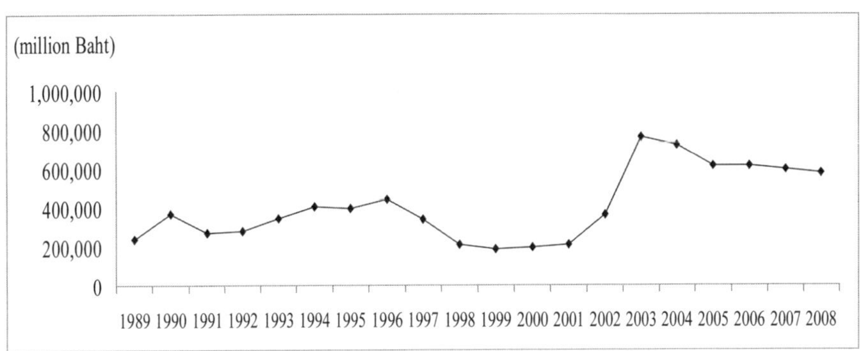

Source: Bank of Thailand, Board of Investment

CONSTRUCTION AREA PERMITTED IN MUNICIPAL ZONE

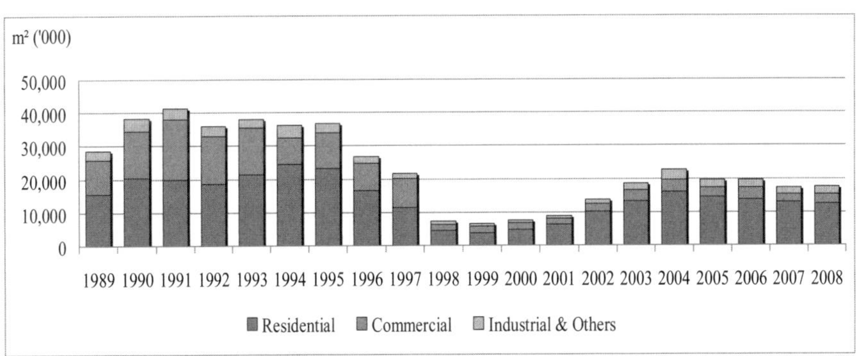

Source: Bank of Thailand, Board of Investment

CONSTRUCTION OUTPUT AT CURRENT PRICES (GDP)

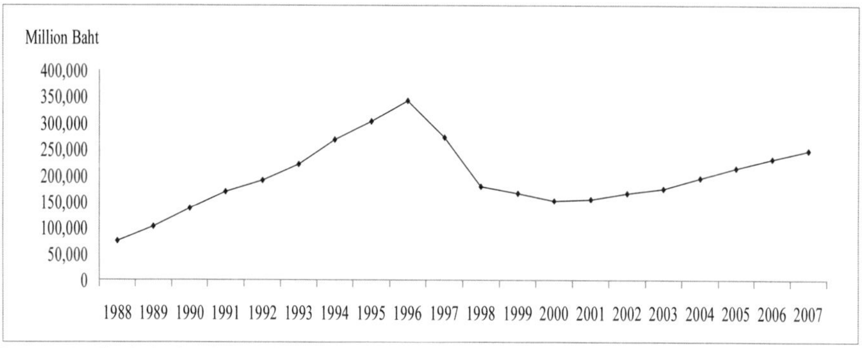

Million Baht
400,000
350,000
300,000
250,000
200,000
150,000
100,000
50,000
0
1988 1989 1990 1991 1992 1993 1994 1995 1996 1997 1998 1999 2000 2001 2002 2003 2004 2005 2006 2007

Source: Bank of Thailand, Board of Investment

The construction industry peaked at 2004. There were signs of slowing down in the last few years due to internal political problems and global recession. The outlook for 2009 is expected to be gloomy but is unlikely to be worst than the 1997 Asian Financial crisis as most of the developers are in better financial shape now.

Characteristics and Structure of the Industry

In the private sector, construction work in Thailand is generally undertaken by various specialist contractors such as piling contractors, main contractors (covering the structure and architectural work with some of them capable of doing mechanical and electrical work too), M&E contractors, aluminium window and curtain wall contractors, interior (ID) works contractors and other specialists contractors.

For large projects, most developers prefer to contract the works directly to specialist contractors under separate contracts instead of nominating sub-contractors under a main contractor. This is partly to avoid paying double taxation on withholding tax.

The main contractor (traditionally undertaking structure and architectural works) will provide attendance and coordinate all other direct contractors on site.

Construction firms are required to be registered to operate business in Thailand. However, there are no specific requirements to obtain a licence to operate as a contractor. For some government projects, only contractors registered with the Council of Contractor are qualified to tender.

Traditionally, architects provide design services only and do not undertake supervision or management functions on the project. The management and supervision of the works is normally undertaken either by the client in-house or by a separately appointed project management firm.

Clients and Finance

The private sector investment accounts for 40% of the construction works and remaining 60% by the public sector.

One of the significant changes in the source of funding in the private sector in the last few years is the presence of property funds. Assets held by property funds have increased from Bt 2 billion to about Bt 45 billion in the last few years.

Selection of Dcsign Consultants

For private sector work, the selection of consultants is normally based on their track record and fee level.

For public sector work, the consultants are normally appointed in a bidding process.

Contractual Arrangements

Construction contract documents may be prepared either in Thai or English. Several versions of the standard form of contract prepared by various departments are used on government projects. For the private sector, no standard form of contract exists although a simplified version of the standard form published by the Joint Contracts Tribunal (JCT) in the UK and FIDIC are commonly adopted.

The traditional lump sum contract is still the most common method of contract procurement. Design-and-build, construction management and Build, Operate and Transfer (BOT) methods are also used occasionally for large projects requiring extensive technical input.

Selection of contractors through competitive bidding is the norm although negotiation with pre-selected contractors and cost-plus arrangements (also negotiated with pre-selected Contractors) are also adopted.

Development Control and Standards

Land titles

There are many types of land title or certificate used as evidence of land ownership, possession rights and other interest in land but only land with Chanote, Nor Sor Sam Gor or Nor Sor Sam can be sold or apply for building approval.

Chanote – Chanote is the only true title deed. The person's name shown on the deed has the legal ownership to the land. The land is accurately survey and plotted with unique numbered marker posts set in the ground.

Nor Sor Sam Gor – Nor Sor Sam Gor certifies that the person named on the certificate has the right to possess the land and use the benefit of the land as an owner. The land is accurately surveyed and the issuance of the title deed is pending.

Nor Sor Sam – Similar to Nor Sor Sam Gor but not all of the formalities to certify the right to use have been performed. The land is not accurately surveyed and may be subject to boundary dispute.

Other forms of land title or rights are Sor Kor Nung, Por Bor Tor 6 and Sor Por Kor 4-01. Land with these type of titles or certificates can neither be transferred nor obtained approval to build on.

Land in Thailand is measured in:

1 Rai 4 Ngan (1600m^2)
1 Ngan 100 Wah (400m^2)
1 Wah 4 m^2
1 Acre is approximately 2.529 Rai
1 Hectare is approximately 6.25 Rai

Generally ownership of land by foreigners is highly restricted. The common option is to set up a majority Thai owned limited company. The other option is by leasing but the maximum land lease which can be registered with the land office is 30 years.

Zoning and building regulations

Construction in Thailand is mainly governed by the Town and City Planning Act and Building Control Act.

The Town and City Planning Act deals with permissible use of land in different zones including Floor Area Ratio (FAR). The zoning regulations also limit the height and size of the building, depending on the width of the frontage road. Each district may have separate zoning restrictions. For example, in Phuket island no construction of any type is allowed on land that is 80 metre above average sea level and within 20 metre from the coastal line.

Annual building inspection

This regulation came into effect since 2007 requiring the owner of buildings of certain size and function to appoint a registered inspector to inspect the building annually.

CONSTRUCTION COST DATA

Cost of Labour

The figures on the next page are typical of labour costs in the Bangkok area as at the fourth quarter of 2008.

	Wage rate (per day) Bt	*Number of hours worked per year*
Site operatives		
Bricklayer	500	2,496
Carpenter	500	2,496
Plumber	600	2,496
Electrician	700	2,496
Structural steel erector	400	2,496
HVAC installer	600	2,496
Semi-skilled worker	450	2,496
Unskilled labourer	350	2,496
Equipment operator	600	2,496
Watchman/security	250	2,496
	(per month)	
Site supervision		
General foreman	25,000	2,496
Trades foreman	20,000	2,496
Clerks of works	25,000	2,496
Resident engineer	35,000	2,496
Contractor's personnel		
Site manager	60,000	2,496
Resident engineer	40,000	2,496
Resident surveyor	40,000	2,496
Junior engineer	30,000	2,496
Junior surveyor	30,000	2,496
Consultants' personnel		
Senior architect	60,000	2,496
Senior engineer	60,000	2,496
Senior surveyor	60,000	2,496
Qualified architect	40,000	2,496
Qualified engineer	40,000	2,496
Qualified surveyor	40,000	2,496

Cost of Materials

The figures that follow are the costs of main construction materials, delivered to site in the Bangkok area, as incurred by contractors in the fourth quarter of 2008. These assume that the materials would be in quantities as required for a medium sized construction project and that the location of the works would be neither constrained nor remote. All the rates in this section exclude value added tax (VAT).

	Unit	*Cost Bt*
Cement and aggregate		
Ordinary portland cement in 50kg bags	tonne	2,600
Coarse aggregates for concrete	m^3	450
Fine aggregates for concrete	m^3	350
Ready mixed concrete (mix Grade 20)	m^3	2,500
Ready mixed concrete (mix Grade 24)	m^3	2,600
Steel		
Mild steel reinforcement	tonne	24,500
High tensile steel reinforcement	tonne	24,500
Structural steel sections	tonne	27,000
Bricks and blocks		
Common bricks (160 x 35 x 70mm)	1,000	1,000
Good quality facing bricks (220 x 65 x 105mm)	1,000	4,500
Hollow concrete blocks (390 x 105 x 65mm)	1,000	4,000
Precast concrete cladding units with exposed aggregate finish	m^2	1,900
Timber and insulation		
Softwood for carpentry	m^3	30,000
Softwood for joinery	m^3	30,000
Hardwood for joinery	m^3	50,500
Exterior quality plywood (20mm)	m^2	820
Plywood for interior joinery (4mm)	m^2	150
Plywood for interior joinery (20mm)	m^2	600
Softwood strip flooring (19mm)	m^2	1,000
Softwood internal door complete with frames and ironmongery	each	6,000
Glass and ceramics		
Float glass (6mm)	m^2	450
Plaster and paint		
Good quality ceramic wall tiles (200 x 200mm)	m^2	600
Plaster in 50kg bags	tonne	1,900
Plasterboard (12mm thick)	m^2	100
Emulsion paint in tins	gallon	350
Gloss oil paint in tins	gallon	590
Tiles and paviors		
Clay floor tiles (100 x 100mm)	m^2	450
Vinyl floor tiles (230 x 230 x 2.0mm)	m^2	300
Precast concrete paving slabs (500 x 500 x 50mm)	m^2	180
Clay roof tiles	m^2	1,200
Precast concrete roof tiles (420 x 330mm)	m^2	800

	Unit	Cost Bt
Drainage		
WC suite complete (medium quality)	each	5,000
Lavatory basin complete (medium quality)	each	3,000
100mm diameter PVC drain pipes	m	200
150mm diameter cast iron drain pipes	m	800

Unit Rates

The descriptions below are generally shortened versions of standard descriptions listed in full in section 4. Where an item has a two digit reference number (e.g. 05 or 33), this relates to the full description against that number in section 4. Where an item has an alphabetic suffix (e.g. 12A or 34B) this indicates that the standard description has been modified. Where a modification is major the complete modified description is included here and the standard description should be ignored; where a modification is minor (e.g. the insertion of a named hardwood) the shortened description has been modified here but, in general, the full description in section 4 prevails.

The unit rates below are for main work items on a typical construction project in the Bangkok area in the fourth quarter of 2008. The rates include all necessary labour, materials and equipment. Allowance of 15% to cover preliminaries and general items and 10% to cover for Contractor's profit and overheads have been included in the unit rates. All the rates in this section exclude value added tax (VAT).

		Unit	Rate Bt
Excavation			
01	Mechanical excavation of foundation trenches including earthwork support	m^3	150
02	Hardcore filling making up levels	m^3	400
Concrete work			
04	Plain in situ concrete in strip foundations in trenches	m^3	2,600
05	Reinforced in situ concrete in beds	m^3	2,900
06	Reinforced in situ concrete in walls	m^3	2,900
07	Reinforced in situ concrete in suspended floors or roof slabs	m^3	2,900
08	Reinforced in situ concrete in columns	m^3	2,900
09	Reinforced in situ concrete in isolated beams	m^3	2,900
10	Precast concrete slabs	m^2	800
Formwork			
11	Softwood formwork to concrete walls	m^2	380
12	Softwood formwork to concrete columns	m^2	380
13	Softwood formwork to horizontal soffits of slabs	m^2	380

		Unit	*Rate Bt*
Reinforcement			
14	Reinforcement in concrete walls	tonne	30,000
15	Reinforcement in suspended concrete slabs	tonne	30,000
16	Fabric reinforcement in concrete beds	m²	95
Steelwork			
17	Fabricate, supply and erect steel framed structure	tonne	55,000
18	Framed structural steelwork in universal joist sections	tonne	55,000
19	Structural steelwork lattice roof trusses	tonne	55,000
Brickwork and blockwork			
21A	Solid (perforated) concrete blocks (70mm thick)	m²	450
23A	Local one brick wall	m²	600
Roofing			
24	Concrete interlocking roof tiles	m²	500
33	Troughed galvanized steel roof cladding	m²	550
Woodwork and metalwork			
36	Single glazed casement window in hardwood, size 650 x 900mm	each	8,000
37	Two panel glazed door in hardwood, size 850 x 2,000mm	each	10,000
38A	Solid core two hours fire resisting hardwood internal flush door, size 800 x 2000mm with ironmongery	each	20,000
41	Hardwood skirtings	m	500
Plumbing			
42A	Light gauge galvanized sheet box gutter 150 x 100mm	m	500
43A	PVC rainwater pipes (100mm diameter) class 8.5	m	540
44A	100mm diameter high pressure polybutylene pipes for cold water supply	m	800
46A	100mm diameter low pressure polybutylene pipes for cold water distribution	m	550
47	UPVC soil and vent pipes (100mm diameter)	m	540
48	White vitreous china WC suite	each	7,000
49	White vitreous china lavatory basin	each	3,500
50	Glazed fireclay shower tray	each	8,000
51	Stainless steel single bowl sink and double drainer	each	4,500

		Unit	*Rate Bt*
Electrical work			
52	PVC insulated and copper sheathed cable	m	80
53A	10 amp unswitched socket outlet	each	200
54	Flush mounted 20 amp, 1 way light switch	each	200
Finishings			
55	2 coats gypsum based plaster on brick walls	m^2	135
56	White glazed tiles on plaster walls	m^2	850
58	Cement and sand screed to concrete floors	m^2	200
60	Mineral fibre tiles on concealed suspension system	m^2	800
Glazing			
61	Glazing to wood	m^2	485
Painting			
62	Emulsion on plaster walls	m^2	90
63	Oil paint on timber	m^2	100

Approximate Estimating

The building costs per unit area given below are averages incurred by building clients for typical buildings in the Bangkok area as at the fourth quarter of 2008. They are based upon the total floor area of all storeys, measured between external walls and without deduction for internal walls.

Approximate estimating costs generally include mechanical and electrical installations but exclude furniture, loose or special equipment, and external works; they also exclude fees for professional services. The costs shown are for specifications and standards appropriate to Thailand and this should be borne in mind when attempting comparisons with similarly described building types in other countries. A discussion of this issue is included in section 2. Comparative data for countries covered in this publication, including construction cost data, are presented in Part Three.

Approximate estimating costs must be treated with caution; they cannot provide more than a rough guide to the probable cost of building. All the rates in this section exclude value added tax (VAT).

	Cost *m^2 Bt*	*Cost* *ft^2 Bt*
Industrial		
Factories for letting	16,600	1,540
Factories for owner occupation (light industrial use)	16,600	1,540
Factories for owner occupation (heavy industrial use)	23,000	2,140
Factory/office (high-tech) for letting (shell and core only)	24,000	2,230

	Cost m^2 Bt	Cost ft^2 Bt
Factory/office (high-tech) for letting (ground floor shell, first floor offices)	26,000	2,420
Factory/office (high-tech) for owner occupation (controlled environment, fully finishes)	26,000	2,420
High tech laboratory workshop centres (air-conditioned)	25,000	2,320
Administrative and commercial buildings		
Civic offices, non air-conditioned	13,000	1,210
Civic offices, fully air-conditioned	15,000	1,390
Offices for letting, 5 to 10 storeys, non air-conditioned	15,000	1,390
Offices for letting, 5 to 10 storeys, air-conditioned	18,000	1,670
Offices for letting, high rise, air-conditioned	21,800	2,030
Offices for owner occupation, 5 to 10 storeys, non air-conditioned	15,000	1,390
Offices for owner occupation, 5 to 10 storeys, air-conditioned	18,000	1,670
Offices for owner occupation, high rise, air-conditioned	21,800	2,030
Prestige/headquarters office, 5 to 10 storeys, air-conditioned	20,000	1,860
Prestige/headquarters office, high rise, air-conditioned	29,600	2,750
Prestige retail/department store	24,300	2,260
Health and education buildings		
General hospitals (excluding specialist equipment and installation) (main hospital)	35,000	3,250
Private hospitals (excluding specialist equipment and installation) (main hospital)	50,000	4,650
Primary/junior schools	15,000	1,390
Secondary/middle schools	15,000	1,390
Recreation and art buildings		
Theatre (over 500 seats) including seating and stage equipment	65,000	6,040
Theatre (less than 500 seats) including seating and stage equipment	70,000	6,500
Concert halls including seating	60,000	5,570
Sports hall including changing and social facilities	50,000	4,650
Swimming pools (international standard) including changing and social facilities	40,000	3,720
National museums including full air-conditioning and standby generator	60,000	5,570
Local museums including air-conditioning	50,000	4,650

	Cost *m^2 Bt*	*Cost* *ft^2 Bt*
Residential buildings		
Social/economic single family housing (multiple units)	18,000	1,670
Private/mass market single family housing 2 storey detached/semi detached (multiple units)	16,000	1,490
Purpose designed single family housing 2 storey detached (single unit)	30,000	2,790
Social/economic apartment housing, low rise (no lifts)	15,000	1,390
Social/economic apartment housing, high rise (with lifts)	16,000	1,490
Private sector apartment building (standard specification)	23,800	2,210
Private sector apartment building (luxury)	35,000	3,250
Student/nurses halls of residence	15,000	1,390
Homes for the elderly (shared accommodation)	20,000	1,860
Homes for the elderly (self contained with shared communal facilities)	18,000	1,670
Hotel, 5 star, city centre (inclusive of FF&E)	59,000	5,480
Hotel, 3 star, city/provincial (inclusive of FF&E)	39,500	3,670
Motel (inclusive of FF&E)	30,000	2,790
Resort hotel, 5 star	68,000	6,320

Value Added Tax (Vat)

The standard rate of value added tax (VAT) is currently 7%.

Regional Variations

The approximate estimating costs are based on projects in the Bangkok area. For other parts of Thailand, adjust these costs by the following factors:

Chiangmai	+10%
Phuket	+12%
Samui	+25%
Pattaya/Cha-Am	+5%

EXCHANGE RATES

The graph on the next page plots the movement of the Thai baht against the sterling, the euro, the US dollar and 100 Japanese yen since 1998. The figures used for the graph are quarterly and the method of calculating these is described and general guidance on the interpretation of the graph provided in section 2. The average exchange rate at the fourth quarter of 2008 was Bt 52.07 to pound sterling, Bt 47.18 to euro, Bt 35.04 to US dollar and Bt 38.38 to 100 Japanese yen.

THE THAI BAHT AGAINST STERLING, EURO, US DOLLAR AND 100 JAPANESE YEN

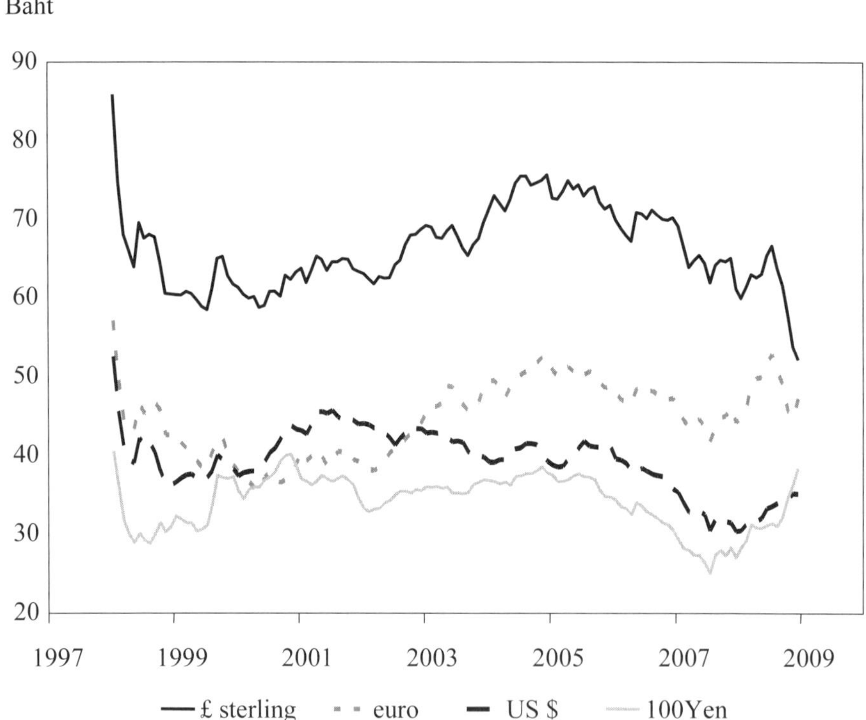

USEFUL ADDRESSES

Public Organizations

Board of Investment
555 Vibhavadi Rangsit Road
Chatuchak
Bangkok 10900
Tel: (66) 2 5378111-55
Fax: (66) 2 5378177
E-mail: head@boi.go.th
Website: www.boi.go.th

Department of Commercial Registration
Ministry of Commerce
44/100 Nonthaburi Road
Amphur Muang
Nonthaburi 11000
Tel: (66) 2 507 8000
Fax: (66) 2 507 7717
E-mail: webmaster@moc.go.th
Website: www.moc.go.th

Department of Town and Country Planning
Ministry of Interior
224 Rama 9 Road, Huay Kwang
Bangkok 10310
Tel: (66) 2 2451420
Fax: (66) 2 2460180
Website: www.dtcp.go.th

Thailand Institute of Scientific and Technological Research
196 Phahonyothin Road
Chatuchak
Bangkok 10900
Tel: (66) 2 5791121-30 / 2 5790160 / 2 5795515
Fax: (66) 2 5614771
E-mail: TISTR@tistr.or.th
Website: www.tistr.or.th

Trade And Professional Associations

The Association of Siamese Architects under Royal Patronage
248/1 Soi Soonvijai 4 (Soi 17)
Rama 9 Road, Bangkapi
Huay Kwang
Bangkok 10310
Tel: (66) 2 3196555
Fax: (66) 2 3196555
E-mail: office@asa.or.th
Website: www.asa.or.th

Council of Architects
Information Technology & Communication Building
Wisutkasat Road
Pranakorn
Bangkok 10200
Tel: (66) 2 280 8880-1
Fax: (66) 2 280 8882
E-mail: office@act.or.th
Website: www.coa.or.th

Council of Engineering
487, Ramkumhaeng 39
Soi Wat Theplela, Wangthonglang
Bangkok 10310
Tel: (66) 2 9356868
Fax: (66) 2 9356697
Website: www.coe.or.th

The Engineering Institute of Thailand
487, Ramkumhaeng 39
Soi Wat Theplela, Wangthonglang
Bangkok 10310
Tel: (66) 2 3192410-3
Fax: (66) 2 3192710-11
E-mail: eit@eit.or.th
Website: www.eit.or.th

Thai Contractors Association
110 Wireless Road, Lumpinee
Pathumwan
Bangkok 10330
Tel: (66) 2 2553991
Fax: (66) 2 2553990
E-mail: thaicom@loxinfo.co.th
Website: www.tca.or.th

UNITED KINGDOM

All data relate to 2008 unless otherwise indicated.

Population	
Population	60.98 mn
Urban population	90%
Population under 16	19%
Population 65 and over	16%
Average annual growth rate (2004 to 2008)	0.5%
Geography	
Land area	244,820 km²
Agricultural area	71%
Capital city	London
Population of capital city	7.6 mn
Economy	
Monetary unit	Pound Sterling (£)
Exchange rate (average fourth quarter 2008)	
the US dollar	£ 0.64
the euro	£ 0.84
the yen x 100	£ 0.67
Average annual inflation – CPI (1999 to 2008)	1.8%
Inflation rate (Dec 2008)	3.1%
Gross Domestic Product (GDP)	£ 1,445 bn
GDP per capita	£ 22,787
Average annual real change in (GDP) (1999 to 2008)	2.6%
Private consumption as a proportion of GDP	63.0%
Public consumption as a proportion of GDP	20.5%
Investment as a proportion of GDP	17.1%
Construction	
Gross value of construction output (current prices, 2007)	£ 122.1 bn
Construction output as a proportion of GDP (2007)	8.4%

THE CONSTRUCTION INDUSTRY

Construction Output

Construction output in the United Kingdom totalled £122.1 billion or approximately US$191 billion in 2007. This represents around 8.4% of GDP. Construction output is estimated to have peaked in 2008 and activity is now set for a prolonged downturn. The breakdown by type of work is shown below:

CONSTRUCTION OUTPUT, 2007 (CURRENT PRICES)

Type of work	*£ billion*	*% of total*
New work		
Residential building	24.3	19.9
of which: Public	4.2	3.4
Private	20.1	16.5
Non-residential building		
Commercial	23.2	19.0
Industrial	5.0	4.1
Other	10.4	8.5
Total	38.6	31.6
Infrastructure	7.0	5.7
Total new work	69.9	57.2
Repair and maintenance		
Public Housing	9.0	7.5
Private Housing	16.9	13.8
Public Non-residential	8.3	6.8
Private Non-residential	18.0	14.7
Total repair and maintenance	52.2	42.8
Total	122.1	100.0

Source: Office for National Statistics (ONS)

After expanding modestly by 1.1% in 2006, UK construction activity accelerated in 2007, with output rising by 2.5%.

CONSTRUCTION OUTPUT % REAL CHANGE, 2006 AND 2007

Type of work	*2006*	*2007*
New work		
Residential building	4.7	1.3
Commercial	13.4	12.6
Industrial	11.1	0.4
Public non-residential building	-5.1	-4.9
Infrastructure	-7.5	2.8
Renovation		
Residential	-3.1	-0.5
Non-Residential	-2.7	1.7
Total	1.1	2.5

Source: Office for National Statistics (ONS)

Industry activity increased further by 2.9% during the first half of 2008 compared to a year earlier. However, since then the reversal in the construction industry's fortunes has been dramatic. In particular, the housing market crash, global credit crunch and increased investor risk aversion have hit the sector. Private investment is retrenching. The fall in the early stages of the current construction downturn has been led by the private housing sector, but the weakness has spread to the industrial and in particular commercial sectors. Since the second half of 2008, the decline has accelerated. Looking ahead, a couple of difficult years lie ahead for the UK construction industry and the risk is that the downturn will last longer than currently anticipated. In particular, the new build sector is expected to decline, primarily due to a drop in private sector construction. In contrast, counter-cyclical measures adopted by the UK government should underpin public investment over the next few years, with priority given to education and health related projects. In addition, work for the 2012 Olympics is set to grow rapidly, as key projects are already under construction or about to start on site.

The general UK housing market began to slow in mid 2007, but the decline has been sharply exacerbated by the credit crunch and deteriorating economic conditions. Mortgage loans and approvals have fallen to historically low levels and house prices are on a sharp downward trend. Credit availability has declined sharply and loans have become more expensive over the course of 2008 and near term, there are little signs that this is likely to improve, as banks remain reluctant to lend. Consequently, housing starts have dropped of dramatically; in fact starts are set to fall to 70,000 - 80,000 in 2009 – levels unseen since the mid 1920s. This is despite government intervention and measures taken by the Bank of England, which cut interest rates to a previously unseen 1% in February 2009, aimed at reviving the housing market. Private housing output is estimated to have fallen 20% in 2008 and a further 30% decline is anticipated for 2009. Recovery is only likely in 2010 if the credit conditions loosen and the economy begins to recover. The concern is that the sharp fall in housing starts coupled with still high latent demand for housing could sow the seeds for renewed escalation in house price inflation once the housing market starts to recover. Housing starts are expected to rebound in 2011 and 2012, but the recovery is not expected to offset the falls between 2008 and 2009, as house builders' loss of capacity may constrain growth in housing starts with critical labour and materials supply having been lost to the industry in the current downturn. The UK Government plans to increase the number of new homes in England to 240,000 p.a. by 2016 and beyond appears out of reach given the current market conditions.

Social housing construction has increased significantly over the course of the last few years, as a result of increased investment in affordable housing. In response to the difficult economic conditions, the government announced £400 million to be brought forward to deliver 5,500 new social homes and the Pre-budget Report of November 2008 announced further £150 million to deliver an extra 2,000 new social homes. A major concern remains the delivery of these programmes. The delivery of social housing has become increasingly intertwined with the private sector through the use of Section 106 agreements. Thus, despite increased funding, Registered Social Landlords (RSLs) are currently finding it difficult to increase the level of social house building activity. Their position is complicated by the fact that they raise a portion of their financing on the private markets and are suffering from the same tight credit conditions as private developers and consumers.

The flow of infrastructure projects in the UK has been erratic over the past decade. After four years of decline, sector activity rebounded modestly in 2007 and accelerated firmly in 2008, with an increase seen across the majority of sub-sectors. Looking ahead, the pipeline of future work, suggests that overall infrastructure work will continue to enjoy positive output over the next few years, though there will be some divergences among the sub-sectors. Capital expenditure by the water & sewerage industry, determined by the regulator OFWAT and set out in 5-year spending plans, is expected to have peaked in 2008 and work is expected to fall back until 2012. Rail-related work is experiencing a renaissance, with activity recovering sharply in 2007 and 2008. Rail work is expected to increase considerably in the years ahead with an array of major projects underway or scheduled to start, including the £16bn Crossrail project. Roads-related work is currently underpinned by a number of motorway schemes, but looking further ahead, funding constraints could limit sector expansion. The 2008 Pre-budget Report set out £700 million to be brought forward from 2010/11 and the majority of this will be focussed on increasing capacity on motorways and major roads, providing additional growth near term at the expense of later years. Construction work on harbours and flood defences is expected to grow significantly due to capacity expansions at three main ports and £2.2bn to be spent on flood defence improvements, £1bn of which identified as construction work.

Private non-residential activity, the main driver of industry growth in recent years, is now at the forefront of the current recession. Commercial construction has remained fairly buoyant in the first half of 2008. However, the sector has been benefiting from high levels of work already under construction in the offices, retail and entertainment sub-sectors. Once these projects are completed, there will be much less coming through, as developers and investors shelve or postpone projects until economic prospects improve. The sector as a whole is expected to see two to three years of consecutive falls, with recovery only expected to occur during the latter half of 2011 at the earliest. The sharp deterioration in the economic outlook, ongoing banking crisis and the sharp fall in the availability of credit have led to a significant re-appraisal of offices development plans. A large number of high profile projects have been delayed or cancelled. Office construction is expected to fall 30% in 2009 with a further similar decline pencilled in for 2010. Recovery is not expected until 2012. The retail sector is also suffering and retail development activity is expected to fall by a fifth in 2009 and – again – no recovery is expected before 2012. The long boom in distribution and logistics-related construction has come to an end in 2007. The sector has been hard hit by the global economic slowdown and drop in business investment. The sector could see a small rebound in 2011 on the back of work developing the hinterlands around major ports projects. PFI financed health and education projects are also classified as commercial work. Although the pipeline of work seems extensive, three factors are currently affecting the viability of projects: the difficulty and cost of obtaining credit; the change in accounting rules so that PFI projects will be brought onto the government's balance sheets; and the fall in the value of the land, often used as collateral. Consequently, a large number of projects previously sought to be delivered via PFI is now expected to rely on traditional government funding. PFI work is expected to fall over the next two years before recovering in 2011 on the back of better credit conditions.

The prospects for activity in the public non-residential sector have brightened significantly. Growth in output in 2008 was robust, almost entirely driven by education construction, a trend that is likely to continue. The *Building Schools for the Future* programme is now finally gathering momentum and should boost sector activity near term. Education-related work is also benefiting from *Building Colleges for the Future*, a £2.3bn capital programme to be delivered by 2010/11 and increased funding for primary schools. Health also remains a key government priority area. However not all of the capital spending will feed through into the construction of new hospitals and GP clinics. A large proportion of funds are allocated to other areas such as IT. The 2008 Pre-budget Report set out £100m to be brought forward for the upgrading of up to 600 GP surgeries. The health sub-sector is expected to grow at 20% during 2009 after an increase of a third in 2008. Public sponsored entertainment work is also benefiting from work related to the 2012 Olympics. However looking further ahead, with government finances extremely tight, spending cuts can be expected from 2011 after the next general elections. The UK construction will hope that by then the private sector will start its recovery.

Public housing repair and maintenance (R&M) output has fallen in recent years. However, a significant increase in funding has led to a rebound in output growth in 2008 and should also ensure growth in 2009. Furthermore if Regional Social Landlords (RSLs) decide to invest their funds from the Housing Corporation in the rehabilitation of existing housing rather than new build, public housing R&M activity could be pushed up further. However, looking further ahead, tight government finances will lead to spending cuts from 2010/11. Private housing repair and maintenance (R&M), dependent on household spending, is set to decline in the two years ahead. One potential benefit to the sector between 2009 and 2011 could be the Carbon Emissions Reduction Target (CERT), which provides support for households who want to improve the energy efficiency of their homes. CERT is estimated to be worth £2.8bn over 3 years. Output in the private non-housing repair and maintenance sector has picked up sharply in 2007 and the first half of 2008. However, the recession, falling asset values, profit margins and cash flow pressures, are likely to have a negative effect on firms' routine and cyclical maintenance expenditure in the two years ahead. Despite a strong performance in 2008, public non-housing repair and maintenance work is expected to fall off near term as local authorities, NHS Trusts and others in the public sector seek to control costs and maintenance expenditure is expected to be one area affected by this.

The table below shows the percentage distribution of population in 2006 and contractors' output in 2007 in the UK regions.

REGIONAL POPULATION AND CONSTRUCTION OUTPUT, 2007

Region	*Population* %	*Output* %
North East	4.3	4.0
Yorkshire and Humberside	8.7	8.8
East Midlands	7.4	6.8
East	9.6	10.1
South East	14.0	14.7
London	12.8	15.4
South West	8.7	8.2
West Midlands	9.1	8.1
North West	11.6	11.0
Wales	5	4.1
Scotland	8.8	8.9
Total	100.0	100.0

Source: Department for Business, Enterprise and Regulatory Reform; National Statistics

Characteristics and Structure of the Industry

The UK construction industry accounts for approximately 8.4% of GDP and has a workforce in excess of 2 million. The industry can be divided into three segments: contractors (including subcontractors), product suppliers and professional service providers.

The bulk of building work in the UK is undertaken by general contractors who traditionally employed their own labour force but now increasingly use labour-only subcontractors. Specialist subcontractors may be nominated by the client's consultant team or employed by the general contractor.

Traditionally, building work in the UK has been administered by professional consultants appointed by the building client. The consultants are responsible for the design and specification of the work, the contractual arrangements and the supervision of the contract. However, integration of design and management of projects has risen over the past decade (see 'Contractual arrangements'), with the growth of design and build contracts. Design and Build has been shown to be the single most prevalent method since 1995. Up until that point, contracts were dominantly Bills of Quantities. According to the Royal Institution of Chartered Surveyors (RICS), smaller projects continue to be dominated by 'plan and specification' procurement routes and lump sum contracts, but larger projects show a preference for Construction Management or a version of Design and Build.

The most obvious characteristic of the UK construction industry is that it is highly fragmented both in terms of firm size and type. It is dominated by Small and Medium Enterprises (SMEs). Employment within the industry is also highly fragmented with 90% of UK firms in the industry employing less than 10 workers and less than 1% employing more than 80 people.

PRIVATE CONTRACTORS

Number of firms, employment and work done by private contractors, 2007

Size of firm by number of employees	*No. of firms*	*Total employment (in thousands)*	*Value of work done (£m)*	*% of total value*
1	74,325	78.9	835	3%
2 - 3	60,313	135.9	1,167	4%
4 - 7	31,814	150.4	1,448	5%
8 - 13	12,669	121.0	1,760	7%
14 - 24	6,860	131.6	2,382	9%
25 - 34	2,128	62.8	1,358	5%
35 - 59	2,129	100.2	2,563	10%
60 - 79	597	42.4	1,264	5%
80 - 114	490	49.1	1,412	5%
115 - 299	595	107.6	3,455	13%
300 - 599	154	68.0	2,342	9%
600 - 1199	65	59.4	1,769	7%
1200+	60	179.0	5,142	19%
Total	192,199	1,286.3	26,898	100%

Source: ONS

The contractor segment comprises around 192,000 firms, the vast majority of which are small, employing 1 to 13 people. Middle-sized firms (up to 80 employees) number 190,000 and large firms (80+ employees) 1,360. In addition, 789,000 people are self-employed, representing nearly two-fifths of firms in the contracting sector, a share well in excess of the UK industry average of approximately 11%. The smallest firms will tend to undertake mainly local work; the larger will have a regional, national or even international focus.

OTHER CONSTRUCTION FIRMS

	Size of firms (no. of employees)					
	1	*2-10*	*11-25*	*26-50*	*Over 50*	*Total*
Architects	4,062 35%	4,857 42%	1,339 12%	1,071 9%	126 1%	11,455 100%
Civil and structural engineers	1,189 20%	2,181 36%	994 16%	1,423 23%	321 5%	6,108 100%
Building services engineers	780 19%	1,688 42%	672 17%	751 19%	110 3%	4,001 100%
Quantity surveyors	550 28%	845 43%	298 15%	242 12%	16 1%	1,951 100%
Other surveyors	619 29%	981 47%	273 13%	205 10%	29 1%	2,107 100%
Managers	255 22%	545 46%	197 17%	147 12%	34 3%	1,178 100%
Others (incl. Planners)	366 32%	464 40%	166 14%	124 11%	27 2%	1,147 100%
Total	7,821	11,561	3,939	3,963	663	27,947

Source: Survey of UK Construction Professionals 2005/06. This survey is undertaken every five years
Note: Figures may not equate exactly due to rounding

The construction products segment turnover around £40 billion, employs 400,000 people within 30,000 companies (15,000 producers spread across 30 different industries and 15,000 suppliers, agents and intermediaries). It too is highly fragmented. 85% of firms in this sector turnover less than £5 million a year. The professional services segment comprises architects, engineers, surveyors, project and facilities managers and planners. Firms tend to be small with the majority employing fewer than ten people.

The *Building m*agazine has for many years published a league table of construction firms by size of turnover. The positions at the top of the table have changed markedly in recent years, with the most recent positions published at the end of 2008 shown in below table. However, this is a still from a rapidly moving film; there has been much restructuring in the industry and the process of change has accelerated in recent years. Some names that were consistently at the top of league have disappeared altogether; others have suddenly appeared, while some have retained their position near the top over many years. Constant change in actual positions at the top is not particularly surprising in an industry that is extremely competitive. The key trends over the past few years have been:

- Consolidation into core businesses of housing or contracting, or specific types of contracting
- Shifts of focus towards services, such as facilities management and maintenance
- Consolidation of the house building sector through a series of takeovers.

MAJOR UK CONTRACTORS

	Major Contractors	*Place in Building's 'Top 200 European Contractors' 2008*
1	Balfour Beatty	11
2	Taylor Wimpey*	18
3	Carillion	21
4	Barratt*	24
5	Laing O'Rourke	25
6	Persimmon*	26
7	Kier Group	36
8	Morgan Sindall	38
9	Newarthill	41
10	Interserve	43
11	Galliford Try	51
12	Bovis Lend Lease	52
13	Amey	53
14	Bellway*	54
15	Miller Group*	55

** Housebuilder*
Sources: Building, January 2009

The UK construction market has seen a large number of mergers and acquisitions in recent years, underpinned by continuing growth in activity. In particular, activity in the construction contracting sector has been driven by the acquisitions of mid-market rivals by high turnover contractors targeting market consolidation and geographical expansion. The house building sector has seen the consolidation of larger companies, with some transactions targeting growth sub-sectors such as retirement homes. The construction products sub-sector has been largely targeting potential market growth and investment returns. A number of market factors have driven these changes. One has been the increase in PFI, as a number of recent takeovers have had as one major objective access to the expertise required to bid for and carry out PFI projects. Another reason has been the very large scale of projects coming forward, which often has been too big for even the largest firm, who have therefore tendered as consortia. Another influence on restructuring has been the increasing influence of major shareholders and City analysts. However this trend has changed over the past year with the event of the global credit and banking crisis. M&A activity has slowed sharply and in some sectors it has dried up completely. The same holds for leveraged buyouts. The M&A market is likely to remain very constraint near term, given the lack of investor confidence and tight credit conditions. In addition, the sharp slowdown and anticipated prolonged construction recession is making the sector less attractive to investors.

In *Building* magazine's 'Top 200 European Contractors' list for 2008 there were 47 UK companies listed overall (42 in 2007) and 29 in the top 100 (24 in 2007). However, no UK firm was listed in the European Top 10 and Balfour Beatty, the UK's biggest player in Europe, ranks 11th. It is more difficult for UK firms to break into foreign markets. Much lower barriers to entry in the UK market compared to France, Germany, Italy and Spain appear to make it harder for UK contractors to break into these markets.

Most architects are members of the Royal Institute of British Architects (RIBA). There are currently a total of 40,500 members, the vast majority of which are corporate members.

Practising civil engineers are normally members of the Institution of Civil Engineers (ICE) whose membership is about 80,000 worldwide. Most structural engineers are members of the Institution of Structural Engineers (ISE), with a membership of about 22,000. Building services engineers are normally members of the Chartered Institution of Building Services Engineers (CIBSE), having a membership of about 15,000. All these institutions have substantially increased their membership in the last few years. The title of Chartered Engineer is registered and protected, either by the professional institution or by the Engineering Council. The idea of construction management as a 'profession' is relatively recent. Chartered Institute of Building (CIOB) has over 30,000 members of whom approximately a third are fully qualified.

Most construction design work in the UK is undertaken by private firms of professional architects or engineers. The amount of in-house work has shrunk considerably in the last years and the remaining contractors' design departments are relatively small, being more concerned with building rather than civil engineering work. Integration of design and management of projects has risen over the past decade, with the growth of design and build contracts. The consulting engineering profession is represented by the Association of Consulting Engineers which supports approximately 800 firms employing over 38,000 qualified personnel.

The surveying profession is very important in the UK. The Royal Institution of Chartered Surveyors (RICS) is an umbrella organization for quantity surveyors and building surveyors as well as a number of other surveying disciplines more concerned with property than the construction industry. RICS membership totals 140,000 members in over 146 countries and there are an additional 34,000 students on 400 accredited degree courses. In the UK, there are approximately 102,300 RICS members across all grades. In contrast to countries not influenced by the British system, the quantity surveyor plays a key role in the UK construction industry. Originally his role was to prepare a bill of quantities and measure work on site. The profession has, however, developed a range of consultancy services for clients and has a full professional status equivalent to that of designers. Nowadays quantity surveyors advise at every stage of the property life-cycle; from raw land, through measurement, planning, funding, design and construction, management, refurbishment and redevelopment. Quantity surveyors work mainly in private practice, but also in the public sector and commercial organizations. Most quantity surveying practices are small but there are a number of very large firms employing several hundred staff. The quantity surveying industry has moved into new fields. Some firms have developed a wider range of expertise and identify for themselves new roles as cost consultants, construction cost advisers or services providers. They are increasingly taking on functions of project and construction managers.

Clients and Finance

Public sector clients are:

- Central Government spending departments, which are directly responsible for much of public sector work, such as infrastructure, health and education.
- Local Government, which are responsible for local roads and other transport facilities, schools, colleges and community buildings.
- Housing Associations, which are responsible for new build social and affordable housing.
- Private Finance Initiative clients (PFI). PFI is one of a new range of public sector procurement methods and its impact has been to alter the whole character of the public sector as a client and to force the industry to approach public sector projects in a new way, leading to structural changes in the industry, such as restructuring of firms, development of consortia and mergers. The PFI is just one of more general Public–Private Partnerships (PPP) with some versions referred to as DBFO (Design, Build, Finance and Operate) schemes.

Private sector clients can be distinguished between:

- Clients for small building work
- Major clients acquiring buildings for their own use, such as British Airports Authority (BAA), British Telecom, large retailers, or water companies
- Property developers
- House buyers

Historically, the UK construction industry maintained a fairly even split between orders from the public and private sectors. However, since the mid 1970s there has been a marked decline in public sector investment. By 1979, 41% of new construction output was for public sector clients; by 1987 this had reduced to 32%, though it rose to 36% in 2007 because of the steep fall in private sector work. Although the decline in public work has been across all types of public construction, it has been most dramatic in the public housing sector and infrastructure – the latter largely as a result of the privatisation of public utilities, such as the water and sewerage industry. By 2007, the share of public sector work including PFI stabilised to around a third of total construction activity.

Non-residential buildings in the private sector may be financed in a number of ways and may be built by owner occupiers or by developers/investors and then let. It is estimated that owner occupiers account for up to 80% of new construction of industrial buildings. However, the amount of other private buildings built and owned by owner occupiers is much less, for example in the offices market. Available statistics suggest that the majority of non-industrial building and non-house building is financed by the banking, pension and insurance sectors or by property developers' own funds.

There has been a marked shift among UK contractors and consultants towards PPP and PFI. The UK is one of the most advanced PFI/PPP markets in Europe, with the market accounting for approximately a tenth of total public sector investment. The PFI/PPP route has been increasingly used for projects in the education and health sectors, as well as transport schemes. The UK Government set to continue its commitment to delivering projects through the PFI route. Future projects that are expected to be (at least partly) delivered through PFI/PPP are the Olympics, the East London Stratford development, retail regeneration schemes in Stevenage, Bracknell and Croydon and the M25 motorway widening proposal. There remains a strong investment appetite for PFI deals in the UK, as new public and private PFI funds are set up, which has resulted in strong activity in the secondary PFI market. The sizes of projects and deals are shaping the market, as only bigger construction companies tend to be involved.

Selection of Design Consultants

In the past, much of the design work of the public sector was done in-house by professional teams; as a result of the privatisation process, some of these teams are increasingly being disbanded, thus providing more work for private design firms.

In the last few years there have been major changes in the method for selection of design consultants in the public sector. Firstly, public clients are now required to select on the basis of a fee competition, although it is not mandatory to accept the lowest tender if greater value for money is achieved by the acceptance of another. And secondly, Public Contract Regulations 2006 which embodied into UK law the revised procurement directive of 31 March 2004 allow for framework agreements whereby the members of the framework will be selected through an advertised procedures, but once selected, the Authority may call off consultants for specific contracts, with or without further competition. In fact most public sector clients interview potential consultants and select on the basis of capability and experience as well as on price.

Nevertheless there is considerable diversity in the approach of the public sector. A large number of substantial and regular public sector clients have their own in-house project managers and they have their own views as to the way in which they will manage a project. In many cases they appoint the architect first and adopt the traditional process except that they themselves are the lead consultants. Others will appoint other professionals first, most commonly a project manager but perhaps a specialized engineer or a quantity surveyor if cost control is especially important.

In the private sector there is more flexibility in the method of selection of consultants with personal attributes of the main player being very important and experience and reputation of the firm of great significance. Fee competition is less usual although it has been used by some and is being considered by others. Work is secured more and more by competitive pitch. It is usual for fees to be negotiated. Regular clients rarely adopt the fee scales recommended by the institutions. In the private sector, there is also great variety in the order in which the consultants are appointed and their responsibilities.

In the last few years the whole ethos of the organization of the construction process has changed from one of a well trodden path to one of choice and flexibility.

Contractual Arrangements

Contractual arrangements for construction projects in the UK have undergone substantial change over the past decade. Procurement routes in the UK are usually divided into 'traditional' and 'non-traditional', with non-traditional including design and build, various forms of management contracting and, more recently, partnering and prime contracts.

Whilst the 'traditional' system of a main contractor appointed by the client remains important, various forms of 'non-traditional' contractual arrangements have become more popular, in particular for larger projects, and the usage of each one depends on the relative bargaining power of the client and other parties to the process, the size, type and complexity of work being undertaken.

In a *'traditional contract'* the employer contracts with an architect or engineer to carry out the design. The architect or engineer, acting as the agent of the employer, supervises the construction of the design, while the contractor enters into a contract with the employer to build that design. The contractor employs both subcontractors and suppliers of services, goods and equipment. Outside the relationship between these parties arises the issue of privity of contract: only parties to a contract can enforce the contract, which means that firstly in the absence of a warranty there is no contractual relationship between the employer and sub-contractors and suppliers and secondly, third parties have no contractual rights. The use of collateral warranties has resolved some of the problems, by providing for contractual relations with third parties. The advantages of the traditional approach are usually seen in its control over design process, the direct reporting of the design team to ensure quality control and that there is no built-in contractor risk premium. However, the main weakness is the division of responsibility for the design and construction.

Various types of 'procurement systems' have evolved to deal with the difficulties perceived within the traditional contract. In particular, *novated 'design-and-build'* contracts have become common. Under this arrangement, the preliminary design is undertaken by the architect or engineer, working to the client's instructions. The design team's original contracts with the employer are rescinded and new contracts are entered into with the contractor. All responsibilities are transferred to the contractor. *Management contracting* refers to an employer engaging the management contractor to partake in the project at an early stage. The management contractor is employed not to undertake the work but to manage the process. All the work is subcontracted to works contractors who carry it out. *Construction management* differs from management contracting in that the employer enters into a direct contract with each specialist contractors. The employer engages the construction manager to act as a 'consultant' to coordinate these specialist contractors. In p*roject management* type of contracts, the project manager is employed to coordinate all the work needed from design to procurement and construction on behalf of the client.

Whatever the method of procurement, the tendering process in the UK is usually based on competitive bidding. To ensure transparency in this process the National Joint Consultative Committee (NJCC), an organisation consisting of the major professional bodies involved with construction has produced codes of procedure. In *open tenders*, the first step is an advertisement calling for expressions of interest, which usually contains a brief description of the location, the type of work being proposed, the scale of the project and the scope of the proposed work. Interested contractors are invited to apply for the details. Local authorities have in the past tended to favour this method of procurement. *Single-stage selective tendering* is, in the NJCC code, considered as suitable for both private and public sector works. This procedure restricts the number of tenderers by pre-selection from either an approved list or on an ad-hoc basis. A limited number are selected on the basis of general skill and experience, financial standing, integrity, proven competence with regard to statutory health and safety requirements, and their approach to quality assurance systems. Thereafter, price alone is the criterion, the lowest tender being selected. In *two-stage selective tenders*, contractors are selected for the first stage on basis of limited scope, e.g. preliminaries, overhead and profit. In the second stage, the full price is negotiated through open book tendering of subcontracts. The NJCC regards this as a suitable method where the early involvement of the main contractor is required before the scheme is fully designed. It enables the design team to make use of the contractor's expertise and the contractor also becomes involved in the planning of the project at an early stage. *Selective tenders*, normally called design and build contracts, include the whole of contractor's proposal including price and design.

The NJCC publishes codes for selective tendering. Prices are based on a lump sum (with or without an activity schedule) or bill of quantities. According to the RICS, the bulk of contracts were based on Joint Contracts Tribunal (JCT) form in 2004, though since then alternative contractual forms such as PPC2000 and the usage of the New Engineering Contract (NEC) have increased significantly.

The 'Joint Contracts Tribunal (JCT) 1998 Standard Form of Contract with quantities' (known as 'JCT98') is still commonly used, in particularly with private sector clients. This contract assumes the use of measured bills of quantities that are normally prepared by the quantity surveyor. The JCT also produces the '1998 Intermediate Form of Building Contract' (known as 'IFC98') for works of simpler content. The tender documentation under both these forms of contract might comprise drawings, specification and bills of quantities, schedules of works or schedules of rates.

Changes are however taking place in these arrangements. In 1994, Sir Michael Latham published a report, entitled *Constructing the Team*. It recommended the increased use of a form of contract known as the NEC. Increasingly, the form of contract used in the UK is the Professional Services Contract (PSC), which is part of the NEC3 suite of contracts (June 2005). The NEC were borne out of an ongoing debate within the ICE, the lead body for the production of the ICE conditions and contract (at that time the standard form used for more civil engineering works in the UK), as to the direction of future contract strategies. The issue was whether the then existing standard forms adequately served the best interests of the parties by focusing on the obligations and responsibilities of the parties rather than on good management. The NEC was drafted with the objective of achieving flexibility, stimulus to good project management, clarity and simplicity. The PSC is intended for use in the appointment of a supplier to provide professional services and can be used for appointing project managers, supervisors, designers, consultants or other suppliers.

Overarching all these is the concept of *partnering*, which can apply to any of the procurement routes. In the partnering scenario, negotiation rather than competitive tender is the key. The rise of partnering in UK construction can be seen as a response to the widely held view that the industry was inherently flawed. After the boom times of the 1980s and subsequent recession of the early 1990s, partnering contracted sharply and a culture of conflict persisted in the industry, with employers and contractors operating in a highly adversarial manner, with contractors taking on greater risks in a fiercely competitive market. In 1998, Sir John Egan launched the publication *Rethinking Construction,* challenging the construction industry to restructure the way it does business and the culture in which such business is conducted. It calls for the reduction of costly and inefficient tendering processes, promoting long-term partnering and sets performance targets on profitability, improvement in construction cost and time, and predictability of projects finishing on time and to budget.

The government recognized that it had to lead from the front in procurement terms and implemented partnering and innovative procurement routes on many public sector projects. The strategy documents such as Achieving Excellence (HM Treasury 1999), Building on Success (OGC 2003), Improving Public Services through Better Construction (NAO 2005) have all assisted in developing the Egan vision. The practice adopted by public sector clients has been to advertise framework agreements in the Official Journal of the European Communities (OJEC), which follow the EC rules for selection and award of the framework. A framework agreement is essentially a template contract for a series of projects to be awarded. The frameworks usually do not last for more than four years and in practice a number of contractors / consultants are selected through the use of pre-qualification questionnaires (PQQ). Framework agreements can be concluded with a single supplier or with several suppliers. Where frameworks are awarded to several suppliers, there are two possible options for awarding call-offs under the framework: Firstly, direct allocation, although this is often limited by the total value of the services being offered, or secondly, hold a mini competition with all those suppliers within the frameworks capable of meeting the particular needs.

In contrast to this, the EU is doing its best to maintain more competition in the public sector. The recent changes to the EU procurement regulations mean that in respect of 'particularly complex projects' public authorities have to conduct a *Competitive Dialogue* with a number of bidders in order to resolve any commercial issues before selecting a preferred contractor. It is too early to say what the effect of this will be. Other factors which impact on the procurement process are health and safety, sustainability and environmental issues.

Building Area Measurement

There are a number of different floor area definitions used for the calculations and appraisal of commercial offices and some other types of building. These are generally in line with the RICS Code of Measurement 6th Edition dated September 2007. Reference to the guide should always be made when considering floor areas in detail; however the following provides a broad classification of the various areas used.

Gross External Area (GEA) is the area of the building measured externally at each floor level and is the basis of measurement for planning applications and approvals, i.e. site coverage.

Gross Internal Area (GIA) is the area of a building measured to the internal face of the perimeter wall at each floor level. That is the brick/block work or plaster coat applied to the brick/block work, not the surface of internal linings installed by the occupier.

Net Internal Area (NIA) is the usable space within a building measured to the internal face of external walls at each floor level, and excludes ancillary and auxiliary spaces such as toilets, lifts, plant rooms, stairwells, corridors and other circulatory areas, internal structural walls and columns, the space occupied by air-conditioning/heating plant, areas with a height of less than 1.5m and parking areas.

Development Control and Standards

The system for planning and control of development was introduced by the Town and Country Planning Act 1947. This has been amended, notably by the Town and Country Planning Act 1971 and then in the Town and Country Planning Act 1990, but the principles remain the same.

The Department for Communities and Local Government (CLG) determines national policies on different aspects of planning and the rules that govern the operation of the system. National planning policies are set out in new-style Planning Policy Statements (PPS), which are gradually replacing Planning Policy Guidance Notes (PPG). These are prepared by the Government after public consultation to explain statutory provisions and provide guidance to local authorities – with whom the main responsibility for planning lies – on planning policy and the operation of the planning system. They also explain the relationship between planning policies and other policies which have an important bearing on issues of development and land use. A development plan for each area must be produced and every development (which is very widely defined) must receive permission from the relevant authority.

The Government has recently introduced a planning reform programme, which includes changes to secondary legislation, reviews of planning policy guidance and a change in culture for the whole of the planning system. The *Planning for a Sustainable Future* White Paper sets out the Government's proposals for reform, building on Kate Barker's recommendations for improving the speed, responsiveness and efficiency in land use planning. It also proposes reforms on how decisions on nationally significant infrastructure projects are taken, including energy, waste, waste-water and transport, responding to the challenges of economic globalisation and climate change. The Government's response to the consultation was published in November 2007 and a *Planning Bill* was introduced, containing a new system for nationally significant infrastructure planning, alongside further reforms to the town and country planning system, aiming to make them more efficient and responsive.

Most building works in the UK, including alterations and/or extensions to existing buildings, are subject to minimum standards of construction in order to safeguard the public interest. The Sustainable Buildings Division (SBD) of Communities and Local Government is responsible for the creation and maintenance of the Building Regulations and associated guidance under the 1984 Building Act. Other built environment responsibilities of the Department include related EU Directives, other non statutory building standards (e.g. the Code for Sustainable Homes) and initiatives (e.g. related to existing buildings) and minor legislation (e.g. Party Wall and Architects Acts). The 1984 Act provides that Building Regulations are made for the purpose of securing the health, safety, welfare and convenience of persons in or about buildings; furthering the conservation of fuel and power; preventing waste, undue consumption, misuse or contamination of water. These purposes have been further extended by the Sustainable and Secure Buildings Act 2004 to provide a power to make regulations (not yet used) relating to furthering the protection or enhancement of the environment, facilitating sustainable development, and furthering the prevention or detection of crime.

The current Building Regulations 2000 (as amended) contain procedural requirements and a broad range of what are termed functional (i.e. performance-based) requirements with which building work must comply. The functional requirements are grouped under fourteen 'parts' (A-P less I and O) and, in essence, they provide a baseline of minimum standards to assure the delivery of 'fit for purpose' new and refurbished buildings.

The majority of the requirements within Building Regulations relate to securing the health and safety of persons in or about buildings (Parts A-D, F-K, N & P). The core requirements from this perspective are structure (A) and fire safety (B). Two parts relate to the welfare and convenience of persons in or about buildings (Parts E and M). The requirements relating to the conservation of fuel and power are contained in Part L. This is the key area to influence the environmental impact of buildings, with energy efficiency seen as the key to the reduction of harmful greenhouse gas emissions (CO_2). This is one of a number of strategic priorities to achieve the Department for the Environment, Food and Rural Affairs (DEFRA) principal aim of sustainable development. The requirements relating to preventing waste, undue consumption and misuse of water are not covered yet. The requirement for preventing contamination of water is covered in

Parts H2 and J6. Any activities in this area need to coordinate with DEFRA, who are responsible for all aspects of water policy in England.

The regulations themselves are concerned with definitions and implementation. What are normally referred to as regulations are requirements. These are extended in a standard set of Approved Documents, which give detailed guidance on how technical requirements of the regulations can be met. Owners and builders are required by law to obtain building control approval, which provides an independent check that the Building Regulations have been complied with. Building control is carried out by local authorities and private sector Approved Inspectors.

Recently, the Government has recognised that there are some issues with the existing system of Building Regulation, including concerns about both compliance and enforcement. There has been some criticism that the pace of review and change to the Building Regulations and associated guidance has been too great and that together with increasing technical complexity is resulting in practitioners failing to understand the requirements, leading to non-compliance. Furthermore, the enforcement bodies have also been unable to adequately keep up with developments and are suffering re-sourcing problems, further weakening compliance levels. Therefore, CLG is currently undertaking a wide ranging review of the principles of and requirements for building standards.

In March 2007, CLG published a report on the *Future of Building Control*, which sets out a package of options that the Government is minded to develop further and invites interested parties to provide suggestions on how the reform should proceed. The paper recognises a number of important shortcomings with the current system, including the lack of a clear future vision for the purpose of Building Control, the current piecemeal approach to regulatory change and the complexity of guidance. Problems with achieving compliance and with effective enforcement are also highlighted as key areas for action.

In December 2006, the *Code for Sustainable Homes* was introduced for consultation, to drive a step-change in sustainable home building practice. April 2007 saw the 'Go Live' where the code became mandatory for Housing Corporation and English Partnership funded schemes The Code measures the sustainability of a new home against categories of sustainable design, rating the 'whole home' as a complete package. The Code uses a 1 to 6 star rating system to communicate the overall sustainability performance of a new home, with Level 3 being the current minimum standard. The design categories included within the Code are energy/CO_2, pollution, water, health and well-being, materials, management, surface water run-off, ecology and waste. The Code sets minimum standards for energy and water use at each level and, within England, replaces the EcoHomes scheme, developed by the Building Research Establishment (BRE). It will form the basis for future developments of the Building Regulations in relation to carbon emissions from, and energy use in homes.

In July 2007, the Government published a consultation, The future of the Code for Sustainable Homes – Making a rating mandatory. This consultation asks about the future of the Code, including whether Lifetime Homes standards should be made mandatory at progressively lower levels of the Code over time. In November 2007, the Government published responses to this consultation and the main message is that the Government will be proceeding with the implementation of mandatory rating against the Code for all new homes.

CONSTRUCTION COST DATA

Cost of Labour

The figures below are typical of labour costs in the London area as at the fourth quarter of 2008, unless otherwise stated. The wage rate is the basis of an employee's income, while the cost of labour indicates the cost to a contractor of employing that employee. The difference between the two covers a variety of mandatory and voluntary contributions – a list of items which could be included is given in section 2.

	Wage rate (per hour) £	*Cost of labour (per hour)* £	*Number of hours worked (per year)*
Site operatives			
Craft operative	10.30	12.96	1,802
Skill rate 1	9.82	12.35	1,802
Skill rate 2	9.46	11.90	1,802
Skill rate 3	8.85	11.13	1,802
Skill rate 4	8.35	10.50	1,802
General operative	7.75	9.74	1,802
Electrician	12.06	15.26	1,688
HVAC installer	9.72	12.19	1,740
Trained plumber	10.39	13.48	1,725

Note: To the above Cost of Labour rates the contractor's typical overhead and profit percentages is

Building	+120%
Electrical	+140%
Mechanical	+140%

	Wage rate (per year) £	*Cost of labour (per year)* £
Site supervision		
General foreman	37,500	52,125
Trades foreman	35,000	48,650
Site agent (BSc 7 years experience)	40,000	55,600

Source: Hays Salary Survey May 2008

	Wage rate (per year) £	*Cost of labour (per year)* £
Contractors' personnel		
Contract manager (MCIOB, 10 years of experience)	55,000	76,450
Engineer (BEng, 3 years of experience)	32000	44,480
Junior engineer (BEng 1 years of experience)	23,000	31,970
Contract QS (BSc/HND/MRICS, 2 years of experience	42,000	58,380
Junior Surveyor (Assistant QS, 1 - 2 of years experience)	32,000	44,480
Planner (BSc/CIOB, 5 years of experience)	43,000	59,770

Source: Hays Salary Survey May 2008

	Wage rate (per year) £	*Cost of labour (per year)* £
Consultants' personnel		
Architect (Associate)	52,000	72,280
Qualified architect (6 years of experience)	42,000	58,380
Surveyor (Associate)	52,000	72,280
Senior surveyor (6 years of experience)	48,000	66,720
Newly qualified surveyor	42,000	58,380
Engineer (Associate)*	55,000	76,450
Senior engineer (6 years of experience)	43,000	59,770
Engineer (2 years of experience)	29,000	40,310

Source: Hays Salary Survey May 2008

Consultants Cost of labour does not include any additional benefits such as medical insurance, car allowances, etc.

Cost of Materials

The figures that follow are the costs of main construction materials, delivered to site in the London area, as incurred by contractors in the fourth quarter of 2008. These assume that the materials would be in quantities as required for a medium sized construction project and that the location of the works would be neither constrained nor remote.

All the costs in this section exclude value added tax (VAT).

	Unit	*Cost £*
Cement and aggregate		
Ordinary Portland Cement	25kg	9.63
Single aggregates for concrete (40mm)	tonne	16.46
Sharp sand for concrete	tonne	19.53
All-in ballast	tonne	23.38
Ready mixed concrete (10N/mm^2) (20mm aggregate)	m^3	85.43
Ready mixed concrete (25N/mm^2) (20mm aggregate)	m^3	97.48
Steel		
Mild steel reinforcement (12mm)	tonne	938.11
Structural steel sections	tonne	1,060.58
A252 fabric reinforcement	m^2	3.38
Bricks and blocks		
Common bricks (215 x 102.5 x 65mm)	1,000	215.94
Good quality facing bricks (215 x 102.5 x 65mm)	1,000	328.60
Hollow concrete blocks (450 x 225 x 140mm)	1,000	1,621.00
Solid concrete blocks (450 x 225 x 140mm)	1,000	1,101.00
Precast concrete cladding units with exposed aggregate finish	m^2	283.07
Timber and insulation		
Softwood sections for carpentry	m^3	249.25
Softwood for joinery	m^3	223.18
Hardwood for joinery (Iroko)	m^3	1,029.00
Exterior plywood (18mm t&g)	m^2	13.99
Plywood for interior joinery (6mm)	m^2	3.69
WS t&g flooring (22mm)	m^2	8.88
Chipboard sheet flooring (18mm t&g)	m^2	2.98
100mm thick quilt insulation	m^2	3.97
100mm thick rigid slab insulation	m^2	11.89
Softwood internal door with frame with steel ironmongery	each	311.80
Glass and ceramics		
Float glass (6mm)	m^2	22.02

	Unit	*Cost £*
Sealed double glazing units (softwood frames)	m²	370.96
Good quality ceramic wall tiles (150 x 150 x 5.5mm), white	m²	17.00
Plaster and paint		
Plaster in 25kg bags (Carlite)	bag	6.66
Plasterboard (9.5mm thick, wallboard)	m²	1.83
Emulsion paint (5 litre tins)	5 litre	17.60
Gloss oil paint (5 litre tins)	5 litre	20.54
Tiles and paviours		
Clay floor tiles (150 x 150 x 12.5mm)	m²	21.95
Vinyl floor tiles (300 x 300 x 2mm)	m²	6.74
Precast concrete paving slabs (200 x 100 x 60mm)	m²	12.54
Clay roof tiles (plain 265 x 165mm)	1,000	300.43
Precast concrete roof tiles (419 x 330mm)	1,000	619.66
Drainage		
WC suite complete	each	274.75
Lavatory basin complete (white)	each	232.65
100mm diameter clay drain pipes	m	9.03
150mm diameter cast iron drain pipes	m	55.08

Unit Rates

The descriptions on the next page are generally shortened versions of standard descriptions listed in full in section 4. Where an item has a two digit reference number (e.g. 05 or 33), this relates to the full description against that number in section 4. Where an item has an alphabetic suffix (e.g. 12A or 34B) this indicates that the standard description has been modified. Where a modification is major the complete modified description is included here and the standard description should be ignored. Where a modification is minor (e.g. the insertion of a named hardwood) the shortened description has been modified here but, in general, the full description in section 4 prevails.

The unit rates on the next page are for main work items on a typical construction project in the Outer London area in the fourth quarter of 2008. The rates include all necessary labour, materials and equipment. Allowances to cover preliminary and general items and contractors' overheads and profit have been included in the rates. All the rates in this section exclude value added tax (VAT).

		Unit	*Rate £*
Excavation			
01	Mechanical excavation of foundation trenches	m^3	7.22
02	Hardcore filling making up levels	m^3	22.96
03	Earthwork support	m^2	1.77
Concrete work			
04A	Plain in situ concrete in strip foundations in trenches 10N/mm²	m^3	99.58
05A	Reinforced in situ concrete in beds 25N/mm² (150 - 450mm thick)	m^3	106.01
06A	Reinforced in situ concrete in walls 25N/mm² (150 - 450mm thick)	m^3	123.72
07A	Reinforced in situ concrete in suspended slabs or roof slabs 20N/mm² (150 - 450mm thick)	m^3	121.74
08A	Reinforced in situ concrete in columns 25N/mm²	m^3	147.95
09A	Reinforced in situ concrete in isolated beams, 25N/mm²	m^3	137.47
10A	Prestressed precast concrete slabs (100mm thick, 1200mm wide)	m^2	33.40
Formwork			
11	Softwood formwork to concrete walls	m^2	35.14
12A	Softwood or metal formwork to concrete isolated columns regular shaped square	m^2	44.71
13A	Softwood or metal formwork to horizontal soffits of slabs; 1.5 - 3m height to soffit not exceeding 200mm thick	m^2	32.70
Reinforcement			
14A	Reinforcement in concrete walls (16mm), straight	tonne	987.00
15A	Reinforcement in suspended concrete slabs (16mm), straight	tonne	987.00
16A	Fabric reinforcement in concrete beds, A252	m^2	4.55
Steelwork			
17	Fabricate, supply and erect steel frame structure	tonne	1,536.00
18	Framed structural steelwork in universal joist sections	tonne	1,443.00
19A	Structural steelwork roof trusses, circular hollow sections	tonne	1,909.00

		Unit	*Rate £*
Brickwork and blockwork			
20A	Precast lightweight hollow concrete block walls (140mm thick)	m²	23.36
21A	Solid concrete blocks (100mm thick)	m²	25.10
22A	Solid (perforated) sand lime bricks half brick thick commons bricks	m²	38.00
23A	Facing bricks (PC £350/1000), half brick thick stretcher bond	m²	50.64
Roofing			
24A	Concrete interlocking roof tiles, 420 x 330mm	m²	18.54
25A	Plain clay roof tiles 265 x 165mm	m²	49.89
26	Fibre cement roof slates 600 x 300mm	m²	28.55
27A	Sawn softwood roof boarding to gutter bottom or sides 19mm thick over 300mm wide	m²	12.53
28A	W.S. board roof covering, 19mm t&g, tanalised sawn softwood	m²	13.35
29A	3 layers polyester based bitumen felt roof covering including chippings	m²	14.23
30A	Bitumen based mastic asphalt roof covering, 20mm thick 2 coats	m²	15.68
31A	Glass fibre mat roof insulation 60mm thick	m²	20.50
32A	Rigid sheet loadbearing roof insulation 100mm thick	m²	46.24
33	Troughed galvanised steel roof cladding	m²	13.25
33A	Plastic drain pipe, 110mm diameter	m	22.55
Woodwork and metalwork			
34A	Preservative treated sawn softwood, 50 x 100mm, flat roof member	m	3.78
35A	Preservative treated sawn softwood, 50 x 150mm, flat roof member	m	4.53
36A	Single glazed casement window in Meranti hardwood, 630 x 900 (including sills and ironmongery)	each	237.46
37A	Two panel glazed door in American White Ash hardwood, 838 x 1981 x 63mm (excluding glazing and ironmongery)	each	164.11
38A	Solid core half hour fire resisting hardwood (American light oak veneer) internal flush door, 826 x 2060 x 44mm thick	each	131.15

		Unit	*Rate £*
39	Aluminium double glazed window, 1200 x 1200mm (including glazing)	each	800.86
40	Aluminium double glazed door 850 x 2100mm (including glazing)	each	1,257.16
41A	Hardwood skirtings Sapele, 25 x 94mm	m	10.16

Plumbing

42A	UPVC half round eaves gutter (112mm)	m	12.74
43A	UPVC rainwater pipes (110mm)	m	17.40
44	Light gauge copper cold water tubing (15mm),capillary fittings	m	7.35
45A	Plastic waste pipes (32mm), polypropylene with "O" fittings	m	5.14
46A	Blue MDPE pipes for cold water distribution (50mm)	m	5.76
47A	UPVC soil & vent pipes (110mm)	m	17.43
48	White vitreous china WC suite	each	289.18
49A	Coloured vitreous china lavatory basin	each	307.57
50A	White glazed fireclay shower tray	each	177.12
51	Stainless steel single bowl sink and double drainer	each	260.74

Electrical work

52A	PVC insulated and PVC sheathed copper cable core and earth (1.5mm²)	m	1.92
53A	13 amp 1 gang unswitched socket outlet	each	15.54
54A	Flush mounted 6 amp, 1 gang 1 way light switch	each	14.98

Finishings

55	2 coats gypsum based plaster on brick walls	m²	13.04
56A	White glazed tiles on plaster walls (150 x 150 x 5.5mm)	m²	30.92
57A	Red clay quarry tiles on concrete floors (150 x 150 x 12.5mm)	m²	36.29
58	Cement and sand screed (1:3) to concrete floors, 50mm thick	m²	10.65
59A	2mm thick vinyl tiles	m²	13.76
60	Mineral fibre tiles on concealed suspension system (suspension system included)	m²	18.21

Glazing

61A	6mm clear float glass, glazing to wood with screwed beads	m²	36.62

		Unit	*Rate £*
Painting			
62A	Emulsion on plaster walls (1 mist, 2 emulsion), to brick or block; internal	m²	3.42
63A	Oil paint on timber (knot, 1 primer, 2 undercoats and 1 finish); internal	m²	5.93

Approximate Estimating

The building costs per unit area given below are averages incurred by building clients for typical buildings in the United Kingdom as at the fourth quarter of 2008. They are based upon the total floor area of all storeys, measured between external walls and without deduction for internal walls.

Approximate estimating costs generally include mechanical and electrical installations but exclude furniture, loose or special equipment, and external works; they also exclude fees for professional services. The costs shown are for specifications and standards appropriate to the United Kingdom and this should be borne in mind when attempting comparisons with similarly described building types in other countries. A discussion of this issue is included in section 2. Comparative data for countries covered in this publication, including construction cost data, are presented in Part Three.

Approximate estimating costs must be treated with caution; they cannot provide more than a rough guide to the probable cost of building. All the rates in this section exclude value added tax (VAT).

	Cost m² £	*Cost ft² £*
Industrial		
Factories for letting (including lighting, power and heating)	503	47
Factories for owner occupation (light industrial use)	689	64
Factories for owner occupation (heavy industrial use)	1,090	101
Factory/office building high technology production for letting (shell & core only)	670	62
Factory/office for owner occupation (controlled environment, fully finished)	1,425	132
High tech laboratory workshop centres, air-conditioned	2,929	272
Industrial building shell with heating to office areas only, 1,000 - 2,000m²	452	42
Industrial building shell including services to production areas, 1,000 - 2,000m²	652	61
Warehouses, high bay, 10 - 15m high, for owner occupation (no heating), 10,000 - 20,000m²	251	23
Cold stores/refrigerated stores	773	72

	Cost m^2 £	Cost ft^2 £
Administrative and commercial protective service facilities		
Civic offices, non air-conditioned	1,364	127
Civic offices, fully air-conditioned	1,648	153
Offices for letting, medium rise, non air-conditioned	1,672	155
Offices for letting, medium rise, air-conditioned	1,765	164
Offices for letting, high rise, air-conditioned	2,142	199
Offices for owner occupation, medium rise, non air-conditioned	2,030	186
Offices for owner occupation, medium rise, air-conditioned	1,965	183
Offices for owner occupation, high rise, air-conditioned	2,295	213
Prestige office, medium rise	2,011	187
Prestige office, high rise	2,933	272
Health and welfare facilities		
District hospitals	1,518	141
Private hospitals	1,588	148
Hospital teaching centres	1,560	145
Health centres	1,220	113
Nursery schools	1,336	124
Primary/junior schools	1,490	138
Secondary/middle schools	1,704	158
University (arts) buildings	1,211	113
University (science) buildings	1,453	135
Management training centres	1,415	131
Recreation and arts facilities		
Theatres (500 seats) including stage equipment	4,074	378
Workshop (less than 500 seats) excluding stage equipment	3,059	284
Concert halls including seating and stage equipment	2,938	273
Sports halls including changing	968	90
Swimming pools (international standard)	3,208	298
Swimming pools (schools standard)	1,187	110
National museum including fully air-conditioning and standby generator	3,180	295
Local museums air-conditioned	1,965	183
City centre libraries	1,737	161
Branch libraries	1,183	110
Residential buildings		
Social/economic single family housing (multiple units)	796	74
Private/mass market single family housing 2 storey detached/semi detached (multiple units)	1,155	107

	Cost m^2 £	*Cost* ft^2 £
Purpose designed single family housing 2 storey detached (single unit)	1,117	104
Social/economic apartment housing, low rise (no lifts)	899	84
Social/economic apartment housing, high rise (with lifts)	987	92
Private sector apartment buildings (standard specification)	1,374	128
Private sector apartment buildings (luxury specification)	1,769	164
Student / nurses halls of residences	1,271	118
Homes for the elderly – residential (self contained with shared communal facilities)	1,020	95
Hotel, 5 star, city centre	2,421	225
Hotel, 3 star, city/provincial	1,867	173
Motel	1,141	106

Regional Variations

The approximate estimating costs are based on average UK rates. Adjust these costs by the following factors for regional variations:

Against UK Mean

Inner London	11%	East Anglia	-3%
Outer London	8%	Yorkshire & Humberside	-5%
South East	3%	North West England	-5%
South West	1%	Scotland	3%
East Midlands	-4%	Wales	-11%
West Midlands	-2%	Northern Ireland	-22%

Value Added Tax (VAT)

In his Pre-Budget Report on 24 November 2008, the Chancellor announced that the standard rate of VAT will be reduced to 15% on 1 December 2008.

This means that for general building work goods or services that take place on or after 1 December 2008, providers should charge VAT at the new rate of 15%.

The 15% rate will remain until 31st December 2009, and from 1 January 2010 it will revert to 17.5%.

EXCHANGE RATES AND INFLATION

The combined effect of exchange rates and inflation on prices within a country and price comparisons between countries is discussed in section 2.

Exchange Rates

The graph below plots the movement of sterling against the euro, the US dollar and 100 Japanese yen since 1998. The values used for the graph are quarterly and the method of calculating these is described and general guidance on the interpretation of the graph provided in section 2. The average exchange rate in the fourth quarter of 2008 was £ 0.84 to euro, £ 0.64 to US dollar and £ 0.67 to 100 Japanese yen.

STERLING AGAINST EURO, US DOLLAR AND 100 JAPANESE YEN

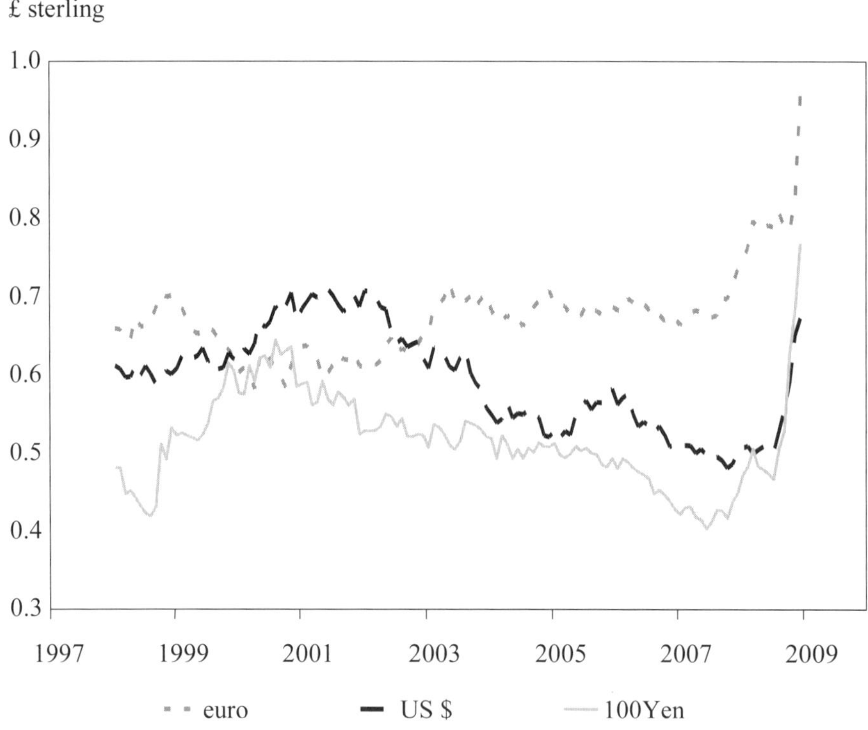

Price Inflation

The following table presents retail price, building cost and tender price inflation in the United Kingdom since 1998. The basis of the first column is the official consumer price index. The other two indices are produced by Davis Langdon LLP: the building cost index provides an index of price movements in general building costs, and the tender price index indicates movements in general building prices in the Greater London area.

The chart clearly indicates the dramatic downturn in UK tender prices which began in the 3rd quarter of 2008 and is set to continue through 2010.

CONSUMER PRICE, BUILDING COST AND BUILDING TENDER PRICE INDICES (1997 TO 2009)

Year	*CPI*	*% Change*	*BCI*	*% Change*	*TPI*	*% Change*
1998	100.0	0.0%	100.0	0.0%	100.0	0.0%
1999	101.3	1.3%	104.8	4.8%	106.1	6.1%
2000	102.2	0.9%	111.4	6.3%	114.7	8.1%
2001	103.4	1.2%	115.5	3.7%	122.7	7.0%
2002	104.7	1.3%	122.1	5.7%	131.6	7.3%
2003	106.1	1.4%	129.2	5.8%	137.1	4.1%
2004	107.6	1.3%	136.7	5.8%	141.2	3.0%
2005	109.8	2.0%	146.0	6.8%	148.2	5.0%
2006	112.3	2.3%	155.4	6.4%	156.9	5.8%
2007	114.9	2.3%	163.1	5.0%	167.7	6.9%
2008*	119.1	3.6%	173.1	5.8%	170.3	1.5%
2009f			181.1	4.6%	153.7	-9.8%

Source: Davis Langdon LLP

* = *estimate*

f = *forecast*

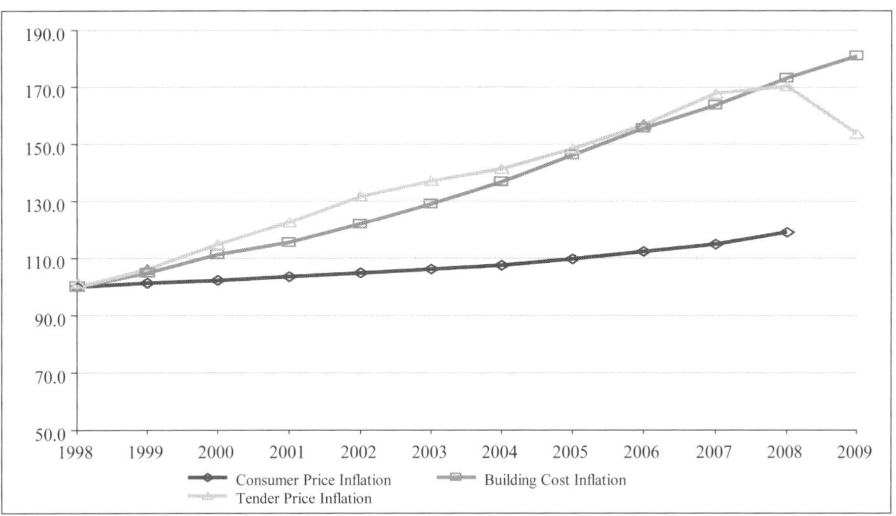

USEFUL ADDRESSES

Government And Public Organizations

British Standards Institution
389 Chiswick High Road
London W4 4AL
Tel: +44 (0) 208 996 9001
Fax: +44 (0)20 8996 7001
E-mail: cservices@bsigroup.com
Website: www.bsi-global.com

Building Research Establishment
Bucknalls Lane
Garston
Watford WD25 9XX
Hertfordshire
Tel: +44 (0) 192 366 4000
Fax: +44 (0) 192 366 4010
E-mail: enquiries@bre.co.uk
Website: www.bre.co.uk

Construction Industry Council
26 Store Street
London WC1E 7BT
Tel: +44 (0) 207 399 7400
Fax: +44 (0) 207 399 7425
E-mail: info@cic.org.uk
Website: www.cic.org.uk

Department for Communities and Local Government
Eland House
Bressenden Place
London SW1E 5DU
Tel: +44 (0) 207 944 4400
Email: contactus@communities.gov.uk
Website: www.communities.gov.uk

Office for National Statistics
1 Myddelton Street
London EC1R 1UW
Tel: +44 (0) 845 601 3034
Fax: +44 (0) 1633 652 747
Website: www.statistics.gov.uk

Trade And Professional Associations

Association for Consultancy and Engineers – ACE
Alliance House
12 Caxton Street
London SW1H 0QL
Tel: +44 (0) 207 222 6557
Fax: +44 (0) 207 222 0750
E-mail: consult@acenet.co.uk
Website: www.acenet.co.uk

Association for Project Management
150 West Wycombe Road
High Wycombe HP12 3AE
Buckinghamshire
Tel: +44 (0) 845 458 1944
Fax: +44 (0) 1494 528 937
E-mail: info@apm.org.uk
Website: www.apm.org.uk

Chartered Institute of Building
Englemere
Kings Ride
Ascot SL5 7TB
Berkshire
Tel: +44 (0) 134 463 0700
Fax: +44 (0) 134 463 0777
E-mail: reception@ciob.org.uk
Website: www.ciob.org.uk

Chartered Institution of Building Services Engineers (CIBSE)
222 Balham High Road
London SW12 9BS
Tel: +44 (0) 208 675 5211
Fax: +44 (0) 208 675 5449
Website: www.cibse.org

Construction Confederation
55 Turfton Street
Westminster
London SW1P 3QL
Tel: +44 (0) 870 898 9090
Fax: +44 (0) 870 898 9095
E-mail: enquiries@thecc.org.uk
Website: www thecc.org.uk

Engineering Council UK
246 High Holborn
London WC1V 7EX
Tel: +44 (0) 203 206 0500
Fax: +44 (0) 203 206 0501
Website: www.engc.org.uk

The House Builders Federation
1st Floor, Bryon House
7-9 St James's Street
London SW1A 1DW
Tel: +44 (0) 207 960 1600
Fax: +44 (0) 207 960 1601
E-mail: info@hbf.co.uk
Website: www.hbf.co.uk

Institution of Civil Engineers
1 Great George Street
Westminster
London SW1P 3AA
Tel: +44 (0) 207 222 7722
Website: www.ice.org.uk

Institution of Structural Engineers
11 Upper Belgrave Street
London SW1X 8BH
Tel: +44 (0) 207 235 4535
Fax: +44 (0) 207 235 4294
Website: www.istructe.org.uk

Royal Institute of British Architects – RIBA
66 Portland Place
London W1B 1AD
Tel: +44 (0) 207 580 5533
Fax: +44 (0) 207 255 1541
E-mail: info@inst.riba.org
Website: www.architecture.com

Royal Institution of Chartered Surveyors – RICS
12 Great George Street
London SW1P 3AD
Tel: +44 (0) 207 222 7000
Fax: +44 (0) 171 334 3811
E-mail: contactrics@rics.org
Website: www.rics.org.uk

Other Organizations

British Board of Agrément – BBA
Bucknalls Lane
Garston
Watford WD25 9BA
Hertfordshire
Tel: +44 (0) 192 366 5300
Fax: +44 (0) 192 366 5301
E-mail: contact@bba.star.co.uk
Website: www.bbacerts.co.uk

The Building Centre
26 Store Street
London WC1E 7BT
Tel: +44 (0) 207 692 4000
Fax: +44 (0) 207 580 9641
Website: www.buildingcentre.co.uk

Construction Industry Research and Information Association – CIRIA
Classic House, 174-180 Old Street
London EC1V 9BP
Tel: +44 (0) 207 549 3300
Fax: +44 (0) 207 253 0523
E-mail: enquiries@ciria.org.uk
Website: www.ciria.org.uk

UNITED STATES OF AMERICA

All data relate to 2008 unless otherwise indicated.

Population	
Population	304.1 mn
Urban population (2000)	79%
Population under 15	20%
Population 65 and over	13%
Average annual growth rate (2004 to 2008)	0.94%
Geography	
Land area	9,161,630 km^2
Agricultural area	18%
Capital city	Washington
Population of capital city	591,833
Economy	
Monetary unit	US Dollar (US$)
Exchange rate (average fourth quarter 2008) to:	
the pound sterling	US$ 1.99
the euro	US$ 1.48
the yen x100	US$ 0.93
Average annual inflation (1998 to 2007)	2.9%
Inflation rate	3.8%
Gross Domestic Product (GDP)	US$ 14,264.6 bn
GDP per capita	US$ 47,395
Average annual real change in (GDP) (1998 to 2008)	6.4%
Private consumption as a proportion of GDP	69.6%
Public consumption as a proportion of GDP	20.4%
Investment as a proportion of GDP	13.7%
Construction	
Gross value of construction output (2007)	US$ 1.2 trillion
Gross value of construction output as a proportion of GDP	8.8%

THE CONSTRUCTION INDUSTRY

Construction Output

The total value of gross output of new work of the US construction industry in 2007 was almost US$1.2 trillion – which makes up 8.8% of the GDP. The table below shows the breakdown of output by type of work for 2007.

OUTPUT OF NEW CONSTRUCTION, 2007 (CURRENT PRICES)

Type of work	*US$ million*	*% of total*
Residential private		
New housing units	352,487	28.0
Improvements	173,028	13.8
Total private residential	525,515	41.8
Non-residential private		
Offices and banks	55,154	4.4
Industrial buildings and warehouses	28,609	2.3
Stores	46,061	3.7
Other	219,459	17.5
Total private non-residential	349,283	27.8
Public buildings		
Housing	7,059	0.6
Office and Industrial	123,921	9.9
Other	120,365	9.6
Total public buildings	251,345	20.0
Public non-buildings		
Highways	76,456	6.1
Sewers and water supply	39,225	3.1
Other	15,733	1.3
Total public non-buildings	131,414	10.4
Total	1,257,557	100.0

Source: US Census Bureau of the Department of Commerce Construction Spending for 2007

Construction output grew by 4.8% in 2006 but fell by 2.6% in 2007. For 2008, the market fell dramatically, led primarily by the collapse in residential construction, but with declines evident across most sectors. Investors' and consumers' confidence has been substantially impacted by disruption of the credit market caused by the collapse of sub-prime mortgages and associated structured debt instruments. This in turn is having a negative effect on the prospects for the construction market, particularly on residential and speculative investment. The correction in the construction markets comes after a period of very strong growth;

over the past ten years, construction output has roughly doubled, representing a sustained growth rate of around 7% per annum. Much of this growth was in the south east and western states, which contrast with states such as Michigan where the market is substantially unchanged over the same period of time. Generally, the biggest losses in volume have occurred in the states which had experienced the greatest growth, most notably California, Nevada and Florida. The evolution of the trend in construction in 2006 and 2007 is shown below.

CONSTRUCTION OUTPUT % REAL CHANGE, 2006 AND 2007

Type of work	*2006*	*2007*
Residential private	-1.9	1.1
Non-residential private		
Offices	18.3	1.9
Commercial	10.8	2.3
Healthcare	16.0	1.8
Total private non-residential	16.2	0.9
Public buildings		
Housing	7.3	4.4
Office	1.1	3.2
Commercial	-5.1	4.4
Healthcare	8.1	2.0
Highways	14.8	1.5
Total public buildings	10.1	0.7
Total	4.8	-2.6

Source: US Census Bureau of the Department of Commerce Construction Spending for 2007

The geographical distribution of construction work in 2008 is shown below.

POPULATION AND CONSTRUCTION EMPLOYMENT, 2008 REGIONAL DISTRIBUTION

Regions	*Population (a) (% of total)*	*Construction Employment (b) (% of total)*
Alabama	1.6	1.5
Alaska	0.2	0.2
Arizona	2.6	2.9
Arkansas	0.8	0.9
California	11.1	12.2
Colorado	2.2	1.6
Connecticut	0.9	1.2
Delaware	0.4	0.3
District of Columbia	0.2	0.2
Florida	7.1	6.1

Regions	*Population (a) (% of total)*	*Construction Employment (b) (% of total)*
Georgia	2.9	3.2
Hawaii	0.5	0.4
Idaho	0.7	0.5
Illinois	3.6	4.3
Indiana	2.0	2.1
Iowa	1.0	1.0
Kansas	0.9	0.9
Kentucky	1.2	1.4
Louisiana	1.9	1.5
Maine	0.4	0.4
Maryland	2.6	1.9
Massachusetts	1.8	2.2
Michigan	2.1	3.3
Minnesota	1.6	1.7
Mississippi	0.8	1.0
Missouri	2.0	2.0
Montana	0.4	0.3
Nebraska	0.7	0.6
Nevada	1.7	0.9
New Hampshire	0.4	0.4
New Jersey	2.3	2.9
New Mexico	0.8	0.7
New York	4.8	6.5
North Carolina	3.5	3.1
North Dakota	0.3	0.2
Ohio	3.0	3.8
Oklahoma	1.0	1.2
Oregon	1.3	1.3
Pennsylvania	3.5	4.1
Rhode Island	0.3	0.3
South Carolina	1.6	1.5
South Dakota	0.3	0.3
Tennessee	1.9	2.1
Texas	9.1	8.1
Utah	1.3	0.9
Vermont	0.2	0.2
Virginia	3.2	2.6
Washington	2.8	2.2
West Virginia	0.5	0.6
Wisconsin	1.7	1.9
Wyoming	0.4	0.2
Total	100	100

Source: (a) US Census Bureau – Department of Commerce, 2008 State Population Estimates
(b) US Department of Labor – Bureau of Labor Statistics, Current Employment Statistics

The work abroad by US contractors present in the *Engineering News Record's Top 225 International Contractors* was valued at US$3.1 trillion and the work is spread through the following regions with a percentage of the total domestic construction revenue in each country:

US CONSTRUCTION WORK ABROAD, 2007

Area	*$ billion*	*% of total*
Europe	9.7	10.0
Africa	2.0	6.8
Middle East	13.5	21.4
Asia	8.2	14.8
Canada	5.5	66.7
Latin America	3.5	16.7
Total	42.4	136.7

Source: Engineering News Record, August 2008

Characteristics and Structure of the Industry

The construction industry has over 700,000 firms employing about 7.3 million people. All states require contractors to be licensed, and licensing laws are generally strongly enforced. About 14% of main contractors and trade contractors use union registered employees and negotiate wages with the unions. The percentage has fallen from 17.5% in 2000. 'Open-shop contracting' has grown significantly over the last 20 years, especially in housebuilding. This growth has moderated the behaviour and wage demands of unions. With the growth of non-union, construction became an expansion of training outside the union sector.

There are a large number of specialist trade contractors and they play an important role. They usually have to organize and manage the work on site with little direction from the main contractor, and they often supply major items of plant and equipment. Labour-only subcontracting is rarely used.

In 2008, there were 35 US contractors listed in *Engineering News Record's Top 225 International Contractors* in terms of revenue and 10 in the top 100. Out of the international regions, US has 11.9% of the total construction market and trails Europe's 64 firms out of the 225 and 31.1% of the revenue. Europe is followed by the Middle East and their 141 firms along with 20.3% of the revenue and lastly followed by Asia/Australia's 155 firms and 17.9% of the revenue. The table on the next page shows the principal US contractors according to *Engineering News Record's Top 225 International Contractors*, 2008 edition.

MAJOR US CONTRACTORS, 2007 AND 2008

Major contractors	*Place in ENR's 2007 Top 225 International Contractors*	*Place in ENR's 2008 Top 225 International Contractors*
Bechtel, California	6	6
Flour Corp, Texas	10	11
KBR, Texas	8	15
Foster Wheeler, New Jersey	27	22
CB&I, Texas	34	29
McDermott International, Texas	36	32
Jacobs, California	28	40
Kiewit Corp, Nebraska	66	56
URS Corp, California	N/A	80

Source: Engineering News Record, August 2008

Architects are required to register in the state in which they practise. The National Council of Architectural Registration Boards grants a certificate to a qualified architect which is usually recognized by states; nevertheless, most states will still require an architect to take additional examinations to practise in that state. There are very few quantity surveyors in the US, as it is the architect who is principally concerned with the cost of projects. However, there are construction cost consultants who may originally have been architects or engineers but are increasingly being augmented by quantity surveyors. Contractors are often prepared to give cost advice to the architect.

There are fewer building engineers than architects in the US. Engineers have to be registered, which generally requires a recognized engineering degree and four years' work experience. There are a number of substantial multidisciplinary practices in the US.

Private Home Market Turmoil and Homeownership

68% of all housing units were owner occupied in the first quarter of 2003, but this fell to 67.8% by the first quarter of 2008. Home ownership varies dramatically depending on the race group in the US. Blacks own homes, as of the first quarter 2008, on average 47%, Whites 75%, Hispanics 48.9% and all other races 58.1%. Private rented property accounts for the bulk of the remainder. Home ownership percentages have crept up over the past 8 years because of access to relatively cheap money from financial institutions. This forced home prices up dramatically over a five year period, with owners using this new found equity for their own consumption habits. Some home buyers selected Adjustable Rate Mortgages (ARMs) with no money down and no interest payments for an initial period of time, only to find that the ARMs, when due, would increase the individual's payment dramatically. This led to a speculative bubble that burst during 2007, sending the credit markets into turmoil after these individuals defaulted on their loans. Because of this, it is becoming more difficult for potential home owners to get loans, as there are more restrictions on money put down for a home, as well as increased interest rates to borrow the money. The overall cost to the housing market has been a dramatic decrease in the value of homes with California and Florida being the hardest hit states in the US.

Selection of Design Consultants

For public sector projects and many large private sector projects, most architects and engineers are selected on the basis of the highest qualification for each project and at a fair and reasonable price. For other private sector projects, architects are often selected with little or no competition. The selection process for public projects typically involves open invitations published for interested architects and engineers (usually only in the state where the project is located) to indicate their interest and to submit detailed specific information on their qualifications. The owner will normally develop a shortlist of three to eight design teams, who are then invited to make presentations before a selection committee, with the preferred team being invited to enter into negotiations with the client. If no agreement is reached, the second candidate enters negotiations, and so on. The process is costly to firms tendering for a project.

Contractual Arrangements

The most common methods of selecting a general contractor are by competitive bidding, by negotiation, or by a combination of the two. Most public sector clients are statutorily required to use open competitive bidding. There are numerous forms of negotiated contracts, but most are of the 'cost-plus-fee' type. Negotiated contracts are normally limited to privately financed work, although there are some statutory exceptions allowing it for some public projects. There is a small but growing trend towards Target Costing and forms of Public/Private partnerships.

'Fixed price' contracts are the most common. Tenders for buildings are customarily prepared on a 'lump sum' basis, whereas engineering projects are generally bid as a series of unit prices. It is standard practice for contractors to prepare their own quantities, which do not form part of the contract. With few exceptions, bids are accompanied by a bid bond guaranteeing that the contractor will enter into a contract if declared successful.

Standard contract conditions have been developed by various bodies, including the American Institute of Architects, the National Society of Professional Engineers, the Associated General Contractors of America and various federal, state and municipal governments. Where a contract provides for arbitration or mediation, most stipulate that it shall be conducted under the auspices of the American Arbitration Association.

There has been an increase in the use of 'management fee' or 'construction management' arrangements for large projects, but often still retaining a guaranteed maximum price. However, this tendency is now less apparent. 'Design-and-build' projects are also becoming more popular, although there are often other provisions for the contractor to offer advice at the design stage.

Specialist trade contractors are usually included in the general contractor's bid. Nomination is virtually unknown.

The lien laws in the US provide a large degree of protection for the contractors and subcontractors working on a project. Under their provisions, a contractor can place a lien on the real property if he has not received payment for goods and services provided. This lien is registered on the title deed of the property and, if not resolved, can be a major impediment for subsequent sale or mortgage financing on the property. The owner is therefore obliged to ensure all payments are properly effected to each supplier of goods or services. In the event that the employer has made a payment to the general contractor, but the general contractor has not paid his subcontractors, then the subcontractors are entitled to place a lien on the property. In this case the employer may have to pay for the works twice to remove the lien, unless he has a labour and materials payment bond in force, in which case he can recover the double payment from the bond company. Standard bond forms are available and in common use throughout the US. Employers and their agents need to monitor payments carefully on projects to avoid lien actions.

Development Control and Standards

The planning process in the US is known as planning control and zoning control. It is very fragmented and every town and county has its own system. There can be hundreds of separate zoning authorities in any one state. There is normally a Zoning Commission Board, a Zoning Board of Appeal and often also a Planning Commission or Board in each jurisdiction. The ease with which development zones of a town can be changed varies according to the attitude of the town or the state. There is usually no statutory period for approval, and projects can be significantly delayed by the planning process. This is particularly true for contentious projects which may even be ultimately decided by public referendum (initiative).

There is no single national building code for the whole of the US. As with planning, there are thousands of jurisdictions, including cities, counties, states and their agencies, federal agencies, and such like. Often there are overlapping jurisdictions, with more than one entity having code enforcement authority over a single project. Nevertheless, various model codes have been prepared. The most widely used is the International Code Council's International Building Code. The most commonly used alternative code is the NFPA 5000, developed by the National Fire Protection Association (NFPA).

There is usually no statutory period for receiving building code approval. During the height of the construction boom, projects could easily take several weeks, if not months, to receive their code approval (building permit). For health care projects in California, approvals could take over a year, although the general slowdown and new procedures are reducing that period. Construction projects are frequently inspected by code officials to ensure compliance with the building permit, and often work cannot proceed until preceding work has been inspected.

Standards are continually referred to in the building codes. They may be mandatory or discretionary. There are some 150 organizations which develop standards, of which perhaps a dozen or so are important. These include the American Society for Testing and Materials (ASTM) and the American National Standards Institute (ANSI).

Liability and Insurance

The contractor is liable for damages caused by his own acts or omissions. He must therefore obtain comprehensive liability insurance to protect himself and his subcontractors.

The liability of designers and contractors varies with the contract used and from state to state. In the US, the architect or engineer has a contractual obligation to check the shop drawings of specialist trade contractors and this affects the liability. Normally, professional liability extends three to four years, but in some circumstances it can extend up to ten years.

Professional indemnity insurance covers the liability of parties involved in design, except that trade contractors may not be covered or, if they are, may be insufficiently so. Professional indemnity insurance is, in any case, very expensive in the USA.

CONSTRUCTION COST DATA

Cost of Labour

The figures below are typical of labour costs in the Washington DC area as at the fourth quarter of 2008. The wage rate is the basis of an employee's income, while the cost of labour indicates the cost to a contractor of employing that employee. The difference between the two covers a variety of mandatory and voluntary contributions – a list of items which could be included is given in section 2.

	Wage rate (per hour) US$	*Cost of labour (per hour) US$*
Site operatives		
Mason/bricklayer	25.90	32.09
Carpenter	24.37	30.52
Plumber	31.52	44.24
Electrician	34.55	45.94
Structural steel erector	25.68	37.68
HVAC installer	31.80	46.50
Semi-skilled worker	19.84	27.74
Unskilled labourer	18.41	23.23
Equipment operator	25.77	32.59
Watchmen/security	18.41	23.23

	Wage rate (per hour) US$	*Cost of labour (per hour) US$*
Site supervision		
General foreman	29.37	37.52
Trades foreman	39.55	52.84
	(per week)	*(per week)*
Clerk of works	1,154	1,673
Contractors' personnel		
Site manager	2,404	3,486
Resident engineer	1,827	2,649
Resident surveyor	1,827	2,649
Junior engineer	1,635	2,371
Junior surveyor	1,635	2,371
Planner	1,250	1,813

Cost of Materials

The figures that follow are the average costs for main construction materials, delivered to site in the Washington DC area, as incurred by contractors in the fourth quarter of 2008. These assume that the materials would be in quantities as required for a medium sized construction project and that the location of the works would be neither constrained nor remote.

	Unit	*Cost US$*
Cement and aggregate		
Ordinary portland cement in 50kg bags	bag	8.50
Coarse aggregates for concrete	tonne	18.00
Fine aggregates for concrete	tonne	22.00
Ready mixed concrete (mix 17MPa)	m^3	117.00
Ready mixed concrete (mix 21MPa)	m^3	125.00
Steel		
Mild steel reinforcement	tonne	930.00
High tensile steel reinforcement	tonne	980.00
Bricks and blocks		
Common bricks (8" x 2.67" x 4")	1,000	420.00
Good quality facing bricks (8" x 2.67" x 4")	1,000	850.00
Hollow concrete blocks (8" x 8" x 16")	1,000	1,500.00
Solid concrete blocks (4" x 8" x 16")	1,000	1,600.00
Precast concrete cladding units with exposed aggregate finish	m^2	350.00

	Unit	*Cost US$*
Timber and insulation		
Exterior quality plywood (13mm)	m^2	12.50
Plywood for interior joinery (6mm)	m^2	8.50
Softwood strip flooring (25 x 102mm)	m^2	45.00
89mm thick unfaced fibreglass blanket	m^2	2.50
100mm thick rigid slab insulation	m^2	16.50
Softwood internal door complete with frames and ironmongery	each	530.00
Glass and ceramics		
Float glass (5mm)	m^2	52.00
Sealed double glazing units (16mm)	m^2	140.00
Good quality ceramic wall tiles	m^2	45.00
Plaster and paint		
Plaster in 36kg bags	bag	15.00
Plasterboard (10mm thick)	m^2	2.40
Emulsion paint in 5 litre tins	gallon	28.00
Gloss oil paint in 5 litre tins	gallon	52.00
Tiles and paviors		
Clay floor tiles (102 x 102 x 13mm)	m^2	39.50
Vinyl floor tiles (305 x 305 x 3mm)	m^2	27.50
Clay roof tiles	m^2	90.00
Precast concrete roof tiles	m^2	24.00
Drainage		
WC suite complete	each	380.00

Unit Rates

The descriptions overleaf are generally shortened versions of standard descriptions listed in section 4. Where an item has a two digit reference number (e.g. 05 or 33), this relates to the full description against that number in section 4. Where an item has an alphabetic suffix (e.g. 12A or 34B) this indicates that the standard description has been modified. Where a modification is major the complete modified description is included here and the standard description should be ignored; where a modification is minor (e.g. the insertion of a named hardwood) the shortened description has been modified here but, in general, the full description in section 4 prevails.

The unit rates overleaf are US national average rates for main work items on a typical construction project as at the fourth quarter of 2008. The rates include all necessary labour, materials, equipment and allowances to cover preliminary and general items and contractors' overheads and profit.

		Unit	Rate US$
Excavation			
01	Mechanical excavation of foundation trenches	m^3	18.00
02	Hardcore filling making up levels	m^3	35.00
03	Earthwork support	m^2	250.00
Concrete work			
04	Plain in situ concrete in strip foundations in trenches	m^3	325.00
05	Reinforced in situ concrete in beds	m^3	295.00
06	Reinforced in situ concrete in walls	m^3	460.00
07	Reinforced in situ concrete in suspended floors or roof slabs	m^2	150.00
08	Reinforced in situ concrete in columns	m^3	825.00
09	Reinforced in situ concrete in isolated beams	m^3	650.00
10	Precast concrete slabs	m^2	350.00
Formwork			
11	Softwood formwork to concrete walls	m^2	150.00
12	Softwood or metal formwork to concrete columns	m^2	180.00
13	Softwood or metal formwork to horizontal soffits of slabs	m^2	120.00
Reinforcement			
14	Reinforcement in concrete walls (16mm)	tonne	2,400.00
15	Reinforcement in suspended concrete slabs	tonne	2,300.00
16	Fabric reinforcement in concrete beds	m^2	32.00
Steelwork			
17	Fabricate, supply and erect steel framed structure	tonne	3,500.00
18	Framed structural steelwork in universal joist sections	tonne	3,500.00
19	Structural steelwork lattice roof trusses	tonne	4,500.00
Brickwork and blockwork			
20	Precast lightweight aggregate hollow concrete block walls	m^2	220.00
21	Solid (perforated) concrete bricks	m^2	260.00
22	Solid (perforated) sand lime bricks	m^2	280.00
23	Facing bricks	m^2	320.00
Roofing			
24	Concrete interlocking roof tiles	m^2	95.00
25	Plain clay roof tiles	m^2	225.00
29	3 layers glass-fibre based bitumen felt roof covering include chipping	m^2	55.00

		Unit	*Rate US$*
30	Bitumen based mastic asphalt lap cement	m^2	4.00
31	Glass-fibre mat roof insulation	m^2	12.00
32	Rigid sheet loadbearing roof insulation 75mm thick	m^2	75.00
33	Troughed galvanized steel roof cladding	m^2	220.00
Woodwork and metalwork			
36	Single glazed casement window in hardwood, size 650 x 900mm	each	800.00
38	Solid core half hour fire resisting hardwood internal flush doors, size 800 x 2000mm	each	1,600.00
39	Aluminum double glazed window, size 1200 x 1200mm	each	1,100.00
40	Aluminum double glazed door, size 850 x 2100mm	each	3,500.00
Plumbing			
42	UPVC half round eaves gutter	m	24.00
43A	Rainwater pipes, 76mm DWC PVC	m	35.00
44A	Type L copper water tubing 13mm	m	115.00
45A	Plastic pipes for cold water distribution, Sch 80 CPVC, 76mm	m	95.00
46A	Plastic pipes for cold water distribution, Sch 40 CPVC, 38mm	m	60.00
47A	Soil and vent pipes, 102mm	m	195.00
48	White vitreous china WC suite	each	850.00
49	White vitreous china lavatory basin	each	650.00
50	White glazed fireclay shower tray	each	600.00
51	Stainless steel single bowl sink	each	600.00
Electrical work			
53	13 amp unswitched socket outlet	each	42.00
54	Flush mounted 20 amp, 1 way light switch	each	42.00
Finishings			
55	2 coats gypsum based plaster on brick walls	m^2	180.00
56	White glazed tiles on plaster walls	m^2	130.00
57	Red clay quarry tiles on concrete floors	m^2	150.00
60	Mineral fibre tiles on concealed suspension system	m^2	45.00
Glazing			
61	Glazing to wood	m^2	120.00

		Unit	Rate US$
Painting			
62	Emulsion on plaster walls	m^2	11.00
63	Oil paint on timber	m^2	35.00

Approximate Estimating

The building costs per unit area below are US national averages incurred by building clients for typical buildings as at the fourth quarter of 2008. They are based upon the total floor area of all storeys, measured between external walls and without deduction for internal walls.

Approximate estimating costs generally include mechanical and electrical installations but exclude furniture, loose or special equipment, and external works; they also exclude fees for professional services.

The costs shown are for specifications and standards appropriate to the United States and this should be borne in mind when attempting comparisons with similarly described building types in other countries. A discussion of this issue is included in section 2. Comparative data for countries covered in this publication, including construction cost data, are presented in Part Three.

Approximate estimating costs must be treated with reserve; they cannot provide more than a rough guide to the probable cost of building.

	Cost m^2 US$	Cost ft^2 US$
Industrial buildings		
Factories for letting (including lighting, power and heating)	1,076	100
Factories for owner occupation (light industrial use)	1,291	120
Factories for owner occupation (heavy industrial use)	1,937	180
Factory/office (high tech) for letting (shell and core only)	1,549	144
Factory/office (high tech) for letting (ground floor shell, first floor offices)	1,893	176
Factory/office (high tech) for owner occupation (controlled environment, fully furnished)	3,873	360
High tech laboratory (air-conditioned)	4,734	440
Warehouses, low bay (6 to 8m high) for letting (no heating)	516	48
Warehouses, low bay for owner occupation (including heating)	689	64
Warehouses, high bay for owner occupation (including heating)	818	76
Cold stores/refrigerated stores	2,152	200

	Cost m² US$	*Cost ft² US$*
Administrative and commercial buildings		
Civic offices, non air-conditioned	2,582	240
Civic offices, fully air-conditioned	2,797	260
Offices for letting, 5 to 10 storeys, non air-conditioned	2,152	200
Offices for letting, 5 to 10 storeys, air-conditioned	2,367	220
Offices for letting, high rise, air-conditioned	3,012	280
Offices for owner occupation 5 to 10 storeys, non air-conditioned	2,668	248
Offices for owner occupation, high rise, air-conditioned	3,529	328
Prestige/headquarters office, 5 to 10 storeys, air conditioned	3,443	320
Prestige/headquarters office, high rise, air-conditioned	3,873	360
Health and education buildings		
General hospitals	2,797	260
Teaching hospitals	3,012	280
Private hospitals	3,012	280
Health centres	2,582	240
Nursery schools	2,066	192
Primary/junior schools	2,152	200
Secondary/middle schools	2,410	224
University (arts) buildings	3,012	280
University (science) buildings	4,303	400
Management training centres	3,529	328
Recreation and arts buildings		
Theatres (over 500 seats) including seating and stage equipment	6,455	600
Theatres (less than 500 seats) including seating and stage equipment	5,164	480
Concert halls including seating and stage equipment	6,993	650
Swimming pools (international standard) including changing facilities	3,615	336
Swimming pools (schools standard) including changing facilities	3,012	280
City centre/central libraries	3,873	360
Branch/local libraries	3,228	300

	Cost m² US$	Cost ft² US$
Residential buildings		
Social/economic single family housing (multiple units)	1,937	180
Private/mass market single family housing 2 storey detached / semi detached (multiple units)	2,066	192
Purpose designed single family housing 2 storey detached (single unit)	2,754	256
Social/economic apartment housing, low rise (no lifts)	1,937	180
Social/economic apartment housing, high rise (with lifts)	2,797	260
Private sector apartment building (standard specification)	3,271	304
Private sector apartment buildings (luxury)	3,615	336
Student/nurses halls of residence	1,893	176
Home for the elderly (shared accommodations)	1,980	184
Homes for the elderly (self contained with shared communal facilities)	2,410	224
Motel	1,205	112

Regional Variations

The approximate estimating costs are based on US national average costs (using costs in Washington DC). Adjust these costs by the following factors for regional variations:

Los Angeles, CA	:	+17%	Providence, RI	:	+6.6%
Hartford, CT	:	+8.5%	Las Vegas, NV	:	+7.5%
Miami, FL	:	-5.7%	Columbia, SC	:	-15.1%
New York, NY	:	+32.1%	Seattle, WA	:	+11.3%
Philadelphia, PA	:	+17.9%	Dallas, TX	:	-15.1%

Source: Marshall & Swift Current Cost Multipliers, Q4 2008

EXCHANGE RATES AND INFLATION

The combined effect of exchange rates and inflation on prices within the United States and on price comparisons between countries is discussed in section 2.

Exchange Rates

The graph below plots the movement of the US dollar against the sterling, the euro, and 100 Japanese yen since 1998. The values used for the graph are quarterly and the method of calculating these is described and general guidance on the interpretation of the graph provided in section 2. The average exchange rate in the fourth quarter of 2008 was US$1.99 to pound sterling and US$1.48 to euro and US$0.93 to 100 Japanese yen.

THE US DOLLAR AGAINST STERLING, EURO AND 100 JAPANESE YEN

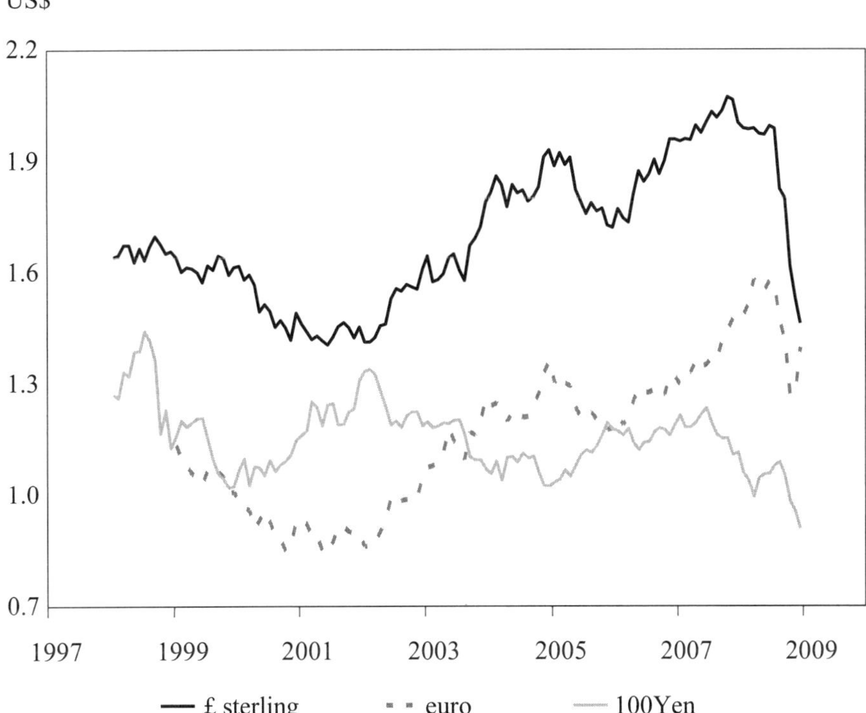

Inflation

The table overleaf presents the indices for consumer price and building cost inflation in the USA since 2000. The indices have been rebased to 2000=100. The annual change is the percentage change between the average index of consecutive years.

CONSUMER PRICE, BUILDING COST AND BUILDING MATERIAL COST INDICES

	Consumer price index(a)		*Building cost index (b)*		*Building material cost index (c)*	
Year	*annual average*	*change %*	*annual average*	*change %*	*annual average*	*change %*
2000	100	3.4	100	2.4	100	1.8
2001	102.8	2.8	101	1.0	97.3	-2.7
2002	104.5	1.6	102.4	1.4	97.9	0.6
2003	106.8	2.3	104.4	1.9	100.3	2.5
2004	109.7	2.7	112.6	7.9	113.2	12.8
2005	113.4	3.4	118.8	5.5	121.6	7.5
2006	117.1	3.2	123.5	3.9	133.1	9.4
2007	120.4	2.9	126.7	2.7	137.2	3.0
2008	125.0	3.8	132.6	4.6	144.9	5.6

Source: (a) US Bureau of Labor Statistics
(b) ENR Building Cost Index
(c) Davis Langdon materials cost index

USEFUL ADDRESSES

Government And Public Organizations

American National Standards Institute
25 West 43rd St
New York NY 10036
Tel: (1) 212 642 4900
Fax: (1) 212 398 0023
Website: www.ansi.org

Army Corps of Engineers
441 G St, NW
Washington DC 20314
Tel: (1) 202 761 0011
Website: www.hq.usace.army.mil

Department of Housing and Urban Development
451 7th Street, SW
Washington DC 20410
Tel: (1) 202 708 1112
Fax: (1) 202 708 1455
Website: www.hud.gov

Department of Transportation
1200 New Jersey Ave SE
Washington DC 20590
Tel: (1) 202 366 4000
Website: www.dot.gov

General Service Administration – GSA
National Capital Region
7th & D Streets, SW
Washington DC 20407
Tel: (1) 202 708 9100
Fax: (1) 202 708 9966
Website: www.gsa.gov

National Academy of Sciences
500 5th St, NW
Washington DC 20001
Tel: (1) 202 334 2000
Website: www.nas.edu

National Institute of Standards and Technology
100 Bureau Drive
Stop 1070
Gaithersburg MD 20899-0001
Tel: (1) 301 975 6478
E-mail: inquiries@nist.gov
Website: www.nist.gov

Small Business Administration
409 3rd Street, SW
Washington DC 20416
Tel: (1) 800 827 5722
E-mail: answerdesk@sba.gov
Website: www.sba.gov

US Census Bureau
4600 Silver Hill Road
Suitland MD 20746
Website: www.census.gov

Trade And Professional Associations

American Association of Cost Engineers International – AACE
209 Prairie Avenue, Suite 100
Morgantown WV 26501-5934
Tel: (1) 304 296-8444
Fax: (1) 304 291-5728
E-mail: info@aacei.org
Website: www.aacei.org

American Institute of Architects – AIA
1735 New York Ave, NW
Washington DC 20006-5292
Tel: (1) 202 626 7300
Fax: (1) 202 626 7547
E-mail: infocentral@aia.org
Website: www.aia.org

American Society for Testing and Materials
100 Barr Harbor Drive
PO Box C700
West Conshohocken PA 19428-2959
Tel: (1) 610 832 9500
Fax: (1) 610 832 9555
E-mail: service@astm.org
Website: www.astm.org

Association of Construction Inspectors
21640 North 19th Avenue, Suite C-2
Phoenix AZ 85027
Tel: (1) 623 580-4646
Fax: (1) 623 580-9656
E-mail: info@aci-assoc.org
Website: www.aci-assoc.org

Associated General Contractors of America
2300 Wilson Blvd, Suite 400
Arlington VA 22201
Tel: (1) 703 548 3118
Fax: (1) 703 548 3119
E-mail: info@agc.org
Website: www.agc.org

Construction Specifications Institute
99 Canal Center Plaza, Suite 300
Alexandria VA 22314-1588
Tel: (1) 800 689 2900
Fax: (1) 703 684 8436
E-mail: csi@csinet.org
Website: www.csinet.org

Design-Build Institute of America
1100 H Street NW, Suite 500
Washington DC 20005-5476
Tel: (1) 202 682 0110
Fax: (1) 202 682 5877
E-mail: dbia@dbia.org
Website: www.dbia.org

National Society of Professional Engineers – NSPE
1420 King Street
Alexandria VA 22314 2794
Tel: (1) 703 684 2800
Fax: (1) 703 836 4875
Website: www.nspe.org

DAVIS LANGDON & SEAH VIETNAM CO. LTD

Davis Langdon & Seah Vietnam Co. Ltd manages client requirements, controls risk, manages cost and maximises value for money, throughout the course of construction projects, always aiming to be – and to deliver – the best.

TYPICAL PROJECT STAGES, INTEGRATED SERVICES AND THEIR EFFECT

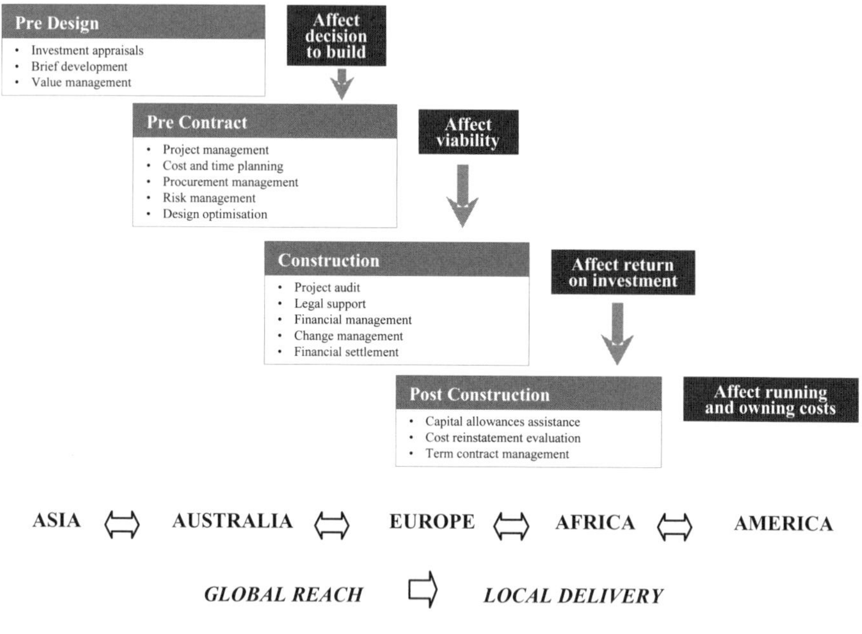

ASIA ⇔ AUSTRALIA ⇔ EUROPE ⇔ AFRICA ⇔ AMERICA

GLOBAL REACH ⇨ LOCAL DELIVERY

HANOI OFFICE
#706 7/F North Star Building
4 Da Tuong Street, Hoan Kiem District
Hanoi
Tel : (844) 942 7525
Fax: (844) 942 7526

HO CHI MINH CITY OFFICE
9th Level, Unit E, OSIC Building
8 Nguyen Hue Street, District 1
Ho Chi Minh City
Tel : (848) 823 8297
Fax: (848) 823 8197

DAVIS LANGDON & SEAH INTERNATIONAL
www.davislangdon.com

VIETNAM

All data relate to 2008 unless otherwise indicated.

Population	
Population	86.16 mn
Urban population	28%
Population under 15	26%
Population 65 and over	7%
Average annual growth rate (2005 to 2008)	1.24%
Geography	
Land area	331,689 km^2
Agricultural area	28%
Capital city	Hanoi
Population of capital city	6.1 mn
Economy	
Monetary unit	Vietnamese Dong (VND)
Exchange rate (average fourth quarter 2008)	
the pound sterling	VND 27,076
the US dollar	VND 17,195
the euro	VND 22,667
the yen x 100	VND 17,912
Average annual inflation (1999 to 2008)	6.4%
Gross Domestic Product (GDP)	VND 1,478,695 bn
GDP per capita	VND 17,162,000
Average annual real change in (GDP) (1999 to 2008)	7.2%
Final consumption as a proportion of GDP	70.9%
Investment as a proportion of GDP	43.1%
Construction	
Gross value of construction output	VND 95,964 bn
Net value of construction output	N/A
Net value of construction output as a proportion of GDP (1999 to 2008)	6.12%

THE CONSTRUCTION INDUSTRY

Construction Outlook

Notwithstanding the sub-prime market concerns of 2007 and their subsequent manifestation into a global financial crisis during 2008, Vietnam has continued to enjoy construction growth during 2007, however growth has been modest during 2008.

It is predicted that government spending will increase during 2009 primarily on the back of Official Development Assistance (ODA) invested projects and as consequence a level of construction growth will be exhibited within the infrastructure sector.

Bank base rates were hiked up by 2.5% during 2008 to a peak of 13.5% and commercial lending rates are currently around 20%+. The base rate was lowered in October to 12% and further reductions are anticipated in 2009. Reductions in the Base lending rate should allow a reduction of commercial lending rates and this should help increase the flow of private domestic capital.

Construction inflation costs have been considerable in 2008 arising largely as a consequence of the increases in international commodity prices (especially steel). Prices of many commodities are now reducing globally and this will reduce prices in the Vietnam Construction market accordingly. Construction costs will be lower in 2009 and this should attract some of the available domestic capital in to the property markets.

The international demand for commodities is not the only significant factor affecting construction inflation, domestic demand can also have a significant effect upon construction inflation. The construction market within Vietnam is relatively small scale compared to its regional neighbours and consequently relatively small shifts in demand can culminate in erratic fluctuations of tender prices which tend to "spike" depending on the timing of the tender period. There are little reliable government hard data or tender price indices available at present and developers looking to receive competitive bids need to consider strategically issuing tender documents at the optimal time but balanced against the perceived market demand for the facility at completion

Construction Output

Development in all sectors has slowed during 2008 due to fiscal measures (namely interest rate increases) which have been implemented to reduce both inflation and the balance of payment deficit. It is further anticipated that Foreign Direct Investment will be slow during 2009 as consequence of the current global liquidity problems. Notwithstanding and as explained above interest rate reductions in 2009 should stimulate some local investment and will lower construction costs. ODA investment should help infrastructure growth.

Construction output data in Vietnam are not reliable due to the nature of the political system; transparency is cited as a major hurdle for the Vietnamese Government to overcome. The three components of construction output are; state funded, local private funded and international investment. There are difficulties in measuring and defining construction output and even in more regulated economies private local investment in construction figures are not discernable. The usual split is one third for each component so US$4 billion (2007) for state projects would equate to circa US$10 billion to US$12 billion for all components of construction output. Interestingly the construction output expressed as a % GDP for more developed nations in South East Asia is 12% to 15% and for Vietnam the figures is circa 6% to 7%.

The construction output figures therefore demonstrate a real growth of 12% in 2007.

Other statistical indicators such as cement and steel production show year-on-year growth of 11% and 5% respectively which are in line with the construction output statistics.

Year	*2000*	*2001*	*2002*	*2003*	*2004*	*2005*	*2006*	*2007*	*2008*
Construction output (US$ billion)	1.63	1.94	2.07	2.39	2.82	3.35	4.09	4.94	6.23
Proportion of GDP (%)	5.35	5.80	5.89	6.05	6.23	6.35	6.62	6.96	6.61
Real Construction output growth (%)	7.51	12.78	10.57	10.59	9.03	10.87	11.05	12.01	0.02

Source: Statistics Directory 2008 – General Statistics Office of Vietnam

The domestic material production industry cannot meet anticipated and expected demand alone leading to the need for imports and stockpiling of key materials. The Vietnam Construction Materials Association has its sights on US$1.0 billion in export earnings by 2010 and aims for 25% to 30% export of all Vietnam produced construction materials with 25% year-on-year growth in export volume. If this is achieved, it will increase the percentage of domestically produced materials incorporated into construction resulting in a reduction in construction costs.

The former "70% imported materials and components" rule of thumb no longer applies in Vietnam with most developers now opting for domestically produced material and components. The architectural and structural trades namely bricks, blocks, tiles, marbles and granites and joinery are all available locally and of acceptable quality for international projects but with limited selection in the domestic market. China sourced products have flooded the market of late with tiles and architectural coverings normally sourced from Thailand and Indonesia.

Cement is produced locally and supply and price are controlled by careful manipulation of supply stock. HOLCIM, a Swiss/Vietnamese joint venture company, is the largest producer of cement in Vietnam. Local supplier such as HA TIEN is also a large producer of cement; supply is currently not an issue in Vietnam.

Hot rolled structural steel sections are not produced in Vietnam and are normally imported from Korea, Japan, Taiwan or China. High grade pre-engineered building steel is also imported in flat plate sections and is fabricated locally to form I-beams and angles (i.e. built-up sections). Reinforcement bar steel is hot rolled in Vietnam and can be purchased with quality "kite-marks" such as JIS. Developers opting for imported reinforcement will incur import tax of 15%.

Sand (used for land filling) is now a very valuable commodity in the low lying delta region of southern Vietnam and the price is now carefully controlled by the formulation of cartels owned and controlled by state agencies which have recently increased prices by circa 20% to 30%.

Aluminum and metal-works are often imported there being only a few architectural metalwork production facilities in Vietnam. The frame extrusions are imported from regional source and fabricated locally. Glass from China and Vietnam is available but in varying standards. Higher-end developments use imported tempered or laminated glass from Japan and Thailand or US.

High-end curtain walling will involve imported frames and glass; lower end will be a hybrid of imported/local frame/glass. High-end curtain walling poses a big problem for Grade 'A' building investors as none have established in Vietnam and there seems to be a reluctance to enter the Vietnam market when order books are full elsewhere.

Vietnam exports over US$7 billion worth of furniture worldwide to US and Europe, accordingly fabricating joinery off-shore and importing to Vietnam is not cost effective except perhaps for the higher-end developments involving highly specialist items.

Mechanical and electrical major plant and equipment for example; gen-sets, chillers, split-units, water-treatment and sewage treatment packaged units, pumps and valves are still imported and regional price adjustments apply. The supply sources are generally G8 countries such as UK, US and Japan and source depends on developer's preference. Cabling and ductwork are produced in Vietnam but are of inferior quality. Imported components are available from stockists or distributorships located in Hanoi and Ho Chi Minh but large quantities are likely to be imported from regional centres such as Malaysia, Thailand China, Japan and Korea.

Characteristics and Structure of the Industry

The Ministry of Construction is responsible for all construction activities in Vietnam. The headquarters is located in Hanoi with a regional office in Ho Chi Minh City. The Ministry is divided into four key sections each under the jurisdiction of a Vice-Minister. These sections are: the Department of Construction Economics and Management; Science and Technology; Urban Management and Development of Construction Materials.

The Ministry is structured into departments, institutes and various construction related companies. Specific responsibilities of the four key sections are:

- Construction Economics and Management – sets guidelines on the costs of labour, materials and plant; to advise on bidding procedures and overhead, profit and taxation allowances for state projects.

- Science and Technology – approves Construction Standards and Regulations with particular attention to the environment.
- Urban Management – responsible for water supply and treatment, waste disposal, construction of low income housing and approval of each province's master plan.
- Development of Construction Materials – promotes and gears foreign investment in domestic resources such as cement, bricks, sanitary ware and roofing products in order to realize the benefits of the forecast increase in demand in products of an international standard both in Vietnam and for the potential export market.

In order to proceed with the construction of any development the investor initially registers for a Construction Investment Licence. The procedures for obtaining the Construction Investment Licence are generally prescribed in Law on Construction for Investment Projects for Construction Works.

- For very large projects (Investment Level not defined) the Investor will need to prepare a Construction Investment Report for approval by the Prime Minister.
- For other projects the Investor will need to ensure that the project complies with General Master Planning requirements accordingly documentation must be submitted to Ministry of Planning and Investment.

Following Construction Investment Approval as discussed above and assuming a Joint Venture structure with the local Land owner (as discussed in detail in the next chapter) the next step in the sequence of development procedures is to ensure the project complies with Municipal Master Plans. The procedure is generally prescribed as follows:

- Planning Permission must be obtained by the Investor from the Municipal Chief Planning Architectural Department.

Once Planning Permission has been obtained then both a Feasibility Study and Basic Design must be prepared and submitted to the Construction Department for approval. It is prudent at this stage to consider the appointment of a Local Architect in order to follow through the design steps required.

- The Feasibility Study now renamed 'Project for Construction of Work' and will comprise:
    - Master plan
    - General Project Explanation
    - Basic Design Drawings
    - Total Investment Cost
- Once Basic Design is approved then Technical Design can be prepared.
- The Technical Design shall be appraised by an independent third party and this shall be organized by the Investor. The Technical Design must also be submitted to the Fire and Police Department for approval.
- Once the Technical Design has been appraised then Detailed Drawings (Construction Drawings) are prepared and again appraised by an independent third party as organized by the Investor.

- The Construction Permit will be issued by the Construction Department following receipt of the Detailed Drawings.
- The tender for construction works may be issued in parallel to obtaining the Construction Permit however works above ground level cannot commence on site until the Construction Permit has been granted.

Other important organizational bodies which need to be considered within Vietnam comprise:

Provincial People's Committee: Each province has a separate People's Committee whose responsibilities include determining the project's compliance with the appropriate development plan, evaluation of the project assets, financial status of the relevant parties, organization of utility supply of the project (if relevant) and determining land use ownership, rights and terms. All new Representative Offices must register with the People's Committee of the province in which the office will be located.

The Chamber of Commerce and Industry (CCI): The CCI is an independent, non-government organization whose functions are to promote trade and investment in Vietnam and abroad; to represent the Vietnamese business community for the promotion and protection of its interests in domestic and international relations; to serve as a forum for exchange of information between investment enterprises and the State on matters concerning the economic activity and business environment in Vietnam.

The Vietnamese Union of Architects (VUA): The VUA is based in Hanoi and has branches throughout the country. The Union is a member of the International Union of Architects and, theoretically, has links with similar institutions around the world. Provincial branches of the Architects' Union are similarly organized and perform similar functions as the National Union in Hanoi.

The Vietnam Consultant Association (VECAS): VECAS is a professional body representing consultants and providing both educational and publication services to consultants throughout Vietnam. VECAS is a Member of International Federation of Consulting Engineers (FIDIC) and published FIDIC Contracts in Vietnamese and English and are a key organization in the development and internationalization of the Vietnam construction industry.

Selection of Contractors

There are four unofficial tiers of contractors available for tendering foreign investment projects.

Tier 1 Foreign contractors from Japan, Korea, Singapore, Hong Kong and Europe Australia
Tier 2 Foreign contractors from China, Russia, Taiwan and Malaysia
Tier 3 Local private contractors or joint stock companies
Tier 4 State contractors

There is a considerable diversity between these tiers in terms of general expertise, technological know-how, human-resource skills and training which leads to a different risk profile for delivery of buildings to time, quality and cost targets.

The lowest tiers use less modern equipment, machinery and system formwork and have much lower human resource costs. However, all tiers are currently experiencing full order books and speculating as to which projects are more viable in terms of their own risk/benefit assessments. This is inevitably leading to higher tender returns. A rigorous pre-qualification procedure is recommended with weighted scoring in line with project objectives e.g. lowest cost objective will have an inevitable impact on quality and time.

A typical pre-qualification assessment might cover (indicative weighting for Vietnam 1 thru 10, 10 highest)

- Vietnam experience (if foreign) : 8
- Project type experience (local and foreign) : 7
- Financial capacity (particularly local) : 8
- Plant and equipment inventory (particularly local) : 7
- Direct labor strength (local and foreign) : 6
- Proposed joint venture / subcontracting arrangements (local and foreign) : 7
- Planning and schedule capability (particularly local) : 8
- Cognizance of international procedures, codes and specifications (local) : 8

Clients and Finance

Up to the end of 2007, Vietnam had 8,684 valid projects with the total registered capital of US$85.05 billion and total implemented capital of US$30 billion. Industry and construction accounted for the highest proportion of 67% of total projects and 60% of total registered capital. Eighty-two countries and territories were investing in Vietnam. Four leading investment countries including Korea, Singapore, Taiwan and Japan made up 55% of total registered capital. The year 2007 witnessed the highest record of Foreign Direct Investment (FDI) flow to Vietnam with the total registered capital of US$21.3 billion.

Since the resumption of ODA to Vietnam in 1993, ODA commitment has been increasing annually. During the period of 1993-2007, total ODA commitment valued at US$ 42,438 million, mostly from large donors such as Japan, World Bank, Asian Development Bank and United Nations agencies. ODA has made significant contribution to the development of Vietnam, accounting for 11% of total social investment and 17% of investment from the budget. The majority of this aid money is to be channeled into infrastructure development, improvement and technical assistance programs. After 2010 when Vietnam reaches the middle income level (GDP per capita over USD 1,000) it will not be further provided with high concessional ODA.

(Source: Ministry of Planning and Investment)

Selection of Consultants

The selection of consultants for public invested projects should comply with Law on Tendering and Decree 16/2005/ND-CP on management of construction investment projects. The Prime Minister also issued Decision 131/2007/QD-TTg to provide Regulations on employing foreign consultants in construction activities in Vietnam. It is important for all foreign investors to be aware of these regulations prior to making final selection of consultants. Foreign clients usually select the design consultants known to them or those who have a reputation for a specific type of building. In recent times these design consultants tended to come from their own countries, however, there appears to be a move to use in-country consultants.

In late 2007, the Government issued Decree 99/2007/ND-CP titled "Government Decree on Cost Management in Civil Engineering Investment" which is in many ways a milestone piece of legislation which acknowledges the need for further market reforms and better cost management within the Government sector. This decree called for the need of suitably qualified Cost Management consultants.

Foreign Banks providing ODA investment also publish their own requirements which must be adhered to and accordingly Contract Documents in such circumstances must be compliant to: The International Standard General Contract Conditions Vietnam Legislation and ODA Guidelines.

Contractual Arrangements

The primary consideration in the choice of procurement strategy is the need to obtain overall value for money during the entire life of the facility and each method has a different risk profile for the employer and contractor. In Vietnam the following are currently prevalent.

- Traditional Lump Sum – high extent
- Management Contracting – low extent
- Construction Management – medium to high extent
- Design and Construct – low extent
- Prime Contracting – low extent
- Framework Agreements – rising extent

Most if not all projects in Vietnam are tendered in competition and the process is covered by the Law of Tendering promulgated in 2006 intended primarily for state projects defined as over 30% total investment capital by a state entity and for Vietnamese Private firms. Foreign Investors do not need to follow the Law of Tendering although it is advisable. The law recognizes open tendering, limited tendering and competitive tendering. For state projects tenders are normally sought using a "two-envelope system" i.e. technical and financial the former being opened first to check for compliance.

Most foreign entities shortlist tendering contractors by a having a robust pre-qualification procedure for checking financial and technical competencies. Tenders are usually open for 90 days.

The FIDIC suite of contracts is widely used for Vietnam construction contracts with the 1999 Red Book being now widely accepted. There are official translations of some of the FIDIC forms and the VECAS is an official member of FIDIC. Most Official Development Aid projects in Vietnam adopt FIDIC also.

Development Control and Standards

The Ministry of Planning and Investment (MPI), the successor to the State Committee for Cooperation and Investment (SCCI), is the body responsible for control of all foreign investments in Vietnam and for issuing investment licences. The MPI's role is to circulate investment licence applications among the various Ministries and relevant People's Committee for the region. The MPI's head office is in Hanoi with a representative office in Ho Chi Minh City. The various departments within the MPI cover Project Evaluation, Investment Promotion, General Office, Information and Legislation.

The Project Evaluation Department's role is to determine whether an investment proposal is in accordance with the best interests of the State. During the evaluation process, which can take up to three months, questions from the various Ministries must be answered within a stipulated time frame which at the time of publishing is 45 days, after which the application is deemed to have lapsed. The MPI provides pro-forma applications and documents and encourages reference to previously successful application to limit abortive time. After the evaluation process an Investment Licence is either issued or the application is rejected.

All foreign investors are required to register their financial accounting practices with the Ministry of Finance. Regular reports on the financial standing of an enterprise must be submitted. The Ministry of Finance also advises the MPI on fiscal matters such as taxation levels, subsidies and incentives for the various forms of foreign investment.

CONSTRUCTION COST DATA

Cost of Labour

The figures below are typical of labour costs in Vietnam as at the fourth quarter of 2008 for joint venture/international projects. The cost of labour indicates the cost to a contractor of employing that employee.

	Cost of labour (per day) US$	*Number of hours worked per year*
Site operatives		
Bricklayer	6.70	2,496
Carpenter	6.70	2,496
Plumber	10.68	2,496
Electrician	10.68	2,496
Structural steel erector	10.68	2,496
Welder	10.68	2,496
Labourer	5.59	2,496
Equipment operator	17.64	2,496

Cost of Materials

The figures that follow are the costs of main construction materials, delivered to site in Vietnam, as incurred by contractors in the fourth quarter of 2008. These assume that the materials would be in quantities as required for a medium sized construction project and that the location of the works would be neither constrained nor remote.

	Unit	*Cost US$*
Cement and aggregate		
Ordinary portland cement in 50kg bags	tonne	68.98
Coarse aggregates for concrete 40mm	m^3	3.02
Fine aggregates for concrete 20mm	m^3	4.08
Ready mixed concrete (Grade 40)	m^3	63.03
Ready mixed concrete (Grade 35)	m^3	59.66
Ready mixed concrete (Grade 20)	m^3	51.26
Steel		
Mild steel reinforcement	tonne	705.18
High tensile steel reinforcement	tonne	715.71
Structural steel sections	tonne	840.34
Bricks		
Hollow concrete blocks (390 x 190 x 100mm)	1,000	315.51
Well burned clay brick unit (80 x 80 x180mm)	1,000	58.82

	Unit	*Cost US$*
Timber and insulation		
Hardwood for joinery	m^3	578.00
Exterior quality plywood (12mm)	m^2	15.75
Plywood for interior joinery (12mm)	m^2	13.94
50mm thick quilt insulation (16kg/m^3)	m^2	1.63
50mm thick rigid slab insulation (60kg/m^3)	m^2	6.10
Hardwood internal door complete with frames and ironmongery	each	350.00
Glass and ceramics		
Float glass (6mm)	m^2	10.43
Laminated glass (10.38mm)	m^2	23.53
Tempered Glass 8mm	m^2	17.65
Plaster and paint		
Good quality ceramic wall tiles (300 x 300 x 8mm)	m^2	12.45
Good quality homogeneous wall tiles (600 x 1200 x 8mm)	m^2	19.89
Good quality marble wall tiles (600 x 1200 x 20mm)	m^2	113.00
Good quality granite wall tiles (600 x 1200 x 20mm)	m^2	106.00
Plasterboard (13mm thick) - gypsum	m^2	11.50
Emulsion paint in 5 litre tins	litre	7.30
Gloss oil paint in 5 litre tins	litre	3.60
Tiles and paviors		
Clay floor tiles (100 x 200 x 8mm)	m^2	5.88
Vinyl floor tiles (300 x 300 x 2mm)	m^2	15.00
Clay roof tiles	1,000	363.64
Precast concrete roof tiles	1,000	588.24
Drainage		
WC suite complete	each	315.64
Wash hand basin complete	each	150.00
100mm diameter PVC pipes	m	5.61
150mm diameter PPR pipes (local product for Cold Water)	m	46.00
150mm diameter cast iron drain pipes (medium grade)	m	44.00

Unit Rates

The descriptions on the next page are generally shortened versions of standard descriptions listed in full in section 4. Where an item has a two digit reference number (e.g. 05 or 3), this relates to the full description against that number in section 4. Where an item has an alphabetic suffix (e.g. 12A or 34B) this indicates that the standard description has been modified. Where a modification is major the complete modified description is included here and the standard description should be ignored; where a modification is minor (e.g. the insertion of a named hardwood) the shortened description has been modified here but, in general, the full description in section 4 prevails.

The unit rates that follow are for main work items on a typical joint venture/international project in the city area in the fourth quarter of 2008. The rates include all necessary labour, materials and equipment. An allowance of 12% has been added to the rates to cover preliminary and general items.

		Unit	*Rate US$*
Excavation			
01	Mechanical excavation of foundation trenches	m^3	3.53
02	Hardcore filling making up levels; 150mm thick	m^3	8.82
Concrete work			
04	Plain in situ concrete in strip foundations in trenches (Grade 20)	m^3	78.00
05A	Reinforced in situ concrete in beds (Grade 35)	m^3	95.00
06A	Reinforced in situ concrete in walls (Grade 35)	m^3	110.00
07A	Reinforced in situ concrete in suspended floors or roof slabs (Grade 35)	m^3	98.00
08A	Reinforced in situ concrete in columns (Grade 35)	m^3	106.00
09A	Reinforced in situ concrete in isolated beams (Grade 35)	m^3	98.00
Formwork			
11A	Waterproof plywood formwork to concrete walls	m^2	8.00
12A	Waterproof plywood formwork to concrete columns	m^2	9.00
13A	Waterproof plywood formwork to horizontal soffits of slabs	m^2	12.50
Reinforcement			
14	Reinforcement in concrete walls	tonne	962.00
15	Reinforcement in suspended concrete slabs	tonne	949.00
16	Fabric reinforcement in concrete beds	m^2	5.16
Steelwork			
17	Fabricate, supply and erect steel framed structure	tonne	1,749.00
19	Structural steelwork lattice roof trusses	tonne	2,098.00

		Unit	*Rate US$*
Brickwork and blockwork			
20	Precast lightweight aggregate hollow concrete block walls	m^2	17.50
21A	Solid (perforated) concrete blocks	m^2	22.00
23	Facing bricks (215mm thick)	m^2	27.00
Roofing			
25	Plain clay roof tiles 260 x 160mm	m^2	10.94
29	3 layers glass-fibre based bitumen felt roof covering	m^2	21.18
33	Troughed galvanized steel roof cladding	m^2	15.13
Woodwork and metalwork			
34	Preservative treated sawn hardwood 50 x 100mm	m	9.00
35	Preservative treated sawn hardwood 50 x 150mm	m	12.00
37	Two panel glazed door in Kapur hardwood, size 850 x 2000mm	each	358.00
38	Solid core half hour fire resisting hardwood internal flush door, size 800 x 2000mm	each	480.00
39	Aluminium double glazed window, size 1200 x 1200mm	each	270.00
41	Hardwood skirtings	m	7.00
Plumbing			
42	UPVC half round eaves gutter	m	19.60
43A	UPVC rainwater pipes; 300mm diameter	m	39.30
44	Light gauge copper cold water tubing	m	14.20
45	High pressure plastic pipes for cold water supply	m	7.70
47	UPVC soil and vent pipes	m	24.80
48	White vitreous china WC suite	each	294.10
49	White vitreous china wash hand basin	each	264.70
51A	Stainless steel double bowl sink and double drainer	each	312.40
Electrical work			
52	PVC insulated and copper sheathed cable	m	4.40
53	13 amp unswitched socket outlet	each	11.00
54	Flush mounted 20 amp, 1 way light switch	each	12.00
Finishing			
55A	2 coats cement and sand (1:4) plaster on brick walls	m^2	2.30
56	White glazed tiles on plaster walls	m^2	12.00
57	Red clay quarry tiles on concrete floors	m^2	20.00
58	Cement and sand screed to concrete floors (20mm thick)	m^2	2.40

		Unit	Rate US$
59	Thermoplastic floor tiles on screed	m²	26.25
60	Mineral fibre tiles on concealed suspension system	m²	25.00
Glazing			
61	Glazing to wood	m²	22.00
Painting			
62	Emulsion on plaster walls	m²	2.60
63	Oil paint on timber	m²	3.00

Approximate Estimating

The building costs per unit area below are averages incurred by building clients for joint venture/international projects in Vietnam as at the fourth quarter of 2008. They are based upon the total floor area of all stories, measured between external walls and without deduction for internal walls.

Approximate estimating costs generally include mechanical and electrical installations but exclude furniture, loose or special equipment, and external works; they also exclude fees for professional services. The costs shown are for specifications and standards appropriate to Vietnam and this should be borne in mind when attempting comparisons with similarly described building types in other countries. A discussion of this issue is included in section 2. Comparative data for countries covered in this publication, including construction cost data, are presented in Part Three.

Approximate estimating costs must be treated with caution; they cannot provide more than a rough guide to the probable cost of building.

	Cost m² US$	Cost ft² US$
Industrial buildings		
Factories for letting	280	26
Factories for owner occupation (light industrial use)	290	27
Factories for owner occupation (heavy industrial use)	430	40
Factory/office (high tech) for owner occupation (controlled environment, fully finished)	430	40
Warehouses, low bay (6 to 8m high) for letting	280	26
Warehouses, low bay for owner occupation	280	26
Warehouses, high bay for owner occupation	340	32
Administrative and commercial buildings		
Offices for letting, 5 to 10 storeys, non air-conditioned	580	54
Offices for letting, 5 to 10 storeys, air-conditioned	650	60
Offices for letting, high rise, air-conditioned	780	72

	Cost m^2 US$	Cost ft^2 US$
Offices for owner occupation high rise, air-conditioned	910	85
Prestige/headquarters office, 5 to 10 storeys, air-conditioned	710	66
Prestige/headquarters office, high rise, air-conditioned	1,010	94
Residential buildings		
Purpose designed single family housing 2 storey detached (single unit)	540	50
Social/economic apartment housing, high rise (with lifts)	500	46
Private sector apartment building (standard specification)	640	59
Private sector apartment buildings (luxury)	840	78
Hotel, 5 star, city centre	1,740	162
Hotel, 3 star, city/provincial	1,460	136

EXCHANGE RATES AND INFLATION

The combined effect of exchange rates and inflation on prices within a country and price comparisons between countries is discussed in section 2.

Exchange Rates

The graph on the next page plots the movement of the Vietnamese Dong against the sterling, the euro, the US dollar and 100 Japanese yen since 1998. The values used for the graph are quarterly and the method of calculating these is described and general guidance on the interpretation of the graph provided in section 2. The average exchange rates in the fourth quarter of 2008 were VND 27,026 to pound sterling, VND 22,667 to euro, VND 17,195 to US dollar and VND 17,912 to 100 Japanese yen.

THE VIETNAM DONG AGAINST STERLING, EURO, US DOLLAR AND 100 JAPANESE YEN

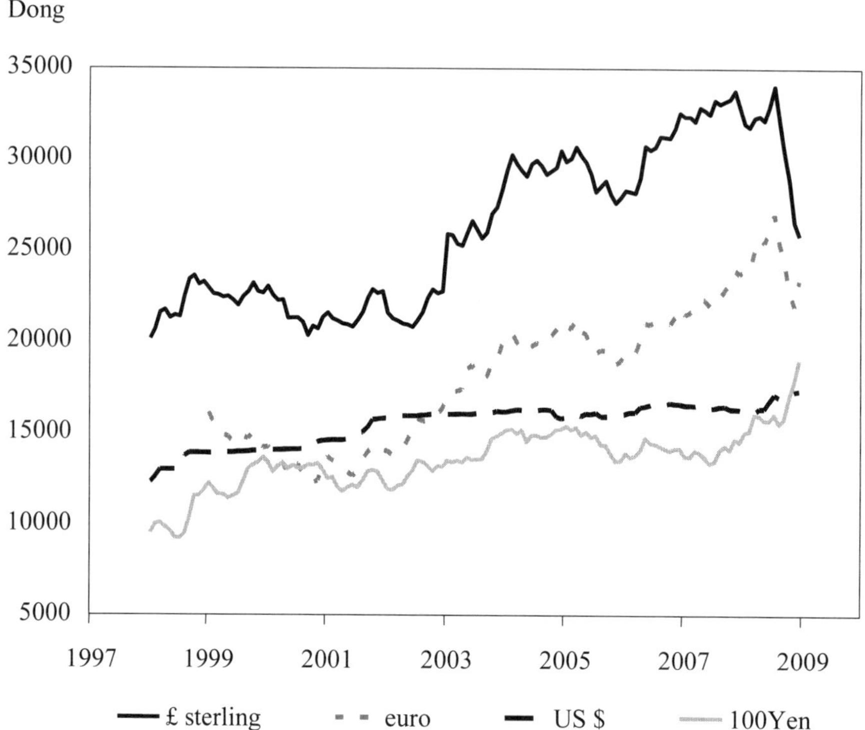

USEFUL ADDRESSES

Ministries

Government Inspectorate
220 Doi Can
Ba Dinh District
Hanoi
Tel: (84) 4 3832 5558 / 3832 5896
Fax: (84) 080 48493
E-mail: ttcp@thanhtra.gov.vn
Website: www.thanhtra.gov.vn

Government Office
1 Hoang Hoa Tham
Ba Dinh District
Hanoi
Tel: (84) 080 43100 / 080 43569
Fax: (84) 080 44130
E-mail: vpcp@chinhphu.vn
Website: vpcp.chinhphu.vn

Ministry of Agricultural and Rural Development
2 Ngoc Ha
Ba Dinh District
Hanoi
Tel: (84) 4 3459 2999 / 3459 2555 / 3846 8161
Fax: (84) 4 459 2888 / 3845 4319
E-mail: icard@agroviet.gov.vn
Website: www.agroviet.gov.vn

Ministry of Construction
37 Le Dai Hanh
Hai Ba Trung District
Hanoi
Tel: (84) 4 3974 0112
Fax: (84) 4 3976 2153
Website: www.moc.gov.vn

Ministry of Culture, Sports and Tourism
51 Ngo Quyen
Hoan Kiem District
Hanoi
Tel: (84) 4 3943 6615 / 3943 9265
Fax: (84) 4 3943 9009 / 3945 4330
Website: www.cinet.gov.vn

Ministry of Education and Training
49 Dai Co Viet
Hai Ba Trung District
Hanoi
Tel: (84) 4 3869 5144 / 3869 7215
Fax: (84) 4 3869 4085
Website: www.moet.gov.vn

Ministry of Finance
28 Tran Hung Dao
Hoan Kiem District
Hanoi
Tel: (84) 4 2220 2828
Fax: (84) 4 2220 8010
E-mail: support@mof.gov.vn
Website: www.mof.gov.vn

Ministry of Foreign Affairs
1 Ton That Dam
Ba Dinh District
Hanoi
Tel: (84) 4 3799 2000
Fax: (84) 4 3823 1872
Website: www.mofa.gov.vn

Ministry of Health
138A Giang Vo
Ba Dinh District
Hanoi
Tel: (84) 4 6273 2273
Fax: (84) 4 3846 4051
E-mail: byt@moh.gov.vn
Website: www.moh.gov.vn

Ministry of Home Affairs
37A Nguyen Binh Khiem
Hai Ba Trung District
Hanoi
Tel: (84) 4 3976 4116 / 976 4278
Fax: (84) 4 3978 1005
E-mail: vucaicachhanhchinh@moha.gov.vn
Website: www.caicachhanhchinh.gov.vn

Ministry of Industry and Trade
54 Hai Ba Trung
Hoan Kiem District
Hanoi
Tel: (84) 4 2220 2101
Fax: (84) 4 2220 2525
E-mail: cenit@moit.gov.vn
Website: www.moit.gov.vn

Ministry of Information and Communications
18 Nguyen Du
Hoan Kiem District
Hanoi
Tel: (84) 4 3943 5602
Fax: (84) 4 3826 3477
E-mail: office@mic.gov.vn
Website: www.mic.gov.vn

Ministry of Justice
56-60 Tran Phu
Ba Dinh
Hanoi District
Tel: (84) 4 3733 8068
Fax: (84) 4 3843 1431
E-mail: botuphap@moj.gov.vn
Website: www.moj.gov.vn

Ministry of Labour, War-Invalids and Social Affairs
12 Ngo Quyen
Hoan Kiem District
Hanoi
Tel: (84) 4 3826 9557 / 3826 9558
Fax: (84) 4 3824 8036
E-mail: lasic@molisa.gov.vn
Website: www.molisa.gov.vn

Ministry of National Defense
7 Nguyen Tri Phuong
Ba Dinh District
Hanoi
Tel: (84) 069 534 223 / 069 531 593
Fax: (84) 069 532 090

Ministry of Natural Resources and Environment
83 Nguyen Chi Thanh
Dong Da District
Hanoi
Tel: (84) 4 3834 3005 / 773 2731
Fax: (84) 4 3835 9221
E-mail: webmaster@nonre.gov.vn
Website: www.monre.gov.vn

Ministry of Planning and Investment
2 Hoang Van Thu
Ba Dinh District
Hanoi
Tel: (84) 4 3845 5298
Fax: (84) 4 3823 4453
Website: www.mpi.gov.vn

Ministry of Public Security
44 Yet Kieu Street
Hoan Kiem District
Hanoi
Tel: (84) 4 069 42545
Fax: (84) 4 3942 0223

Ministry of Science and Technology
39 Tran Hung Dao
Hoan Kiem District
Hanoi
Tel: (84) 4 3943 9731 / 3943 7056
Fax: (84) 4 3943 9733
E-mail: ttth@most.gov.vn
Website: www.most.gov.vn

Ministry of Transport
80 Tran Hung Dao
Hoan Kiem District
Hanoi
Tel: (84) 4 3942 4015 / 3942 4085
Fax: (84) 4 3942 3291
Website: www.mt.gov.vn

State Bank
49 Ly Thai To
Hoan Kiem District
Hanoi
Tel: (84) 4 3825 4845 / 3826 8779
Fax: (84) 4 3934 9569
Website: www.sbv.gov.vn

People's Committees

General Statistical Office
2 Hoang Van Thu
Ba Dinh District
Hanoi
Tel: (84) 4 3733 2997
Fax: (84) 4 3864 4921
E-mail: banbientap@gso.gov.vn
Website: www.gso.gov.vn

People's Committee of Hanoi
79 Dinh Tien Hoang
Hoan Kiem Street
Hanoi
Tel: (84) 4 3825 3536
Fax: (84) 4 3824 3126
E-mail: vanthu@hanoi.gov.vn
Website: www.hanoi.gov.vn

People's Committee of Ho Chi Minh City
86 Le Thanh Ton
District 1
Ho Chi Minh City
Tel: (84) 8 3829 1036 / 3829 2030 / 3829 6999
Fax: (84) 8 3829 5675 / 3829 6988
E-mail: ubnd@tphcm.gov.vn
Website: www.hochiminhcity.gov.vn

Vietnam Chamber of Commerce and Industry
International Trade Centre
4 Floor, 9 Dao Duy Anh
Dong Da District
Hanoi
Tel: (84) 4 3574 2022
Fax: (84) 4 3574 2622
E-mail: vcci@fmail.vnn.vn
Website: www.vcci.com.vn

4. Amplified descriptions of construction items

EXCAVATION

(Assume excavation in firm soil)

1 Mechanical excavation of foundation trenches; starting from ground level (including removal of excavation material from site); over 0.30m wide, not exceeding 2.00m deep.

2 Hardcore filling in making up levels; hard brick, broken stone (or sand where appropriate); crushed to pass a 100mm ring 150mm deep.

3 Earthwork support; sides of trench excavation; distance between opposing faces not exceeding 2.00m; maximum depth 2.00m.

CONCRETE WORK

(Formwork and reinforcement measured separately)

4 Plain in situ concrete in strip foundations in trenches 20N/mm^2; ordinary portland cement, 20mm coarse aggregate; size 500mm wide x 300mm thick.

5 Reinforced in situ concrete in beds 20N/mm^2; ordinary portland cement, 20mm coarse aggregate; 200mm thick.

6 Reinforced in situ concrete in walls 20N/mm^2; ordinary portland cement, 20mm coarse aggregate; 200mm thick.

7 Reinforced in situ concrete in suspended floor or roof slabs 20N/mm^2; ordinary portland cement, 20mm coarse aggregate; 150mm thick.

8 Reinforced in situ concrete in columns 20N/mm^2; ordinary portland cement, 20mm coarse aggregate; size 400 x 400mm.

9 Reinforced in situ concrete in isolated beams 20N/mm^2; ordinary portland cement, 20mm coarse aggregate; size 400 x 600mm deep.

10 Precast concrete slabs (including reinforcement as necessary); contractor designed for total loading of 3N/mm^2; 5.00m span.

FORMWORK

(Assume a simple repetitive design which allows three uses of formwork)

11 Softwood or metal formwork to concrete walls; basic finish (one side only).

12 Softwood or metal formwork to concrete columns; basic finish; columns 1600mm girth.

13 Softwood or metal formwork to horizontal soffits of slabs; basic finish; slabs 150mm thick, not exceeding 3.50m high.

REINFORCEMENT

14 Reinforcement in concrete walls; hot rolled high tensile bars cut, bent and laid, 16mm diameter.

15 Reinforcement in suspended concrete slabs; hot rolled high tensile bars cut, bent and laid, 25mm diameter.

16 Fabric (mat) reinforcement in concrete beds (measured separately); weight approximately 3.0 kg/m^2; laid in position with 150mm side and end laps.

STEELWORK

17 Fabricate, supply and erect steel framed structure; including painting all steel with one coat primer.

18 Framed structural steelwork in universal joist sections; bolted or welded connections, including erecting on site and painting one coat at works.

19 Structural steelwork lattice roof trusses; bolted or welded connections, including erecting on site and painting one coat at works.

BRICKWORK AND BLOCKWORK

(Assume a notional thickness of 100mm for bricks and blocks. Rates should be for the nearest standard size to 100mm)

20 Precast lightweight aggregate hollow concrete block walls; gauged mortar; 100mm thick.

21 Solid (perforated) clay or concrete common bricks (priced at .... per m² delivered to site); gauged mortar; 100mm thick walls.

22 Solid (perforated) sand lime bricks (priced at .... per m² delivered to site); gauged mortar; 100mm thick walls.

23 Facing bricks (priced at .... per m² delivered to site); gauged mortar, flush pointed as work proceeds; half brick thick walls.

ROOFING

24 Concrete interlocking roof tiles 430 x 380mm (or nearest equivalent); on and including battens and underfelt; laid to 355mm gauge with 75mm laps (excluding eaves fittings or ridge tiles).

25 Plain clay roof tiles 260 x 160mm (or nearest equivalent); on and including battens and underfelt; laid to 100mm lap (excluding eaves fittings or ridge tiles).

26 Fibre cement roof slates 600 x 300mm (or nearest equivalent); on and including battens and underfelt; laid flat or to fall as coverings for roofs.

27 Sawn softwood roof boarding, preservative treated 25mm thick; laid flat or to fall.

28 Particle board roof coverings with tongued and grooved joints 25mm thick; laid flat or to fall.

29 Three layers glass-fibre based bitumen felt roof covering; finished with limestone chippings in hot bitumen; to flat roofs.

30 Bitumen based mastic asphalt roof covering in two layers; on and including sheathing felt underlay, with white chippings finish; to flat roofs.

31 Glass-fibre mat roof insulation 160mm thick; laid flat between ceiling joists.

32 Rigid sheet resin-bonded loadbearing glass-fibre roof insulation 75mm thick; laid on flat roofs.

33 0.8mm troughed galvanized steel roof cladding in single spans of 3.00m with loading of 0.75 KN/m^2; fixed to steel roof trusses with bolts; to pitched roofs.

WOODWORK AND METALWORK

(Hardwood should be assumed to be of reasonable exterior quality)

34 Preservative treated sawn softwood; size 50 x 100mm; framed in partitions.

35 Preservative treated sawn softwood; size 50 x 150mm; pitched roof members.

36 Single glazed casement window in (.............) hardwood including hardwood frame and sill; including steel butts and anodized aluminium espagnolette bolt; size approximately 650 x 900mm with 38 x 100mm frame and 75 x 125mm sill.

37 Two panel door with panels open for glass in (............) hardwood including hardwood frame and sill; including glazing with 6mm wired polished plate security glass fixed with hardwood beads and including steel butts, anodized handles and push plates and security locks; size approximately 850 x 2000mm with 38 x 100mm frame and 38 x 150mm sill.

38 Solid core half hour fire resisting hardwood internal flush door lipped on all edges; unpainted, including steel butts, anodized handles and push plates and mortice lock; size approximately 800 x 2000mm.

39 Aluminium double glazed window and hardwood sub-frame; standard anodized horizontally sliding double glazed in (..............) hardwood sub-frame and sill; including double glazing with 4mm glass, including all ironmongery; size approximately 1200 x 1200mm with 38 x 100mm sub-frame and 75 x 125mm sill.

40 Aluminium double glazed door set and hardwood sub-frame; standard anodized aluminium, double glazed in (..............) hardwood sub-frame and sill; including double glazing with 4mm glass, including all ironmongery; size approximately 850 x 2100mm with 38 x 100mm sub-frame and 75 x 125mm sill.

41 Hardwood skirtings. Wrought (...............) hardwood; fixed on softwood grounds; size 20 x 100mm.

PLUMBING

(Sizes of sanitary installations and pipes are indicative)

42 UPVC half round eaves gutter; screwed to softwood at 1.00m centres; 110mm external diameter (excluding bends, outlets etc.).

43 UPVC rainwater pipes with pushfit joints; screwed to brickwork at 1.50m centres; 100mm external diameter (excluding bends, outlets etc.).

44 Light gauge copper cold water tubing with compression or capillary fittings; screwed to brickwork horizontally at 1.00m centres; 15mm external diameter.

45 High pressure polypropylene, polythene or UPVC (as appropriate) pipes for cold water supply; fixed horizontally to brick walls at 1.00m centres; 15mm external diameter, complete with fittings.

46 Low pressure polypropylene, polythene or UPVC (as appropriate) pipes for cold water distribution; with plastic compression fittings 20mm external diameter, laid in trenches.

47 UPVC soil and vent pipes with solvent welded or ring seal joints; fixed vertically to brickwork with brackets at 1.50m centres; 100mm external diameter.

48 White vitreous china WC suite with black plastic seat and cover and plastic low level cistern, 9 litre capacity; complete with ball valve and float and flush pipe to WC suite; fixed to concrete.

49 White vitreous china lavatory basin with 2 No. chrome plated taps (or medium quality chrome plated mixer taps); including plug, overflow and waste connections (excluding trap); size approximately 560 x 400mm, fixed to brickwork with concealed brackets.

50 Glazed fireclay shower tray; including overflow and waste (excluding trap); size approximately 750 x 750 x 175mm, fixed to concrete.

51 Stainless steel single bowl sink and double drainer (excluding taps); including plug, overflow and connections (excluding trap); size approximately 1500 x 600mm, fixed to softwood sink unit (excluding sink base).

ELECTRICAL WORK

52 PVC insulated and copper sheathed cable, 450/750 volt grade, twin core and ECC $6mm^2$ cross section area; fixed to timber with clips.

53 13 amp, 2 gang flush mounted white, unswitched socket outlet; including 6.0m of $2.5mm^2$ concealed PVC insulated copper cable (excluding conduit); flush mounted to brickwork including all fittings and fixing as necessary.

54 Flush mounted 20 amp, 2 gang, 1 way white light switch; including 6.0m of $1.5mm^2$ concealed mineral insulated copper cable (excluding conduit); flush mounted to brickwork including all fittings and fixings as necessary.

FINISHINGS

55 Two coats gypsum based plaster on brick walls 13mm thick; floated finish.

56 White glazed tiles on plaster walls size 100 x 100 x 4mm; fixed with adhesive and grouted between tiles.

57 Red clay quarry tiles on concrete floors size 150 x 150 x 16mm; bedded and jointed in mortar.

58 Floor screed; cement and sand screed to concrete floors 1:3 mix; 50mm thick; floated finish.

59 Thermoplastic floor tiles on screed 2.5mm thick; fixed with adhesive.

60 Suspended ceiling system; fissured mineral fibre tiles size 300 x 300 x 15mm; on galvanized steel concealed suspension system; fixed to concrete soffits with 500mm drop (excluding lamp fittings).

GLAZING

61 Glazing to wood; ordinary quality 4mm glass; softwood beads.

PAINTING

62 Emulsion on plaster walls; one coat diluted sealer coat and two coats full vinyl emulsion paint.

63 Oil paint on timber; one coat primer and two coats oil based paint.

PART THREE: COMPARATIVE DATA

5. Introductory notes

Part Three brings together data from a variety of sources but mainly Part Two, and presents them in the form of tables to allow rapid comparison among the countries included in the book. This also helps place countries, their main statistical indicators and their construction costs in an international context.

There are nineteen tables derived from Part Two arranged in three sections:

Key national indicators

- Population
- The economy
- Geography

Construction output indicators

- Construction output
- Construction output per capita

Construction cost data

- Mason/bricklayer and unskilled labour costs
- Site manager and qualified architect labour costs
- Material costs – Cement and concrete aggregates
- Material costs – Ready mixed concrete and reinforcement steel
- Material costs – Common bricks and hollow concrete blocks
- Material costs – Softwood for joinery and quilt insulation
- Material costs – Sheet glass and plasterboard
- Material costs – Emulsion paint and vinyl floor tiles
- Approximate estimating – Factories and warehouses
- Approximate estimating – Offices
- Approximate estimating – Housing
- Approximate estimating – Hospitals and schools
- Approximate estimating – Theatres and sports halls
- Approximate estimating – Hotels

The first five tables are based on the Key data sheets at the beginning of each country section, the remainder are drawn from the Construction cost data in each country section. Each table is prefaced by explanatory notes. There are inherent dangers in attempting to compare international data, particularly where two sets of data are used (e.g. construction output and population) and, even more so, when exchange rates are used. While these tables can provide useful initial comparisons between countries they should, nevertheless, be used with caution.

6. Key national indicators

POPULATION

The table below summarizes population statistics for all twenty countries included in this book. The table highlights not only the differences in total population among the countries but also variations in the distribution of population between age groups within countries, in population growth rates and in the proportion of the population living in urban areas.

The table includes the most populous country in the world (China) and four others from the top ten most populous countries (India, USA, Indonesia and Pakistan). The developed countries generally have high rates of urbanisation though so also do the two city states of Hong Kong and Singapore. The developed countries also have relatively low proportions of population under 15 and relatively high populations over 65.

Population growth rates vary from less than 1% per annum in the China, Hong Kong, Japan, South Korea, Thailand, UK and USA to over 2% in Brunei, Malaysia, Singapore and Taiwan.

	Population				
Country	*Total (mn)*	*Urban %*	*Under 15 %*	*Over 65 %*	*Growth % pa*
Australia	21.2	75	20	13	1.5
Brunei	0.4	72	32	3	2.7
Cambodia	14.4	18	35	4	1.9
China	1,321.0	45	18	9	0.6
Hong Kong	7.0	94	13	13	0.8
India	1,149.0	29	32	5	1.6
Indonesia	225.6	44	28	5	1.3
Japan	127.7	65	15	21	-0.1
Malaysia	27.2	65	32	5	2.1
New Zealand	4.3	72	22	13	1.4
Pakistan	164.7	35	37	4	2.0
Philippines	88.6	30	32	3	2.0
Singapore	4.6	100	19	9	2.8
South Korea	48.6	82	18	10	0.4
Sri Lanka	20.0	21	28	2	1.1
Taiwan	23.0	55	17	10	3.8
Thailand	66.0	33	23	11	0.8
UK	61.0	90	19	16	0.5
USA	304.1	79	20	13	0.9
Vietnam	86.2	28	26	7	1.2

THE ECONOMY

This table summarizes economic data for the countries included in this book. In the country sections Gross Domestic Product (GDP) figures are given in national currencies; here they have been converted to US dollars using the average exchange rate for the appropriate year – usually 2007. The table contains the two wealthiest nations in the world – USA and Japan – and some of the poorest. As with population density, GDP per capita is a more helpful measure of national wealth than total GDP. UK and USA have amongst the highest GDPs per capita in the world. Brunei has one of the highest GDP per capita in the world while total GDP is less than a tenth of neighbouring Malaysia.

The GDP growth rates are perhaps more interesting indicators of potential wealth. The growth rates are real, that is the effects of inflation are excluded. With the exception of Indonesia and Sri Lanka, average annual inflation 1998-2007 in all countries is below 10%, often well below.

	2007		*1998-2007*	
Country	*GDP US$ bn*	*GDP per capita US$*	*GDP Growth (real) % pa*	*Inflation average % pa*
Australia	712.93	33,628.95	3.6	3.3
Brunei	12.77	32,735.86	1.9	1.8
Cambodia	8.49	591.37	9.4	4.3
China	3,681.99	2,772.18	10.0	1.3
Hong Kong	208.52	30,110.71	2.5	-0.7
India	3,100.00	2,600.00	8.7	6.0
Indonesia	360.52	1,410.48	17.0	15.3
Japan	5,804.37	32,250.97	0.1	0.1
Malaysia	133.26	6,446.91	4.4	2.4
New Zealand	78.50	18,280.92	3.4	2.3
Pakistan	143.65	2,600.00	5.6	5.9
Philippines	154.78	1,746.94	6.1	5.6
Singapore	153.76	35,566.44	5.3	0.7
South Korea	801.47	18,250.66	4.9	2.8
Sri Lanka	32.58	1,629.94	4.0	10.2
Taiwan	374.13	16,238.28	3.7	2.1
Thailand	242.38	3,670.29	4.5	2.8
UK	2,257.81	35,604.69	2.6	1.8
USA	14,264.60	47,395.00	6.4	2.9
Vietnam	86.00	998.08	7.2	6.4

GEOGRAPHY

The table below summarizes geographical statistics for the countries included in this book. As with population the table highlights the differences between countries. It includes two of the largest countries in the world (China and USA); it also includes two of the smallest (Hong Kong and Singapore).

The figures for population density and the percentage of national population in the largest city are perhaps more helpful indicators of land use than total area. As might be expected the table shows that the country with one of the largest areas in the world (Australia) has almost the lowest population density, while Singapore with over 6,000 persons per km² has one of the highest population densities in the world. The percentage of national population in the largest city gives an indication of the relative importance of that city – usually the capital.

	Land area		*Population*	*Largest city*	
Country	*Total*	*Agriculture*			
	000 km²	*Area %*	*per km²*	*000's*	*% of total*
Australia	7,686.85	60	2.76	4,100	19.3
Brunei	5.77	3	67.65	46	11.9
Cambodia	181.04	20	79.54	2,000	13.9
China	9,425.29	60	140.15	16,300	1.2
Hong Kong	1.08	6	6,441.15	n.a.	n.a.
India	3,287.59	61	349.50	17,000	1.5
Indonesia	1,890.75	24	119.32	9,130	4.0
Japan	377.84	13	337.98	12,870	10.1
Malaysia	330.25	18	82.36	1,600	5.9
New Zealand	268.67	60	15.97	379	8.8
Pakistan	803.94	34	204.92	800	0.5
Philippines	300.00	41	295.33	1,350	1.5
Singapore	0.71	2	6,506.36	n.a.	n.a.
South Korea	99.68	22	487.57	9,700	20.0
Sri Lanka	65.61	36	304.83	2,200	11.0
Taiwan	36.19	23	636.66	2,620	11.4
Thailand	514.00	35	128.48	10,000	15.1
UK	244.82	71	249.08	7,600	12.5
USA	9,161.63	18	33.19	592	0.2
Vietnam	331.69	28	259.76	6,100	7.1

7. Construction output indicators

CONSTRUCTION OUTPUT

The table below summarizes construction output statistics from the country Key data sheets. On the Key data sheet for each country, an output figure is given in national currency and the year to which it relates is noted.

In this summary table, figures in national currency are listed and, in addition, in order to facilitate (crude) comparisons, US dollar, pound sterling and yen equivalents are presented for each figure. The currency conversions have been carried out using appropriate exchange rates. As noted earlier, construction statistics, including those for construction output, are notoriously unreliable and, in addition, national definitions of construction output vary widely. It would therefore, be unwise to draw too many conclusions from this table.

Country	*National unit of currency*	*Construction output*			*Billions*
		National currency	*UK£*	*US$*	*Yen*
Australia	A$	n.a.			
Brunei	B$	n.a.			
Cambodia	Riels	2,338.00	0.38	0.57	51.73
China	Rmb	1,401.00	130.93	205.12	19,594.41
Hong Kong	HK$	37.90	3.11	4.89	468.48
India	Rs	n.a.			
Indonesia	Rp	67,318.00	3.91	6.13	587.62
Japan	¥	n.a.			
Malaysia	RM	7.45	1.33	2.09	200.73
New Zealand	NZ$	6.29*	2.31	3.64	351.40
Pakistan	Rs	n.a.			
Philippines	Php	256.50*	3.36	5.30	508.02
Singapore	S$	8.40	3.59	5.64	541.94
South Korea	Won	161,257.30	80.11	146.33	14,986.74
Sri Lanka	SLR	143.00	0.83	1.30	124.89
Taiwan	NT$	269.80	5.39	8.16	758.72
Thailand	Bt	247.00	4.74	7.05	643.56
UK	£	n.a.			
USA	US$	n.a.			
Vietnam	Dong	n.a.			

* *2008*

CONSTRUCTION OUTPUT PER CAPITA

This table is based on the previous one, but has each figure for construction output divided by the population of that country. Despite the uncertainty of both construction and population data and the limitations of exchange rates, the table reveals some useful indicators of construction activity. South Korea has by far the highest construction output per capita – almost 110 times greater than the lowest in the table (Indonesia). Like Cambodia, Indonesia, Malaysia, Philippines and Sri Lanka have output per capita of less than US$100. The remainder of the Asia Pacific countries are spread over the range from Thailand (US$107) to South Korea (US$3,011).

Country	*National unit of currency*	*Construction output per capita*			
		National currency	*UK£*	*US$*	*Yen*
Australia	A$	n.a.			
Brunei	B$	n.a.			
Cambodia	Riels	162,361	26.49	39.35	3,592
China	Rmb	1,061	99.12	155.28	14,833
Hong Kong	HK$	5,453	447.35	703.64	67,407
India	Rs	n.a.			
Indonesia	Rp	298,395	17.33	27.18	2,605
Japan	¥	n.a.			
Malaysia	RM	274	48.89	76.91	7,380
New Zealand	NZ$	1,466*	539.04	847.51	81,911
Pakistan	Rs	n.a.			
Philippines	Php	2,895*	37.97	59.77	5,734
Singapore	S$	1,826	780.38	1,225.56	117,812
South Korea	Won	3,318,051	1,648.31	3,010.94	308,369
Sri Lanka	SLR	7,150	41.40	65.11	6,245
Taiwan	NT$	11,710	233.92	354.31	32,930
Thailand	Bt	3,740	71.83	106.74	9,745
UK	£	n.a.			
USA	US$	n.a.			
Vietnam	Dong	n.a.			

** 2008*

8. Construction cost data

MASON/BRICKLAYER AND UNSKILLED LABOUR COSTS

This table summarizes hourly labour costs for a mason/bricklayer and for unskilled labour in each country as at the fourth quarter of 2008. The figures in national currency are taken from each country's construction cost data and have been converted into pound sterling, US dollar and yen equivalents using the fourth quarter 2008 exchange rates. As indicated earlier, the cost of labour is the cost to a contractor of employing that employee; it is based on the employee's income but also includes allowances for a range of mandatory and voluntary contributions which vary from country to country.

It is probable that the definitions of skilled and unskilled and what is included in labour costs varies between countries, thus these figures should not be taken as strictly comparable. The ranking and relative level of labour costs are broadly similar to the GDP per capita figures though there are interesting detailed differences in ranking.

	Mason/bricklayer			*hour*	*Unskilled labour*			*hour*
Country	*National Currency*	*UK£*	*US$*	*Yen*	*National Currency*	*UK£*	*US$*	*Yen*
Australia	85.00	35.86	55.92	5,380	80.00	33.76	52.63	5,063
Brunei	6.25	2.10	4.31	492	5.94	1.99	4.09	468
Cambodia	2,579.00*	0.42	0.63	57	n.a.			
China	7.21*	0.67	1.06	101	4.62*	0.43	0.68	65
Hong Kong	110.63	9.08	14.27	1,367	77.50	6.36	10.00	958
India	40.00**	0.52	0.83	79	n.a.			
Indonesia	8,750.00	0.51	0.80	76	7,563.00	0.44	0.69	66
Japan	2,587.50*	17.25	26.95	-	1,337.50*	8.92	13.93	-
Malaysia	11.25*	2.01	3.16	303	6.25*	1.12	1.76	168
New Zealand	39.00*	14.34	22.54	2,179	24.00*	8.82	13.87	1,341
Pakistan	75.00	0.63	0.95	85	43.75	0.37	0.55	50
Philippines	73.13	0.96	1.51	145	68.25	0.89	1.41	135
Singapore	6.63	2.83	4.45	427	n.a.			
South Korea	12,012.50	5.97	10.90	1,116	8,412.50	4.18	7.63	782
Sri Lanka	162.50	0.94	1.48	142	121.88	0.71	1.11	106
Taiwan	362.50	7.24	10.97	1,019	287.50	5.74	8.70	808
Thailand	62.50*	1.20	1.78	163	43.75*	0.84	1.25	114
UK	n.a.				n.a.			
USA	32.09	16.13	-	3,451	23.23	11.67	-	2,498
Vietnam	14,401.00	0.53	0.84	80	n.a.			

* *Wage rate*

** *3Q08*

SITE MANAGER AND QUALIFIED ARCHITECT LABOUR RATES

This table is from the same source as the previous and is presented in the same way. Site managers and qualified architects are representative of staff rather than site labour.

	Site manager			*hour*	*Qualified architect*			*hour*
Country	*National Currency*	*UK£*	*US$*	*Yen*	*National Currency*	*UK£*	*US$*	*Yen*
Australia	140.00	59.07	92.11	8,861	170.00	71.73	111.84	10,759
Brunei	46.12	15.48	31.81	3,631	32.35	10.86	22.31	2,547
Cambodia	11,902.00*	1.94	2.88	263	n.a.			
China	36.49*	3.41	5.34	510	33.32*	3.11	4.88	466
Hong Kong	300.00	24.61	38.71	3,708	240.00	19.69	30.97	2,967
India	450.00**	5.87	9.31	894	225.00**	2.94	4.66	447
Indonesia	n.a.				n.a.			
Japan	5,650.00*	37.67	58.85	-	4,787.50*	31.92	49.87	-
Malaysia	41.67*	7.44	11.70	1,123	33.85*	6.05	9.51	913
New Zealand	46.00*	16.91	26.59	2,570	29.00*	10.66	16.76	1,620
Pakistan	n.a.				n.a.			
Philippines	192.25	2.52	3.97	381	180.25	2.36	3.72	357
Singapore	35.98	15.38	24.15	2,321	31.62	13.51	21.22	2,040
South Korea	32,527.47	16.16	29.52	3,023	40,437.50	20.09	36.69	3,758
Sri Lanka	777.78	4.50	7.08	679	468.75	2.71	4.27	409
Taiwan	572.73	11.44	17.33	1,611	1,153.85	23.05	34.91	3,245
Thailand	288.46*	5.54	8.23	752	192.31*	3.69	5.49	501
UK	42.43	-	66.29	6,332	32.40	-	50.62	4,835
USA	87.15	43.79	-	9,371	n.a.			
Vietnam	n.a.				n.a.			

** Wage rate*

*** 3Q08*

MATERIALS COSTS – CEMENT AND CONCRETE AGGREGATES

The table below summarizes costs per tonne for cement and costs per m^3 for concrete aggregates as at the fourth quarter of 2008. The figures in national currency are taken from each country's construction cost data and converted into pound sterling, US dollar and yen equivalents using the fourth quarter 2008 exchange rates.

Costs are as delivered to site in a major – usually the capital – city. It is assumed that the materials are in quantities as required for a medium sized construction project and that the location of the works would be neither constrained or remote. Material costs generally exclude value added tax or other similar taxes.

	Cement			*tonne*	*Aggregate for concrete*			m^3
Country	*National Currency*	*UK£*	*US$*	*Yen*	*National Currency*	*UK£*	*US$*	*Yen*
Australia	n.a.				n.a.			
Brunei	155	52.01	106.90	12,205	33.00	11.07	22.76	2,598
Cambodia	334,206	54.52	81.00	7,394	61,890	10.10	15.00	1,369
China	300	28.04	43.92	4,196	120	11.18	17.51	1,673
Hong Kong	560	45.94	72.26	6,922	62.00	5.09	8.00	766
India	4,440	57.96	91.91	8,818	1,783	23.27	36.90	3,540
Indonesia	700,000	40.66	63.77	6,110	140,000	8.13	12.75	1,222
Japan	40,000	266.67	416.67	-	3,450	23.00	35.94	-
Malaysia	290	51.79	81.46	7,817	40.00	7.14	11.24	1,078
New Zealand	355	130.51	205.20	19,832	55.00	20.22	31.79	3,073
Pakistan	6,740	56.61	85.42	7,659	825	6.93	10.46	938
Philippines	4,625	60.65	95.48	9,160	750	9.83	15.48	1,485
Singapore	123	52.56	82.55	7,935	59.80	25.56	40.13	3,858
South Korea	84,250	41.85	76.45	7,830	17,500	8.69	15.88	1,626
Sri Lanka	15,000	86.86	136.59	13,100	19,000	110.02	173.01	16,594
Taiwan	3,500	69.92	105.90	9,843	700	13.98	21.18	1,969
Thailand	2,600	49.93	74.20	6,774	450	8.64	12.84	1,172
UK	385	-	601.88	57,493	37.86	-	59.15	5,650
USA	170	85.43	-	18,280	41.40	20.80	-	4,452
Vietnam	1,186,111	43.81	68.98	6,622	51,929	1.92	3.02	290

MATERIALS COSTS – READY MIXED CONCRETE AND REINFORCEMENT STEEL

The table below summarizes costs per m^3 for ready mixed concrete and costs per tonne for reinforcement steel as at the fourth quarter of 2008. The figures in national currency are taken from each country's construction cost data and converted into pound sterling, US dollar and yen equivalents using the fourth quarter 2008 exchange rates.

Costs are as delivered to site in a major – usually the capital – city. It is assumed that the materials are in quantities as required for a medium sized construction project and that the location of the works would be neither constrained or remote. Material costs generally exclude value added tax or other similar taxes.

	Ready mixed concrete			*m^3*	*Mild steel reinforcement*			*tonne*
Country	*National Currency*	*UK£*	*US$*	*Yen*	*National Currency*	*UK£*	*US$*	*Yen*
Australia	159	67.00	104.61	10,063	1,250	527.43	822.37	79,114
Brunei	119	39.93	82.07	9,370	1,150	385.91	793.10	90,551
Cambodia	206,300	33.65	50.00	4,564	4,002,220	652.89	970.00	88,545
China	n.a.				4,550	425.23	666.18	63,636
Hong Kong	700	57.42	90.32	8,653	6,000	492.21	774.19	74,166
India	3,500	45.69	72.45	6,951	36,000	469.91	745.19	71,500
Indonesia	460,000	26.72	41.91	4,015	9,250,000	537.35	842.67	80,744
Japan	11,750	78.33	122.40	-	109,000	726.67	1,135.42	-
Malaysia	215	38.39	60.39	5,795	2,200	392.86	617.98	59,299
New Zealand	184	67.65	106.36	10,279	1,900	698.53	1,098.27	106,145
Pakistan	5,400	45.36	68.44	6,136	70,500	592.14	893.54	80,114
Philippines	4,500	59.01	92.90	8,913	33,000	432.73	681.26	65,359
Singapore	121	51.71	81.21	7,806	1,091	466.24	732.21	70,387
South Korea	49,080	24.38	44.54	4,561	593,000	294.59	538.11	55,112
Sri Lanka	12,300	71.22	112.00	10,742	100,000	579.04	910.58	87,336
Taiwan	2,700	53.94	81.69	7,593	16,800	335.60	508.32	47,244
Thailand	2,600	49.93	74.20	6,774	24,500	470.52	699.20	63,835
UK	97	-	152.31	14,549	938	-	1,465.80	140,016
USA	125	62.81	-	13,441	930	467.34	-	100,000
Vietnam	1,025,854	37.89	59.66	5,727	12,125,570	447.83	705.18	67,695

MATERIALS COSTS – COMMON BRICKS AND HOLLOW CONCRETE BLOCKS

The table below summarizes costs per 1,000 pieces for bricks and blocks as at the fourth quarter of 2008. The figures in national currency are taken from each country's construction cost data and converted to pound sterling, US dollar and yen equivalents using the fourth quarter 2008 exchange rates.

Costs are as delivered to site in a major – usually the capital – city. It is assumed that the materials are in quantities as required for a medium sized construction project and that the location of the works would be neither constrained or remote. Material costs generally exclude value added tax or other similar taxes.

The costs of bricks and blocks vary by the availability of raw materials and the national practices in walling construction. Where brick-making clays are not readily available, for example, the cost of bricks may be relatively high. It is probably reasonable to assume that brick dimensions are broadly similar; the dimensions of concrete blocks, however, can and do vary widely.

	Common bricks			*1,000 pcs*	*Hollow concrete blocks*			*1,000 pcs*
Country	*National Currency*	*UK£*	*US$*	*Yen*	*National Currency*	*UK£*	*US$*	*Yen*
Australia	515	217.30	338.82	32,595	158	66.67	103.95	10,000
Brunei	150	50.34	103.45	11,811	450	151.01	310.34	35,433
Cambodia	3,300,800	538.47	800.00	73,027	n.a.			
China	n.a.				1,880	175.70	275.26	26,294
Hong Kong	1,200	98.44	154.84	14,833	2,000	164.07	258.06	24,722
India	2,500	32.63	51.75	4,965	28,000	365.49	579.59	55,611
Indonesia	360,000	20.91	32.80	3,142	n.a.			
Japan	257,000	1,713.33	2,677.08	-	215,000	1,433.33	2,239.58	-
Malaysia	300	53.57	84.27	8,086	2,300	410.71	646.07	61,995
New Zealand	1,498	550.74	865.90	83,687	345	126.84	199.42	19,274
Pakistan	n.a.				19,000	159.58	240.81	21,591
Philippines	n.a.				12,000	157.36	247.73	23,767
Singapore	220	94.02	147.65	14,194	700	299.15	469.80	45,161
South Korea	45,000	22.35	40.83	4,182	600,000	298.06	544.46	55,762
Sri Lanka	7,500	43.43	68.29	6,550	35,700	206.72	325.08	31,179
Taiwan	2,400	47.94	72.62	6,749	47,000	938.87	1,422.09	132,171
Thailand	1,000	19.20	28.54	2,606	4,000	76.82	114.16	10,422
UK	216	-	337.41	32,230	1,621	-	2,532.81	241,940
USA	420	211.06	-	45,161	1,500	753.77	-	161,290
Vietnam	n.a.				5,425,194	200.37	315.51	30,288

MATERIALS COSTS – SOFTWOOD FOR JOINERY AND QUILT INSULATION

The table below summarizes costs per m^3 for softwood for joinery and costs per m^2 for 100mm thick quilt insulation as at the fourth quarter of 2008. The figures in national currency are taken from each country's construction cost data and converted to pound sterling, US dollar and yen equivalents using the fourth quarter 2008 exchange rates.

Costs are as delivered to site in a major – usually the capital – city. It is assumed that the materials are in quantities as required for a medium sized construction project and that the location of the works would be neither constrained or remote. Material costs generally exclude value added tax or other similar taxes.

	Softwood for joinery			*m^3*	*Quilt insulation 100mm*			*m^2*
Country	*National Currency*	*UK£*	*US$*	*Yen*	*National Currency*	*UK£*	*US$*	*Yen*
Australia	1,450	611.81	953.95	91,772	9.00	3.80	5.92	570
Brunei	1,283	430.65	885.06	101,050	6.00	2.01	4.14	472
Cambodia	2,269,300	370.20	550.00	50,206	n.a.			
China	1,950	182.24	285.51	27,273	35.00	3.27	5.12	490
Hong Kong	3,300	270.71	425.81	40,791	55.00	4.51	7.10	680
India	n.a.				730	9.53	15.11	1,450
Indonesia	5,000,000	290.46	455.50	43,645	13,000	7.55	11.84	1,135
Japan	80,000	533.33	833.33	-	n.a.			
Malaysia	n.a.				n.a.			
New Zealand	2,300	845.59	1,329.48	128,492	13.00	4.78	7.51	726
Pakistan	n.a.				n.a.			
Philippines	21,500	281.93	443.85	42,583	650	8.52	13.42	1,287
Singapore	667	284.90	447.43	43,011	7.00	2.99	4.70	452
South Korea	1,047,900	520.57	950.91	97,388	8,680	4.31	7.88	807
Sri Lanka	n.a.				n.a.			
Taiwan	n.a.				10,000	199.76	302.57	28,121
Thailand	30,000	576.15	856.16	78,166	n.a.			
UK	223	-	348.72	33,310	3.97	-	6.20	593
USA	1,417	711.89	-	152,330	n.a.			
Vietnam	19,974,858	737.73	1,161.67	111,517	56,056	2.07	3.26	313

MATERIALS COSTS – SHEET GLASS AND PLASTERBOARD

The table below summarizes costs per m^2 for sheet or float glass and costs per m^2 for 9-12mm thick plasterboard as at the fourth quarter of 2008. The figures in national currency are taken from each country's construction cost data and converted to pound sterling, US dollar and yen equivalents using the fourth quarter 2008 exchange rates.

Costs are as delivered to site in a major – usually the capital – city. It is assumed that the materials are in quantities as required for a medium sized construction project and that the location of the works would be neither constrained or remote. Material costs generally exclude value added tax or other similar taxes.

	Sheet/float glass			*m²*	*Plasterboard 9 - 12mm*			*m²*
Country	*National Currency*	*UK£*	*US$*	*Yen*	*National Currency*	*UK£*	*US$*	*Yen*
Australia	49.00	20.68	32.24	3,101	8.00	3.38	5.26	506
Brunei	40.00	13.42	27.59	3,150	12.00	4.03	8.28	945
Cambodia	78,394	12.79	19.00	1,734	n.a.			
China	35.00	3.27	5.12	490	26.00	2.43	3.81	364
Hong Kong	99.00	8.12	12.77	1,224	140	11.48	18.06	1,731
India	1,200	15.66	24.84	2,383	175	2.28	3.62	348
Indonesia	85,000	4.94	7.74	742	16,667	0.97	1.52	145
Japan	1,330	8.87	13.85	-	145	0.97	1.51	-
Malaysia	40.00	7.14	11.24	1,078	18.00	3.21	5.06	485
New Zealand	200	73.53	115.61	11,173	15.00	5.51	8.67	838
Pakistan	n.a.				n.a.			
Philippines	850	11.15	17.55	1,684	250	3.28	5.16	495
Singapore	40.00	17.09	26.85	2,581	5.00	2.14	3.36	323
South Korea	4,770	2.37	4.33	443	1,920	0.95	1.74	178
Sri Lanka	550	3.18	5.01	480	n.a.			
Taiwan	n.a.				450	8.99	13.62	1,265
Thailand	450	8.64	12.84	1,172	100	1.92	2.85	261
UK	22.02	-	34.41	3,287	1.83	-	2.86	273
USA	52.00	26.13	-	5,591	2.40	1.21	-	258
Vietnam	179,344	6.62	10.43	1,001	197,743	7.30	11.50	1,104

MATERIALS COSTS – EMULSION PAINT AND VINYL FLOOR TILES

The table below summarizes costs per litre for emulsion paint and costs per m² for vinyl floor tiles as at the fourth quarter of 2008. The figures in national currency are taken from each country's construction cost data and converted to pound sterling, US dollar and yen equivalents using the fourth quarter 2008 exchange rates.

Costs are as delivered to site in a major – usually the capital – city. It is assumed that the materials are in quantities as required for a medium sized construction project and that the location of the works would be neither constrained or remote. Material costs generally exclude value added tax or other similar taxes.

	Emulsion paint			*litre*	*Vinyl floor tiles*			*m²*
Country	*National Currency*	*UK£*	*US$*	*Yen*	*National Currency*	*UK£*	*US$*	*Yen*
Australia	n.a.				20.00	8.44	13.16	1,266
Brunei	9.50	3.19	6.55	748	18.00	6.04	12.41	1,417
Cambodia	n.a.				n.a.			
China	14.00*	1.31	2.05	196	40.00	3.74	5.86	559
Hong Kong	39.00	3.20	5.03	482	70.00	5.74	9.03	865
India	170	2.22	3.52	338	300	3.92	6.21	596
Indonesia	21,150*	1.23	1.93	185	95,000	5.52	8.65	829
Japan	310*	2.07	3.23	-	700	4.67	7.29	-
Malaysia	17.50	3.13	4.92	472	32.00	5.71	8.99	863
New Zealand	7.50**	2.76	4.34	419	38.00	13.97	21.97	2,123
Pakistan	255	2.14	3.23	290	n.a.			
Philippines	550***	7.21	11.35	1,089	550	7.21	11.35	1,089
Singapore	3.50	1.50	2.35	226	20.00	8.55	13.42	1,290
South Korea	4,300	2.14	3.90	400	6,400	3.18	5.81	595
Sri Lanka	575	3.33	5.24	502	n.a.			
Taiwan	120	2.40	3.63	337	345	6.89	10.44	970
Thailand	350***	6.72	9.99	912	300	5.76	8.56	782
UK	3.52	-	5.50	525	6.74	-	10.53	1,006
USA	28.00***	14.07	-	3,011	27.50	13.82	-	2,957
Vietnam	125,524	4.64	7.30	701	257,925	9.53	15.00	1,440

* *kg*

** *m²*

*** *gallon*

APPROXIMATE ESTIMATING – FACTORIES AND WAREHOUSES

This table summarizes approximate estimating costs per m² for factories and warehouses. Approximate estimating costs are averages as incurred by building clients for typical buildings in major – usually capital – cities in the fourth quarter of 2008. They are based upon the total floor area of all storeys, measured between external walls and without deduction for internal walls. Approximate estimating costs generally include mechanical and electrical installations but exclude furniture, loose or special equipment, and external works; they also exclude fees for professional services. Where a range of costs has been given, the mid point is shown. The figures in national currency are taken from each country's construction cost data and converted to pound sterling, US dollar, and yen equivalents using the fourth quarter 2008 exchange rates.

It must be borne in mind that even where costs are given under the same description in one or more countries, this is not to say that they are identical, or even physically similar. Approximate estimating costs for a particular country are for the normal standards prevailing in that country. Quality and technical standards vary widely and there are differences between countries in what is, and is not, included. The table, therefore, should be used with care.

	Factories for owner occupation (light industrial use)			*m²*	*Warehouse, low bay (6 - 8m high) for letting (no heating)*			*m²*
Country	*National Currency*	*UK£*	*US$*	*Yen*	*National Currency*	*UK£*	*US$*	*Yen*
Australia	500	210.97	328.95	31,646	450	189.87	296.05	28,481
Brunei	675	226.51	465.52	53,150	650	218.12	448.28	51,181
Cambodia	n.a.				n.a.			
China	4,300	401.87	629.58	60,140	n.a.			
Hong Kong	8,500	697.29	1,096.77	105,068	9,200	754.72	1,187.10	113,721
India	13,450	175.56	278.41	26,713	8,650	112.91	179.05	17,180
Indonesia	3,567,525	207.25	325.00	31,141	2,908,905	168.98	265.00	25,392
Japan	160,000	1,066.67	1,666.67	-	100,000	666.67	1,041.67	-
Malaysia	1,500	267.86	421.35	40,431	n.a.			
New Zealand	750	275.74	433.53	41,899	600	220.59	346.82	33,520
Pakistan	16,678	140.08	211.38	18,952	n.a.			
Philippines	17,300	226.86	357.14	34,264	14,100	184.89	291.08	27,926
Singapore	1,600	683.76	1,073.83	103,226	1,300	555.56	872.48	83,871
South Korea	760,000	377.55	689.66	70,632	610,000	303.03	553.54	56,691
Sri Lanka	24,300	140.71	221.27	21,223	n.a.			
Taiwan	27,200	543.35	823.00	76,490	n.a.			
Thailand	16,600	318.80	473.74	43,252	n.a.			
UK	689	-	1,076.56	102,836	n.a.			
USA	1,291	648.74	-	138,817	516	259.30	-	55,484
Vietnam	4,986,550	184.17	290.00	27,839	4,814,600	177.82	280.00	26,879

APPROXIMATE ESTIMATING – OFFICES

This table summarizes approximate estimating costs per m^2 for two different types of office buildings. Approximate estimating costs are averages as incurred by building clients for typical buildings in major – usually capital – cities in the fourth quarter of 2008. They are based upon the total floor area of all storeys, measured between external walls and without deduction for internal walls. Approximate estimating costs generally include mechanical and electrical installations but exclude furniture, loose or special equipment, and external works; they also exclude fees for professional services. Where a range of costs has been given, the mid point is shown. The figures in national currency are taken from each country's construction cost data and converted to pound sterling, US dollar and yen equivalents using the fourth quarter 2008 exchange rates.

It must be borne in mind that even where costs are given under the same description in one or more countries, this is not to say that they are identical, or even physically similar. Approximate estimating costs for a particular country are for the normal standards prevailing in that country. Quality and technical standards vary widely and there are differences between countries in what is, and is not, included. The table, therefore, should be used with care.

	Offices for letting, 5 - 10 storeys air-conditioned			*m^2*	*Prestige/headquarters office high rise air-conditioned*			*m^2*
Country	*National Currency*	*UK£*	*US$*	*Yen*	*National Currency*	*UK£*	*US$*	*Yen*
Australia	2,400	1,012.66	1,578.95	151,899	3,200	1,350.21	2,105.26	202,532
Brunei	1,100	369.13	758.62	86,614	n.a.			
Cambodia	n.a.				2,888,200	471.16	700.00	63,898
China	5,700	532.71	834.55	79,720	7,500	700.93	1,098.10	104,895
Hong Kong	14,300	1,173.09	1,845.16	176,671	19,800	1,624.28	2,554.84	244,747
India	19,900	259.76	411.92	39,523	23,700	309.36	490.58	47,071
Indonesia	3,293,100	191.30	300.00	28,746	8,836,485	513.33	805.00	77,134
Japan	230,000	1,533.33	2,395.83	-	350,000	2,333.33	3,645.83	-
Malaysia	1,780	317.86	500.00	47,978	3,760	671.43	1,056.18	101,348
New Zealand	1,900	698.53	1,098.27	106,145	3,000	1,102.94	1,734.10	167,598
Pakistan	36,046	302.75	456.86	40,961	69,402	582.92	879.62	78,866
Philippines	28,200	369.79	582.16	55,853	48,000	629.43	990.92	95,068
Singapore	2,300	982.91	1,543.62	148,387	3,050	1,303.42	2,046.98	196,774
South Korea	1,210,000	601.09	1,098.00	112,454	1,820,000	904.12	1,651.54	169,145
Sri Lanka	n.a.				145,300	841.34	1,323.07	126,900
Taiwan	36,300	725.13	1,098.34	102,081	54,450	1,087.69	1,647.50	153,121
Thailand	18,000	345.69	513.70	46,899	29,600	568.47	844.75	77,124
UK	1,765	-	2,757.81	263,433	2,933	-	4,582.81	437,761
USA	2,367	1,189.45	-	254,516	3,873	1,946.23	-	416,452
Vietnam	11,176,750	412.79	650.00	62,398	17,366,950	641.41	1,010.00	96,957

APPROXIMATE ESTIMATING – HOUSING

This table summarizes approximate estimating costs per m² for two different types of housing. In countries where housing types did not match exactly these descriptions, the nearest equivalent has been taken.

Approximate estimating costs are averages as incurred by building clients for typical buildings in major – usually capital – cities in the fourth quarter of 2008. They are based upon the total floor area of all storeys, measured between external walls and without deduction for internal walls. Approximate estimating costs generally include mechanical and electrical installations but exclude furniture, loose or special equipment, and external works; they also exclude fees for professional services. Where a range of costs has been given, the mid point is shown. The figures in national currency are taken from each country's construction cost data and converted to pound sterling, US dollar and yen equivalents using the fourth quarter 2008 exchange rates.

It must be borne in mind that even where costs are given under the same description in one or more countries, this is not to say that they are identical, or even physically similar. Approximate estimating costs for a particular country are for the normal standards prevailing in that country. Quality and technical standards vary widely and there are differences between countries in what is, and is not, included. The table, therefore, should be used with care.

	Single family housing private, detached, semi detached			*m²*	*Private sector apartment building (standard specification)*			*m²*
Country	*National Currency*	*UK£*	*US$*	*Yen*	*National Currency*	*UK£*	*US$*	*Yen*
Australia	1,300	548.52	855.26	82,278	2,700	1,139.24	1,776.32	170,886
Brunei	675	226.51	465.52	53,150	1,100	369.13	758.62	86,614
Cambodia	n.a.				2,063,000	336.54	500.00	45,642
China	7,500	700.93	1,098.10	104,895	3,500	327.10	512.45	48,951
Hong Kong	18,200	1,493.03	2,348.39	224,969	11,400	935.19	1,470.97	140,915
India	14,000	182.74	289.80	27,805	16,800	219.29	347.75	33,366
Indonesia	n.a.				6,311,775	366.67	575.00	55,096
Japan	190,000	1,266.67	1,979.17	-	210,000	1,400.00	2,187.50	-
Malaysia	2,430	433.93	682.58	65,499	1,670	298.21	469.10	45,013
New Zealand	1,400	514.71	809.25	78,212	2,400	882.35	1,387.28	134,078
Pakistan	15,064	126.52	190.93	17,118	37,660	316.31	477.31	42,795
Philippines	14,000	183.58	289.02	27,728	29,200	382.90	602.81	57,833
Singapore	3,600	1,538.46	2,416.11	232,258	3,250	1,388.89	2,181.21	209,677
South Korea	1,200,000	596.13	1,088.93	111,524	1,490,000	740.19	1,352.09	138,476
Sri Lanka	30,300	175.45	275.91	26,463	43,000	248.99	391.55	37,555
Taiwan	n.a.				36,300	725.13	1,098.34	102,081
Thailand	16,000	307.28	456.62	41,688	23,800	457.08	679.22	62,011
UK	1,155	-	1,804.69	172,388	1,374	-	2,146.88	205,075
USA	2,066	1,038.19	-	222,151	3,271	1,643.72	-	351,720
Vietnam	n.a.				11,004,800	406.44	640.00	61,438

APPROXIMATE ESTIMATING – HOSPITALS AND SCHOOLS

This table summarizes approximate estimating costs per m² for general hospitals and secondary or middle schools. Approximate estimating costs are averages as incurred by building clients for typical buildings in major – usually capital – cities in the fourth quarter of 2008. They are based upon the total floor area of all storeys, measured between external walls and without deduction for internal walls. Approximate estimating costs generally include mechanical and electrical installations but exclude furniture, loose or special equipment, and external works; they also exclude fees for professional services. Where a range of costs has been given, the mid point is shown. The figures in national currency are taken from each country's construction cost data and converted to pound sterling, US dollar and yen equivalents using the fourth quarter 2008 exchange rates.

It must be borne in mind that even where costs are given under the same description in one or more countries, this is not to say that they are identical, or even physically similar. Approximate estimating costs for a particular country are for the normal standards prevailing in that country. Quality and technical standards vary widely and there are differences between countries in what is, and is not, included. The table, therefore, should be used with care.

	General hospitals			*m²*	*Secondary/middle schools*			*m²*
Country	*National Currency*	*UK£*	*US$*	*Yen*	*National Currency*	*UK£*	*US$*	*Yen*
Australia	3,500	1,476.79	2,302.63	221,519	2,000	843.88	1,315.79	126,582
Brunei	1,575	528.52	1,086.21	124,016	1,100	369.13	758.62	86,614
Cambodia	n.a.				n.a.			
China	8,000	747.66	1,171.30	111,888	3,600	336.45	527.09	50,350
Hong Kong	22,000	1,804.76	2,838.71	271,941	9,900	812.14	1,277.42	122,373
India	16,200	211.46	335.33	32,175	7,000	91.37	144.90	13,903
Indonesia	8,232,750	478.26	750.00	71,864	5,762,925	334.78	525.00	50,305
Japan	300,000	2,000.00	3,125.00	-	220,000	1,466.67	2,291.67	-
Malaysia	n.a.				1,160	207.14	325.84	31,267
New Zealand	4,000	1,470.59	2,312.14	223,464	2,200	808.82	1,271.68	122,905
Pakistan	75,858	637.14	961.44	86,202	39,274	329.87	479.77	44,630
Philippines	43,500	570.42	898.02	86,156	n.a.			
Singapore	3,550	1,517.09	2,382.55	229,032	1,700	726.50	1,140.94	109,677
South Korea	1,750,000	869.35	1,588.02	162,639	970,000	481.87	880.22	90,149
Sri Lanka	41,200	238.56	375.16	35,983	n.a.			
Taiwan	75,625	1,510.69	2,288.20	212,669	36,300	725.13	1,098.34	102,081
Thailand	35,000	672.17	998.86	91,193	15,000	288.07	428.08	39,083
UK	1,518	-	2,371.88	226,567	1,704	-	2,662.50	254,328
USA	2,797	1,405.53	-	300,753	2,410	1,211.06	-	259,140
Vietnam	n.a.				n.a.			

APPROXIMATE ESTIMATING – THEATRES AND SPORTS HALLS

This table summarizes approximate estimating costs per m² for theatres and sports halls. Approximate estimating costs are averages as incurred by building clients for typical buildings in major – usually capital – cities in the fourth quarter of 2008. They are based upon the total floor area of all storeys, measured between external walls and without deduction for internal walls. Approximate estimating costs generally include mechanical and electrical installations but exclude furniture, loose or special equipment, and external works; they also exclude fees for professional services. Where a range of costs has been given, the mid point is shown. The figures in national currency are taken from each country's construction cost data and converted to pound sterling, US dollar and yen equivalents using the fourth quarter 2008 exchange rates.

It must be borne in mind that even where costs are given under the same description in one or more countries, this is not to say that they are identical, or even physically similar. Approximate estimating costs for a particular country are for the normal standards prevailing in that country. Quality and technical standards vary widely and there are differences between countries in what is, and is not, included. The table, therefore, should be used with care.

	Theatres including seating and stage equipment, over 500 seats			*m²*	*Sports halls including changing and social facilities*			*m²*
Country	*National Currency*	*UK£*	*US$*	*Yen*	*National Currency*	*UK£*	*US$*	*Yen*
Australia	3,200	1,350.21	2,105.26	202,532	2,100	886.08	1,381.58	132,911
Brunei	3,400	1,140.94	2,344.83	267,717	1,600	536.91	1,103.45	125,984
Cambodia	n.a.				n.a.			
China	11,000	1,028.04	1,610.54	153,846	n.a.			
Hong Kong	27,500	2,255.95	3,548.39	339,926	n.a.			
India	n.a.				5,400	70.49	111.78	10,725
Indonesia	n.a.				n.a.			
Japan	400,000	2,666.67	4,166.67	-	220,000	1,466.67	2,291.67	-
Malaysia	n.a.				1,880	335.71	528.09	50,674
New Zealand	3,000	1,102.94	1,734.10	167,598	2,500	919.12	1,445.09	139,665
Pakistan	n.a.				38,736	325.35	490.95	44,018
Philippines	58,000	760.56	1,197.36	114,874	42,000	550.75	867.05	83,185
Singapore	n.a.				2,200	940.17	1,476.51	141,935
South Korea	2,150,000	1,068.06	1,951.00	199,814	1,900,000	943.86	1,724.14	176,580
Sri Lanka	n.a.				n.a.			
Taiwan	56,332	1,125.29	1,704.45	158,414	n.a.			
Thailand	65,000	1,248.32	1,855.02	169,359	50,000	960.25	1,426.94	130,276
UK	4,074	-	6,365.63	608,060	968	-	1,512.50	144,478
USA	6,455	3,243.72	-	694,086	n.a.			
Vietnam	n.a.				n.a.			

APPROXIMATE ESTIMATING – HOTELS

This table summarizes approximate estimating costs per m² for two types of hotels. Approximate estimating costs are averages as incurred by building clients for typical buildings in major – usually capital – cities in the fourth quarter of 2008. They are based upon the total floor area of all storeys, measured between external walls and without deduction for internal walls. Approximate estimating costs generally include mechanical and electrical installations but exclude furniture, loose or special equipment, and external works; they also exclude fees for professional services. Where a range of costs has been given, the mid point is shown. The figures in national currency are taken from each country's construction cost data and converted to pound sterling, US dollar and yen equivalents using the fourth quarter 2008 exchange rates.

It must be borne in mind that even where costs are given under the same description in one or more countries, this is not to say that they are identical, or even physically similar. Approximate estimating costs for a particular country are for the normal standards prevailing in that country. Quality and technical standards vary widely and there are differences between countries in what is, and is not, included. The table, therefore, should be used with care.

Country	*Hotel, 5 star, city centre*			*m²*	*Hotel, 3 star city/provincial*			*m²*
	National Currency	*UK£*	*US$*	*Yen*	*National Currency*	*UK£*	*US$*	*Yen*
Australia	3,450	1,455.70	2,269.74	218,354	2,800	1,181.43	1,842.11	177,215
Brunei	3,000	1,006.71	2,068.97	236,220	2,500	838.93	1,724.14	196,850
Cambodia	6,189,000	1,009.62	1,500.00	136,925	4,538,600	740.39	1,100.00	100,412
China	10,400	971.96	1,522.69	145,455	6,700	626.17	980.97	93,706
Hong Kong	23,400	1,919.61	3,019.35	289,246	18,400	1,509.43	2,374.19	227,441
India	48,450	632.42	1,002.90	96,226	37,700	492.10	780.38	74,876
Indonesia	16,465,500	956.52	1,500.00	143,728	9,879,300	573.91	900.00	86,237
Japan	400,000	2,666.67	4,166.67	-	300,000	2,000.00	3,125.00	-
Malaysia	7,700	1,375.00	2,162.92	207,547	5,520	985.71	1,550.56	148,787
New Zealand	3,900	1,433.82	2,254.34	217,877	3,020	1,110.29	1,745.66	168,715
Pakistan	129,120	1,084.50	1,636.50	146,727	85,004	713.96	1,077.36	96,595
Philippines	66,000	865.46	1,362.51	130,719	50,500	662.21	1,042.53	100,020
Singapore	4,550	1,944.44	3,053.69	293,548	3,350	1,431.62	2,248.32	216,129
South Korea	2,600,000	1,291.60	2,359.35	241,636	2,000,000	993.54	1,814.88	185,874
Sri Lanka	142,000	822.24	1,293.02	124,017	101,800	589.46	926.97	88,908
Taiwan	65,800	1,314.42	1,990.92	185,039	n.a.			
Thailand	59,000	1,133.09	1,683.79	153,726	39,500	758.59	1,127.28	102,918
UK	2,421	-	3,782.81	361,343	1,867	-	2,917.19	278,657
USA	n.a.				n.a.			
Vietnam	29,919,300	1,105.01	1,740.00	167,035	25,104,700	927.19	1,460.00	140,156

Index